F V

ANALOG ELECTRONIC CIRCUITS

ANALOG ELECTRONIC CIRCUITS

Glenn M. Glasford

Professor of Electrical and Computer Engineering
Syracuse University

PRENTICE-HALL
Englewood Cliffs, New Jersey 07632

Library of Congress Cataloging-in-Publication Data

GLASFORD, GLENN M. (date)
  Analog electronic circuits.

  Includes bibliographies and index.
  1. Analog electronic systems. I. Title.
TK7870.G539 1986   621.381   85-19239
ISBN   0-13-032699-2

Editorial/production supervision and
  interior design: Reynold Rieger
Cover design: Wanda Lubelska Design
Manufacturing buyer: Rhett Conklin

Printed in the United States of America

10 9 8 7 6 5 4 3 2 1

ISBN 0-13-032699-2 025

PRENTICE-HALL INTERNATIONAL (UK) LIMITED, *London*
PRENTICE-HALL OF AUSTRALIA PTY. LIMITED, *Sydney*
PRENTICE-HALL CANADA INC., *Toronto*
PRENTICE-HALL HISPANOAMERICANA, S.A., *Mexico*
PRENTICE-HALL OF INDIA PRIVATE LIMITED, *New Delhi*
PRENTICE-HALL OF JAPAN, INC., *Tokyo*
PRENTICE-HALL OF SOUTHEAST ASIA PTE. LTD., *Singapore*
EDITORA PRENTICE-HALL DO BRASIL, LTDA., *Rio de Janeiro*
WHITEHALL BOOKS LIMITED, *Wellington, New Zealand*

To my wife
Ethel

Contents

Preface   *xiii*

Chapter 1

Semiconductors, Junctions, and the Junction Diode   *1*

       Introduction,   1
  1.1 Semiconductor Properties,   2
  1.2 Semiconductor Junction Properties,   7
  1.3 Voltage–Current Relationships at Semiconductor Junctions,   13
  1.4 Metal–Semiconductor Contacts,   19
  1.5 *PN* Junction Diode,   21
  1.6 *PN* Junction Diode Dynamics,   27
  1.7 Diode Switching Characteristics,   33
  1.8 Approximate Temperature Characteristics of *PN* Junctions,   38
  1.9 Noise in Semiconductor Materials and Junctions,   41
       Problems,   46
       References,   49

Chapter 2

Bipolar Junction Transistors: Mathematical and Circuit Models   *50*

       Introduction,   50
  2.1 Structures, Terminology, and General Characteristics,   51

2.2 Basic Mathematical Model for the *NPN* Transistor,   52

2.3 *PNP* Transistor,   57

2.4 Common-Emitter Connection,   58

2.5 Extraction of Basic Parameters from Transistor Measurements,   62

2.6 Linear and Nonlinear Analytical Models for Transistors in the Normal Active Region,   64

2.7 Incrementally Linear (Small-Signal) Circuit Models,   68

2.8 High-Frequency Circuit Models,   75

2.9 Variational Nature of Incremental Parameters,   77

2.10 Structural Parameters and Nonlinearities,   80

2.11 Effects of Temperature Variations on Transistor Parameters,   87

2.12 Noise Sources and Models,   90

Problems,   91

References,   96

Chapter 3

Junction and Insulated Gate Field-Effect Transistors   **98**

Introduction,   98

3.1 Basic *n*-Channel Junction Gate FET Structure,   98

3.2 Basic Equations for Symmetrical Structure,   99

3.3 Semiempirical Models for the Saturation Region,   106

3.4 Incremental (Small-Signal) Circuit Models,   107

3.5 *P*-Channel Junction Field-Effect Transistor,   109

3.6 High-Frequency Incremental Models,   110

3.7 Insulated Gate Field-Effect Transistor,   112

3.8 Basic Equations for IGFET Operation in the Low-Voltage (Triode) Region,   114

3.9 Approximate IGFET Models for the Saturation Region,   120

3.10 IGFET (MOSFET) Symbols,   122

3.11 Temperature Dependence of Field-Effect Transistors,   123

3.12 Noise Sources in Field-Effect Transistors,   123

Problems,   125

References,   129

Chapter 4

Basic Properties of Single-Device Amplifier Structures   **130**

Introduction,   130

4.1 Amplifier Characterizations and Definitions,   131

4.2 Bipolar Transistors as Voltage Amplifiers,   134

4.3 Common-Emitter Amplifier,   135

4.4 Common-Collector Amplifier: Emitter Follower,   137

4.5   Common-Base Amplifier,   140

4.6   Two-Input Transistor Amplifier,   142

4.7   Field-Effect Transistor Amplifiers,   142

4.8   Common-Source and Common-Drain Amplifier,   143

4.9   Common-Gate Amplifier,   146

4.10 Dual-Input FET Amplifier,   148

4.11 Introduction to Frequency- and Time-Domain Responses,   148

4.12 Frequency Response of Single-Device Inverting Amplifiers,   154

4.13 Frequency Response of Noninverting Amplifiers,   163

4.14 Noise Models for Single-Device Amplifiers,   168

      Problems,   173

      References,   179

Chapter 5

Composite Amplifiers: Structure, Performance, and Biasing   *180*

      Introduction,   180

5.1   Approximate Equations for Single-Device Amplifiers,   180

5.2   Cascode Amplifier,   182

5.3   Composite Voltage Followers,   188

5.4   Darlington Amplifier,   191

5.5   Single-Input, Single-Output Emitter-Coupled Amplifier,   194

5.6   Emitter-Coupled Differential Amplifier Pair,   195

5.7   Effects of Internal Device Feedback on Balance,   197

5.8   High-Impedance (Current) Source for Common Emitters,   198

5.9   The Differential-Cascode Amplifier,   199

5.10 Composite-Compound BJT Amplifiers at Low Current Levels: Another
      Viewpoint,   200

5.11 Differential Pair FET Amplifier,   201

5.12 Frequency Response of Voltage Follower Driving a Noninverting Gain
      Stage,   203

5.13 Complementary-Pair Transistor Amplifiers,   203

5.14 Current Sources and Loads for BJTs,   209

5.15 FETs as Sources and Loads,   220

      Problems,   225

      References,   232

Chapter 6

Properties of Amplifiers with Feedback   *233*

      Introduction,   233

6.1   Amplifier Represented as a Gain Block; Some Matters of
      Terminology,   233

6.2   Gain Equations for Amplifiers with Feedback,   235

6.3   Properties of Voltage Feedback: Specific Examples,   237

6.4   Properties of Current Feedback: Feedback Derived in Series with Load,   242

6.5   Summary of Feedback Classification and Properties,   245

6.6   Feedback in Differential Form,   247

6.7   Frequency Characteristics of Negative Feedback Amplifiers,   249

6.8   Frequency Characteristics for Negative Shunt–Shunt Feedback,   250

6.9   Frequency Characteristics of the Differential Feedback Amplifier,   254

6.10 Feedback Amplifiers with Multiple Pole Responses,   256

6.11 Formal Expressions of Stability Criteria for Negative Feedback Amplifiers,   266

6.12 Reduction of Amplifier Distortion with Negative Feedback,   267

6.13 Feedback Amplifiers as Sinusoidal Oscillators,   269

      Problems,   271

      References,   278

Chapter 7

Operational Amplifiers: Specifications, Analysis, and Applications   *279*

      Introduction,   279

7.1   General Dc and Low-Frequency Properties and Specifications,   280

7.2   Frequency- and Time-Domain Response of Operational Amplifiers,   285

7.3   Evolution of Operational-Amplifier Structures,   290

7.4   Approaches to High-Gain Extended Bandwidth Op-Amps,   298

7.5   IGFET (MOSFET) Monolithic Operational Amplifiers,   305

7.6   Operational Transconductance Amplifier,   309

7.7   Gain-Stabilized Transconductance Amplifier,   313

7.8   Micropower Operational Amplifiers,   314

7.9   Operational Amplifiers as Voltage Comparators,   316

7.10 Programmable Operational Amplifiers,   318

7.11 Active Filters Incorporating Operational Amplifiers,   318

7.12 Synthesis of Impedance Elements,   332

      Problems,   334

      References,   338

Chapter 8

Nonlinear Distortion in Devices and Applications of Nonlinear Characteristics *340*

      Introduction,   340

8.1   Nonlinear Characteristics of Field-Effect Transistors,   341

8.2   Nonlinear Representation of Bipolar Transistors,   346

8.3 Transistor Power Amplifier Distortions, 350

8.4 Complementary-Pair Transistor Power Amplifiers, 355

8.5 Transformer Coupling of Power Output Stages, 357

8.6 Utilization of Nonlinear Amplifier Properties, 358

8.7 Analog Multiplier Realizations, 361

8.8 Analog Multiplier Applications, 365

8.9 Logarithmic Amplifiers and Applications, 367

Problems, 370

References, 372

Chapter 9

Analog Switching Circuits, Transmission Channel Time Sharing, Function Generators, and Phase-Locked Loops   *374*

Introduction, 374

9.1 Limiting, Clipping, and Clamping Circuits, 375

9.2 Transmission Gates, Analog Switches, and Elementary Applications, 380

9.3 Simple Bistable, Monostable, and Astable Circuits, 384

9.4 Voltage-Controlled Oscillators, 395

9.5 Periodic Voltage Waveform Generators, 399

9.6 Sawtooth Current Generators, 405

9.7 Phase-Locked Loops, 409

9.8 Specific Applications of Phase-Locked Loops, 415

Problems, 423

References, 437

Chapter 10

Analog-to-Digital and Digital-to-Analog Conversion Fundamentals   *439*

Introduction, 439

10.1 Sampling Rate and Bandwidth Requirements, 440

10.2 Quantizing and Digitizing Analog Information, 444

10.3 Simple Implementation of A/D Converters using Parallel Comparators, 448

10.4 Counter-Ramp A/D Converter, 453

10.5 Digital-to-Analog Conversion Processes and Techniques, 455

10.6 Analog-to-Digital Converters Using Digital-to-Analog Converters, 462

10.7 Concluding Comments on A/D and D/A Converters, 464

Problems, 464

References, 471

Index   *473*

Preface

The study of electronic circuits encompasses all aspects of the theory and application of devices, including the engineering of systems involving both discrete devices and integrated structures. It is usually divided arbitrarily although somewhat logically into two separate but related parts, analog circuits and digital circuits. A third part, nonlinear and switching operations, particularly as they relate to the techniques used for analog–digital–analog conversion processes, may be included with either or both parts, or may be treated separately, as, indeed they would be for a comprehensive treatment, or they may be neglected entirely, as unfortunately they often are.

Analog circuits and digital circuits can be treated pedagogically in either order following an appropriate treatment of device theory and circuit modeling, which may be more or less common on both parts. This philosophy of mutual independence is carried out in this volume and in a companion volume, *Digital Electronic Circuits.*

This volume, *Analog Electronic Circuits,* begins with three chapters covering theory and circuit modeling of semiconductor devices, diodes, bipolar junction transistors, and field-effect transistors with an emphasis on structural parameters and the incremental circuit models that can be derived from them, followed by a comprehensive sequence of chapters on analog circuits and systems including nonlinear and switching operations, with a final chapter on the analog-to-digital and digital-to-analog conversion processes.

The rapid progress in creative electronic circuit design mirrors the progress in the evolution of electronic devices upon which electronic system design is based, from the vacuum tube to the *npn* and *pnp* bipolar transistor to the *p*- and *n*-channel field-effect transistor with depletion- and enhancement-mode variations; and from handwiring to printed-circuit wiring, to the full monolithic integrated circuit. The full use of the variety of discrete devices and integrated structures now available permits innovative designs not dreamed of in the days of the vacuum tube with its very limited flexibility.

Creative circuit designs of professional quality can be carried out only by those individuals who have acquired a thorough understanding of the variations of device characteristics available or that can be created, and only those who also have acquired this understanding will be able to use in an optimum manner in electronic systems, those individual circuit designs that have been created by others. It is the goal of this volume to organize appropriate material pertinent to analog circuits in a way that will help the individual designer not only to learn the factual material, but to present it in such a way that his or her own creative processes will be stimulated, which although seemingly a contradiction can best be accomplished by presenting background material in a carefully structured and logical manner. With these objectives in mind the material presented in this volume is structured in a step-by-step building-block approach that (1) begins with the detailed modeling of single devices and their use as linear amplifiers, (2) progresses through the evolution of composite device structures to the monolithic operational amplifier and its myriad of programmable functions, (3) is followed by the modeling of nonlinear phenomena using the nonlinear characteristics of the basic building blocks, and (4) characteristics of analog switches and the evolution of analog subsystems such as the phase-locked loop and analog-to-digital conversions and their inverse.

Specifically, the first three chapters are devoted to the creation of appropriate circuit and mathematical models of semiconductor devices, principally the junction diode, the bipolar junction transistor, and the junction and insulated gate field-effect transistor. There are two complementary approaches to such modeling: (1) the creation of linear and nonlinear mathematical and circuit models derived from mathematical representation of the internal physical processes and use of the structural parameters inherent in these processes to model various aspects of these processes and (2) derivation of electrical circuit models from an analysis of the terminal characteristics of devices expressed in graphical form without reference to the physical processes involved. Models, sometimes referred to as linear, incremental, or small-signal, may be derived using either approach, with the external device parameters linked to the physical processes through a comparison of two methods. For example the incremental characteristics of bipolar junction transistors, which are usually represented by alternative circuit models known as the h-parameter model, the hybrid-π model, and the T-model can be derived by either method. Each configuration reveals certain aspects of performance better than the others, and the relationships among the models through the various structural parameters is an important aspect of the insight required for creative circuit design that is often neglected in text or reference books but forms a cornerstone of this volume, as these relationships are exploited extensively in later chapters.

Chapter 4 explores the basic properties of single-device amplifiers, using devices in each of their important circuit configurations, through the properties of voltage gain, input impedance, and output impedance, expressed first as general equations that are then simplified to model the most important characteristics including frequency and time-domain responses.

Chapter 5 treats analytically the interconnections of two or more devices in various configurations to form what may be called composite or compound structures whose properties can readily be expressed in terms of the gain, input impedance, and output impedance of the single devices that comprise them, which in turn leads to a much simplified analytical approach, by treating each device as a source or load for the other.

Most amplifiers exhibit the properties of feedback, internal and external, a fact that is stressed repeatedly in Chapters 4 and 5. Feedback in addition, is often introduced to modify drastically overall amplifier characteristics. Therefore it is appropriate to make a thorough and somewhat formal study of feedback characteristics before progressing to the next stage of amplifier development. This is done in Chapter 6 where all of the important properties of feedback are thoroughly explored.

The integrated operational amplifier is created essentially from the basic building blocks of single and composite devices to form a higher-order building block and is treated in Chapter 7 on this basis. Principles of design of op amps for various applications are studied. These include the conventional op amp with restricted open loop bandwidth, the broad-band, or video amplifier, the operational transconductance amplifier, and the micropower op amp, all of which are discussed in detail including applications to active filters, and sinusoidal oscillators as well as the realization of specific circuit elements such as gyrators and impedance converters.

Thus, through Chapter 7, the subject matter of the book deals principally with linear characteristics, and the principal tool of analysis has been linear circuit theory with incremental device models. However, beginning with Chapter 8, nonlinear characteristics and applications dominate the subject matter. Chapter 8 itself analyzes nonlinearities in devices and amplifiers as an extension of the linear amplifier on a power series or harmonic basis. It treats first the undesirable qualities of nonlinearity as distortion and how it may be corrected by balanced circuits and negative feedback; then it treats the exploitation of nonlinear characteristics as a basis for the electronic multiplier and how it in turn forms the basis for nonlinear analog computations; also how it is involved in frequency conversion and modulation and demodulation processes.

Nonlinearity is carried to the ultimate extreme in Chapter 9 where the analog switch is used in waveform generators and in voltage controlled oscillators, which in turn become the essential elements of phase-locked loops. Theory and applications of phase-locked loops are discussed in detail. Also in this chapter, signal sampling techniques and sample-and-hold circuits are applied to the quantization of signal amplitude in preparation for applications to analog-to-digital conversion processes. Properties of bistable circuits important in such processes are also reviewed.

Analog-to-digital and digital-to-analog conversion processes are discussed in Chapter 10 with an emphasis on sampling, quantizing and digitizing of analog data based on specific circuits discussed in the previous chapter, although the reverse processes are covered as well. However the strictly digital signal processing techniques involving the details of logic gates, flip-flops, registers, and memories are not covered in this volume, since they form the core of the companion volume, *Digital Electronic Circuits*. In that volume the same subject of A/D and D/A conversion techniques is also discussed but more from the digital designer's point of view. Thus the final chapters in both volumes are complementary to each other.

This volume may be used in a variety of modes ranging from classroom instruction to self-study and reference use, with users ranging from upper division undergraduate and graduate level students in electrical and computer engineering to practicing engineers and scientists who need more detailed information on various aspects of analog circuits than they may be able to obtain from other sources.

Ideally for instructional use, the students might be either senior or beginning graduate students who have a prior background in semiconductor device theory, linear circuit theory, including steady-state and transient analysis, and at least an elementary survey course in general electronic circuits. For such students the first three chapters might not be covered in detail but referred back to on a need-to-know basis. Used in this way most of the remainder of the material could be covered in a one-semester course. For students less well prepared, the first three chapters may form a substantial portion of the course. In this case it would be reasonable to structure a one-semester course around the first six chapters, with Chapter 7 given a more brief treatment. This would leave Chapters 7, 8, 9, and 10, which includes the details of operational amplifiers, nonlinear and switching circuits, and A/D and D/A converters, to a separate course. However, it is also possible to extend the coverage to the entire book for the students not so well prepared in all areas by a careful selection of topics to be emphasized.

In summary, this volume is intended to present a comprehensive and pedagogically sound analysis of the most important aspects of electronic circuit analysis and design, of use both to the student of various levels of technical maturity and to practicing engineers who need more detailed information on various aspects of analog circuits than they may be able to obtain from other sources.

The author is grateful to many colleagues throughout his professional career with whom the sharing of information and experiences has contributed to his knowledge and his point of view and thus contributed substantially to this book. There are far too many to cite individually. Thanks are also due to editors at Prentice-Hall, Bernard Goodwin, Tim Bozik, and Reynold Rieger, for their positive support and dedicated efforts through the many stages of writing and production. They have been a pleasure to work with.

Particular thanks are due to Bertha Fancher whose skill in typing the final manuscript as well as her patience in carrying some of the material through several stages of preparation has been extremely helpful.

This book could not have been written at all without the advice, patience, understanding, and continuous encouragement of my wife, Ethel. For these things, I express my greatest appreciation.

GLENN M. GLASFORD

Chapter 1

Semiconductors, Junctions, and the Junction Diode

INTRODUCTION

The majority of semiconductor devices of importance in solid-state electronic circuits make use of the electrical properties of junctions of two semiconductor materials with different impurity concentrations or of a metal and a semiconductor. Therefore, as background for the analysis or synthesis of solid-state circuits, it is necessary to understand quantitatively the properties of pure semiconductors and how such properties are modified by the introduction of specific types of impurities. A semiconductor so modified to be made to conduct current primarily by the flow of excess electrons by the addition of a particular type of impurity is referred to as an *n*-type semiconductor, whereas one that is modified by creating electron deficiencies, referred to as holes, and through which conduction takes place by the movement of such holes, which appears like the movement of positive charges, is referred to as a *p*-type semiconductor.

A properly processed *p-n* junction that will conduct readily in one direction with one applied voltage polarity and almost not at all in the other direction with the opposite voltage polarity is referred to as a *rectifying junction*. The control of the movement of charge carriers (electrons or holes) across *p-n* junctions is fundamental to the operation of most semiconductor devices.

The purposes of this chapter may be stated briefly as follows: (1) to review the basic properties of semiconductor materials as based on principles of solid-state physics, (2) to analyze thoroughly the essential properties of semiconductors, and (3) to define analytically the characteristics of the semiconductor diode, which is a specific device making use of the rectifying junction.

The properties of the *p-n* junction and the junction diode are the basis of much more complex devices involving two or more junctions in discrete and integrated circuits. Thus this chapter is an introduction to much of what follows in succeeding chapters, and the relationships developed in it will be referred to extensively.

1.1 SEMICONDUCTOR PROPERTIES

A semiconductor is defined as a material that has an electrical conductivity somewhere between that of materials (usually metals) normally classified as conductors, such as aluminum and copper, and insulators, such as glass and various ceramics.

A metal conducts electric current primarily as a result of the movement of electrons in the conduction band freed from the outer shells of their atoms under the influence of an applied electric field. In a bar or rod, the current density along its axis is given by

$$J = (nq)v \tag{1.1}$$

where n is the electron density (electrons per unit volume), q is the charge of an electron (1.602×10^{-19} coulombs), and v is the velocity of charge carriers, referred to as drift velocity. The (nq) product is the charge density. The drift velocity is proportional to the applied electric field intensity and is given by $v_d = \mu_n\mathscr{E}$, where μ_n is defined as the electron mobility. Then the current density is

$$J = n\mu_n q\mathscr{E} = \sigma\mathscr{E} \tag{1.2}$$

where σ is defined as the conductivity.

In Eq. 1.2, if $\mathscr{E}$ is in units of volts per meter (V/m), q is in coulombs (C), n is electrons per cubic meter (m^3), and μ_n in square meters per volt-second (m^2/V-s), the current density is in units of coulombs per square meter (C/m^2) $\times$ (1/t) or amperes per square meter (A/m^2).

The total current through a cross section of area A in square meters is $I = JA$ A. These relationships lead to Ohm's law:

$$I = \frac{V}{R} = \frac{\sigma AV}{L} \tag{1.3}$$

where L is the length of the conducting element and R is defined as its resistance in ohms (Ω), given by $R = L/\sigma A$.

For a sheet of material of length L and area $A = WT$, where W is the width and T its thickness, the resistance can be expressed as

$$R = \frac{L}{W}\left(\frac{1}{\sigma T}\right)$$

The quantity $1/\sigma T$ is referred to as the *sheet resistance*, which is the resistance of any size of square sheet of the material and is designated by $R_\square$, usually expressed in units of *ohms per square*.

In a semiconductor, electron current is also given by Eq. 1.2. However, the mobility will be found to be a much smaller number. Typical semiconductors are germanium and silicon, each having four valence electrons. Each valence electron in each atom is shared by one of its four adjacent atoms, and the structure of the material is represented conceptually as a tinker-toy-like array called a lattice of what are called covalent bonds, with the entire structure referred to as a crystal. At temperatures near absolute zero, all electrons remain tightly bound and conductivity is near zero. As the temperature is increased, some of the covalent bonds are broken because of thermal energy supplied to the crystal, and electrons acquire sufficient energy to move from the valence band to the conducting band and, thus freed, can move through the crystal, which accounts for a finite mobility. In addition, for each freed electron from an atom, there exists an electron deficiency called a *hole,* which gives the atom a net positive charge. Thus as an electron may move from an adjacent atom to fill the hole, another hole is created by the departed electron. This movement of holes is equivalent to the movement of positive charges, which accounts for an additional component of current referred to as hole conduction. Hence the total current density at any point in a semiconductor is given by

$$J = \mu_n nq\mathcal{E} + \mu_p pq\mathcal{E} \tag{1.4}$$

where p is the hole density and μ_p is defined as the mobility of holes, which is in general considerably smaller than the corresponding electron mobility. In a pure semiconductor, there are an equal number of holes and electrons, each referred to as the intrinsic carrier concentration $n = p = n_i$, which is usually expressed as

$$n_i^2 = pn \tag{1.5}$$

The intrinsic carrier concentration is an increasing function of temperature in a complex manner and is proportional to $T^{3/2}$. It is often specified at a common reference temperature. For example, silicon at 300 K has an intrinsic carrier concentration of $n_i \cong 1.45 \times 10^{10}/\text{cm}^3$.

Impurities in Semiconductors

The electrical properties of semiconductors are modified drastically by atoms of other materials, referred to as *impurities,* introduced into the structure, a process called *doping.* For example, adding only a very small percentage of material having more outer shell electrons contributes additional electrons to increase the electrical conductivity. Usually, pentavalent atoms such as arsenic or antimony are introduced into silicon or germanium, which contributes one free electron for each atom introduced. Such atoms are called *donors,* and the electron density of such donors is symbolized as N_D. Usually, $N_D \gg n_i$ and $n \cong N_D$. For example, silicon contains approximately 5×10^{22} atoms/cm³ and hence, with $n_i = 1.45 \times 10^{10}/\text{cm}^3$, contains approximately 2.9×10^{-13} free carriers per atom, while each impurity atom contributes one free carrier. The introduction of pentavalent atoms of $N_D = 5 \times 10^{16}$ atoms/cm³, which is $(5 \times 10^{16})/(5 \times 10^{22})$ or one impurity atom per 10^6 silicon atoms, yields $N_D \cong 3.45 \times 10^6\ n_i$.

Adding trivalent atoms such as boron or indium creates electron deficiencies or holes into which electrons from the parent atoms are free to move, thus drastically increasing hole conduction. Such impurity atoms are called *acceptors,* and the acceptor density is N_A. Usually, $N_A \gg n_i$; hence $p \cong N_A$.

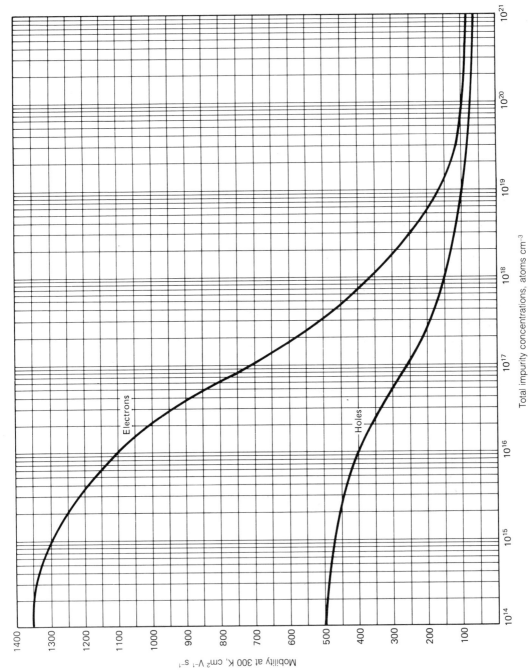

Figure 1.1 Electron and hole mobilities in doped silicon semiconductors.

A doped semiconductor in which the impurities are predominately acceptor atoms is called a p-type semiconductor, and one in which donor atoms predominate is called an n-type semiconductor.

From the energy band theory based on Fermi–Dirac statistics, the laws of statistical mechanics, and the density of state functions derived from quantum mechanics, it can be shown that Eq. 1.5 is equally valid for doped semiconductors.

In the intrinisic semiconductor, the numbers of holes and electrons are equal, and the atoms are on the average electrically neutral. In a doped semiconductor, electrical neutrality is still preserved in the aggregate and, in general, assuming that the acceptors and donors are fully ionized, that is, a negative charge on all acceptors and a positive charge on all donors,

$$(N_A + n) = (N_D + p)$$

which, using $p = n_i^2/n$, can be written as

$$n = \frac{(N_D - N_A) + \sqrt{(N_D - N_A)^2 + 4n_i^2}}{2}$$

For an n-type semiconductor, $N_D >> N_A$ and $N_D >> n_i$; so $n \cong N_D$. Then, for an n-type semiconductor,

$$p \cong \frac{n_i^2}{N_D} \tag{1.6}$$

Similarly, for a p-type semiconductor,

$$n \cong \frac{n_i^2}{N_A} \tag{1.7}$$

The conductivity of a semiconductor does not increase in proportion to the increase in free carriers as a result of doping because carrier mobility is a decreasing function of impurity concentration as a result of random scattering of free electrons due to lattice vibrations, which result from energy imported at elevated temperatures, which impedes the motion of free carriers. The mobilities of holes and electrons at 300 K are shown in Fig. 1.1 for doped silicon.

The impurity concentration in Fig. 1.1 is total impurity concentration $N_D + N_A$. However, usually only donors or acceptors are present, and the electron or hole mobilities are as shown for whichever impurities are present (i.e., electron mobility is given for free electrons in either a p- or n-type semiconductor).

Diffusion Current

Drift current in a semiconductor due to an applied electric field as given by Eq. 1.4 is not the only component of current if there is a gradient of free carriers dn/dx or dp/dx along the axis of conduction, as indicated in Fig. 1.2. The additional components are generated by the diffusion process and are referred to as diffusion currents. The total current densities along the axis are defined by the equations

$$J_{px} = q\mu_p p \mathscr{E}_x - qD_p \frac{dp}{dx} \qquad (1.8)$$

$$J_{nx} = \underbrace{q\mu_n n \mathscr{E}_x}_{\text{drift}} + \underbrace{qD_n \frac{dn}{dx}}_{\text{diffusion}} \qquad (1.9)$$

where D_n and D_p are called diffusion constants for electrons and holes, respectively, and $\mathscr{E}_x$ is the component of the electric field in the positive x direction.

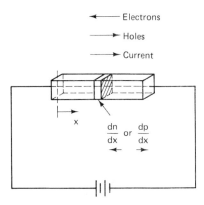

Figure 1.2 Charge carrier flow in a semiconductor bar.

Diffusion constants are related to mobilities by the Einstein relationship, as derived from statistical mechanics, given by

$$\frac{D_n}{\mu_n} = \frac{D_p}{\mu_p} = \frac{kT}{q} \qquad (1.10)$$

where T = temperature in kelvins (K), $q = 1.602 \times 10^{-19}$ C, and k is Boltzmann's constant, given by 1.38×10^{-23} J/K. This equation permits Eqs. 1.8 and 1.9 to be expressed alternatively in terms of mobility or diffusion constant alone.

The potential ϕ at a distance x relative to $x = 0$ is defined by

$$\phi = -\int_0^x \mathscr{E}_x d_x \qquad (1.11)$$

or conversely

$$\mathscr{E}_x = -\frac{d\phi}{dx} \qquad (1.12)$$

An important equation used in the determination of voltage–current relationships in semiconductors is Gauss's law, which is expressed in one-dimensional form as

$$\frac{d\mathscr{E}}{dx} = \frac{\rho_x}{\varepsilon_o \varepsilon_r} \qquad (1.13)$$

where ρ_x is the charge density at the point, ε_o is the permittivity of vacuum (8.854×10^{-14} F/cm), and ε_r is the relative dielectric constant for the particular semiconductor material (11.7 for silicon).

An alternative version known as Poisson's equation and obtained from Eqs. 1.13 and 1.12 is

$$\frac{d^2\phi}{dx^2} = \frac{-\rho_x}{\varepsilon_o \varepsilon_r} \tag{1.14}$$

Another important relationship is the *continuity equation,* which relates the time rate of change of charge carriers passing through an element area as indicated in Fig. 1.2 to the current density flow and the charge carriers lost through a process called recombination. For hole conduction,

$$q \frac{\partial p}{\partial t} = -q \frac{p - p_o}{\tau_p} - \frac{\partial J_p}{\partial x} \tag{1.15}$$

The time constant τ_p is defined as hole lifetime, which is the average time a charge carrier exists in a region without being lost by recombination with an electron or "trapped" within the lattice structure.

The term $p - p_o = p'$ is the *excess carrier concentration* at the point that is above the equilibrium value and that arises from externally induced processes.

A similar equation for negative charge carriers (electrons) is

$$q \frac{\partial n}{\partial t} = q \frac{n - n_o}{\tau_n} + \frac{\partial J_n}{\partial x} \tag{1.16}$$

where τ_n is the electron lifetime.

The lifetime τ is a somewhat indeterminate quantity given by an equation of the general form

$$\tau = \frac{1}{\sigma v_{th} N_T}$$

where σ is referred to as the capture cross section, N_T is the density of recombination centers, and v_{th} is the thermal velocity, given approximately by $v_{th} = \sqrt{3kT/m}$ (which is approximately 10^7 cm/s at 300 K); hence the lifetime is roughly proportional to $T^{-1/2}$.

The foregoing equations are special cases of electrostatic field equations where current flow is confined to a single direction. The electric field-potential relationship, Gauss's law, Poisson's equation, and the continuity equation are all vector quantities, which can be used in their more general form for multidimensional solutions.

1.2 SEMICONDUCTOR JUNCTION PROPERTIES

If p- and n-type semiconductors are joined by an almost perfect contact that preserves the integrity of the crystal lattice structure across the boundary (except for occasional defects), the result is called a *pn* junction, as indicated symbolically in Fig. 1.3(a).

It might be expected that an electric field would exist in the vicinity of the junction as a result of the diffusion of electrons across the boundary into the p region and holes across the boundary into the *n* region. Since an excess of holes would recombine with electrons on the *n* side, it would be expected that on the *n* side there would be left atoms with a net positive charge, and, similarly, by recombination of electrons with holes on the p side, atoms would be left on the *p* side with a net negative charge, as indicated in Fig. 1.3(b). These charged immobile atoms or ions are sometimes referred to as *uncovered charges* since they are not neutralized by free carriers. Thus a narrow region in the vicinity of the junction is essentially depleted of mobile holes and electrons and is referred to as the depletion region, space charge region, or sometimes the transition region.

Whatever the exact charge distribution, with no external excitation no current can flow (a situation referred to as equilibrium conditions), and from Eqs. 1.8 and 1.9, along with Eq. 1.10, the electric field across the junction can be expressed as

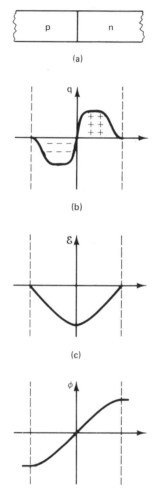

(a)

(b)

(c)

(d)

Figure 1.3 (a) Electrical properties at a semiconductor junction. (b) Charge distribution. (c) Electrical field. (d) Potential distribution.

$$\mathcal{E}_x = -\frac{kT}{q}\frac{1}{n_o}\frac{dn_o}{dx} \tag{1.17}$$

or

$$\mathcal{E}_x = \frac{kT}{q}\frac{1}{p_o}\frac{dp_o}{dx} \tag{1.18}$$

where n_o and p_o identify equilibrium values of carrier concentrations.

This field exists only in the vicinity of the region of uncovered charges that are immediately adjacent to the junction, as indicated in Fig. 1.3(c). As a result, a potential difference exists across the junction as indicated by Eq. 1.11 and shown in Fig. 1.3(d). In general, the potential difference between two points transverse to the junction using Eq. 1.17 or 1.18 in Eq. 1.11 can be written as

$$\phi_x - \phi_{-x} = \frac{kT}{q}\int_{-x}^{x}\frac{1}{n_o}\left(\frac{dn_o}{dx}\right)dx = \frac{kT}{q}\ln n_o\,\Big|_{-x}^{x} \tag{1.19}$$

or

$$\phi_x - \phi_{-x} = -\frac{kT}{q}\int_{-x}^{x}\frac{1}{p_o}\left(\frac{dp_o}{dx}\right)dx = \frac{kT}{q}\ln p_o\,\Big|_{-x}^{x} \tag{1.20}$$

The total potential across the junction incorporating all uncovered charges is

$$\phi_B = \phi_n - \phi_p = \frac{kT}{q}\int_{p}^{n}\frac{dn_o}{n_o} = \frac{kT}{q}\ln n_o\,\Big|_{p}^{n} = \frac{kT}{q}\ln\frac{n_{no}}{n_{po}} \tag{1.21}$$

or

$$\phi_B = \phi_n - \phi_p = -\frac{kT}{q}\int_{p}^{n}\frac{dp_o}{p_o} = -\frac{kT}{q}\ln p_o\,\Big|_{p}^{n} = \frac{kT}{q}\ln\frac{p_{po}}{p_{no}} \tag{1.22}$$

The solution of these equations for ϕ and an explicit determination of depletion layer width depends on certain simplifying assumptions. Two specific cases will be considered as follows.

Uniform Impurity Distribution: Abrupt Junction, $N_D(x) = N_D$, $N_A(-x) = N_A$

For this condition, as indicated in Fig. 1.4(a), the respective charge distributions are assumed to be almost uniform to a distance d_n in the n region and d_p in the p region, as indicated in Fig. 1.4(b), with the charge densities being approximately the density of the impurity atoms in the regions (i.e., $p_{no} \cong N_D$ and $n_{po} \cong N_A$). The corresponding electric field varies almost linearly, as indicated in Fig. 1.4(c). Then, inasmuch as a charge balance must exist on either side of the junction, the condition should be met that

$$qAN_Dd_n = qAN_Ad_p \tag{1.23}$$

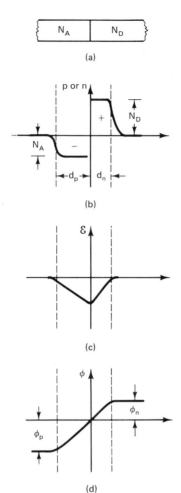

Figure 1.4 Electrical characteristic at a *p-n* junction. (a) Uniform impurity concentrations. (b) Depletion layer charge. (c) Electric field. (d) Built-in potential.

This relationship, known as the *depletion approximation*, assumes that $n \simeq N_D$ and $p \simeq N_A$ for $N \gg n_i$. Then, using the relationship $pn = n_i^2$ together with Eq. 1.21 or 1.22, the total built-in voltage can be obtained as

$$\phi_B = \frac{kT}{q} \ln \frac{N_A N_D}{n_i^2} \tag{1.24}$$

or

$$\phi_B = \phi_n - \phi_p = \frac{kT}{q} \ln \frac{N_A}{n_i} + \frac{kT}{q} \ln \frac{N_D}{n_i} \tag{1.25}$$

which defines the potentials in the *n* and *p* regions separately, as indicated in Fig. 1.4(d).

The relationship between ϕ_B and depletion layer widths d_n and d_p may be determined separately. Two successive integrations of Poisson's equation (Eq. 1.14) using N_A and

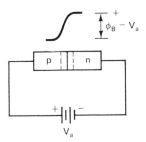

Figure 1.5 Junction potential distribution with an applied potential.

N_D for charge density ρ, noting that $d\phi/dx = 0$ at d_p and d_n and $\phi_B = 0$ at $x = 0$, and using Eq. 1.23 permits ϕ_B to be expressed in terms of depletion layer widths as

$$\phi_B = \phi_n - \phi_p = \frac{q}{2\varepsilon_o\varepsilon_r}(N_A d_p^2 + N_D d_n^2) \tag{1.26}$$

This relationship may be extended to include the application of an external voltage, V_a, as indicated in Fig. 1.5. Since there is no electric field outside the depletion region, the total net voltage appears across the depletion region; that is, the junction potential is given by

$$V_j = \phi_B - v_a \tag{1.27}$$

If v_a is positive, $V_j \rightarrow 0$ as $v_a \rightarrow \phi_B$, and for v_a negative, $V_j = \phi_B + |v_a|$. Thus, with an applied voltage, Eq. 1.26 is still valid using $\phi_B - v_a$ instead of ϕ_B alone. Then, using Eq. 1.23,

$$d_n = \left(\frac{N_A}{N_A N_D + N_D^2}\frac{2\varepsilon_o\varepsilon_r}{q}\right)^{1/2}(\phi_B - v_a)^{1/2} \tag{1.28}$$

$$d_p = \left(\frac{N_D}{N_A N_D + N_A^2}\frac{2\varepsilon_o\varepsilon_r}{q}\right)^{1/2}(\phi_B - v_a)^{1/2} \tag{1.29}$$

with the total depletion width given by

$$d = d_n + d_p = \left[\frac{1}{N_D} + \frac{1}{N_A}\right]^{1/2}\left(\frac{2\varepsilon_o\varepsilon_r}{q}\right)^{1/2}(\phi_B - v_a)^{1/2} \tag{1.30}$$

Nonuniform Impurity Distribution: Linearly Graded Junction

The depletion approximation that led to separate equations for junction potential and depletion layer widths, while not exact for the ideal abrupt junction, is even less accurate when there is a graded change of impurities across the junction, which, however, is a more structurally realizable situation. Practical diffusion processes of junction formation lead to an approximation of a linearly graded junction as indicated in Fig. 1.6, where it is assumed that the gradation extends beyond the widths for which the region is depleted of mobile carriers.

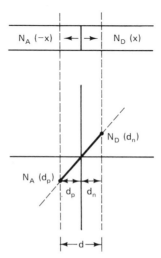

Figure 1.6 Depletion region of a linearly graded junction.

The region outside the depletion layer may or may not be uniform, which is a separate issue to be discussed later.

The slope of the impurity gradation is defined by $m = N(x)/x$ in the vicinity of the junction, the charge density $\rho = qmx$, and

$$\frac{d^2\phi}{dx^2} = -\frac{qmx}{\varepsilon_o\varepsilon_r} \tag{1.31}$$

which, from two successive integrations using appropriate constants of integration, is

$$\phi_B = \frac{qmd^3}{12\varepsilon_o\varepsilon_r} \tag{1.32}$$

or

$$d = \left(\frac{12\varepsilon_o\varepsilon_r\phi_B}{qm}\right)^{1/3} \tag{1.33}$$

This equation is valid as well when an applied voltage is included and may be written as

$$d = \left(\frac{12\varepsilon_o\varepsilon_r(\phi_B - v_a)}{qm}\right)^{1/3} \tag{1.34}$$

Going back to Eq. 1.19 or 1.20 using $n_{po} = N_A(-x) = -mx$ and $p_{no} = N_D(x) = +mx$, along with $pn = n_i^2$, another equation for ϕ_B is

$$\phi_B = \frac{kT}{q} \ln\left(\frac{md_o}{2n_i}\right)^2 \tag{1.35}$$

which may be written as

$$d_o = \frac{2n_i}{m} e^{(q\phi_B/2kT)} \tag{1.36}$$

where d_o is the zero-bias value for depletion layer width.

Since Eqs. 1.33 and 1.36 are independent expressions, they can be solved simultaneously for ϕ_B and d_o separately (see Problem 1.6).

1.3 VOLTAGE–CURRENT RELATIONSHIPS AT SEMICONDUCTOR JUNCTIONS

If a voltage is applied to the *pn* junction having *p* and *n* regions of finite width, as indicated in Fig. 1.7(a), the depletion layer width varies with the resultant junction voltage as discussed in the foregoing section, and the effective widths of the *p* and *n* regions are

$$W'_p = W_p - d_p$$

$$W'_n = W_n - d_n$$

It would be expected under forward-bias voltage (i.e., V_a positive) that with the junction voltage $V_j < \phi_B$ a flow of charge carriers across the boundary would be aided, which would result in a continuous current flow in the direction shown and dependent on v_a. The solution for the resultant voltage–current relationship would be expected to be a function of the impurity distribution throughout the W'_p and W'_n regions, as well as through the depletion region, *d*. However, some general relationships can be established. It would be expected that equations like Eq. 1.21 and 1.22 could be written as

$$\phi_B - v'_a = \frac{kT}{q} \ln \frac{n_n}{n_p} \tag{1.37}$$

$$\phi_B - v'_a = \frac{kT}{q} \ln \frac{p_p}{p_n} \tag{1.38}$$

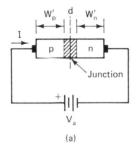

(a)

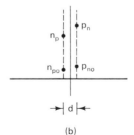

(b)

Figure 1.7 Charge carriers at depletion layer boundaries. (a) Applied voltage. (b) Excess carriers at depletion layer boundaries.

where $\phi_B - v'_a$ is the total potential across the depletion region, n_n and n_p are total negative charge concentrations at the depletion boundary of the n and p regions, respectively, and p_p and p_n are positive charge concentration at the respective depletion boundaries. These are total values rather than equilibrium values, as indicated in Fig. 1.7(b). If it is assumed that no potential variation exists in the region outside the depletion layer or at the external contacts, $v_a = v'_a$, and Eq. 1.37 can be written as

$$n_p = n_n e^{-q(\phi_B/kT)} e^{q(v_a/kT)} \tag{1.39}$$

But for $v_a = 0$,

$$n_p = n_{po} = n_n e^{-q(\phi_B/kT)}$$

where n_{po} is the equilibrium value of the negative charge concentration at the edge of the p region; hence

$$n_p = n_{po} e^{qv_a/kT} \tag{1.40}$$

In addition, we may define the excess carrier concentration at the boundary as

$$n'_p = n_p - n_{po} \tag{1.41}$$

Then from Eqs. 1.40 and 1.41

$$n'_p = n_{po}[e^{q(v_a/kT)} - 1] \tag{1.42}$$

Similarly, starting with Eq. 1.38,

$$p'_n = p_{no}[e^{q(v_a/kT)} - 1] \tag{1.43}$$

where $p'_n = p_n - p_{no}$ defined at the depletion boundary of the n region.

Solution for Uniform Impurity Distribution

For the case of $N_D(x) = N_D$ and $N_A(-x) = N_A$, it may be noted that, in Eq. 1.42, $n_{po} \cong n_i^2/N_A$ and, in Eq. 1.43, $p_{no} \cong n_i^2/N_D$. Hence

$$n'_p = \frac{n_i^2}{N_A} [e^{q(v_a/kT)} - 1] \tag{1.44}$$

and

$$p'_n = \frac{n_i^2}{N_D} [e^{q(v_a/kT)} - 1] \tag{1.45}$$

Electron current flows in the p region and hole current in the n region as a result of the excess carriers n'_p and p'_n produced at the depletion layer boundaries. These currents are pure diffusion currents in the p and n regions outside the depletion layer; hence the electric field $\mathscr{E} = 0$ except through the depletion layer. Thus, from Eqs. 1.8 and 1.9,

$$I_{pn} = -qAD_{pn} \frac{dp'_n(x')}{dx'_n}, \qquad x'_n = x - d_n \tag{1.46}$$

and

$$I_{np} = +qAD_{np} \frac{dn'_p(x')}{dx'_p}, \qquad x'_n = x + d_p \tag{1.47}$$

These are hole currents in the n region and electric currents in the p region, respectively, D_{np} and D_{pn} are diffusion constants for electrons in the p region and holes in the n region, respectively, and $n'_p(x)$ and $p'_n(x)$ are excess carrier concentrations through the p and n regions.

Since the currents are constant through the regions, that is,

$$\frac{dI_n\,(x')}{dx'_p} = 0 \quad \text{and} \quad \frac{dI_p\,(x')}{dx'_n} = 0$$

therefore

$$qAD_{pn}\frac{d^2 p'_n\,(x)}{dx'^2_n} = 0 \quad \text{and} \quad qAD_{np}\frac{d^2 n'_p\,(x')}{dx'^2_p} = 0$$

These conditions and the requirement that $p'_n\,(x') = 0$ at W'_n and $n'_p\,(-x') = 0$ at W'_p yield a solution for excess carrier concentrations through the regions.

$$p'_n\,(x') = p'_n\,(0)\left(1 - \frac{x'}{W'_n}\right) \tag{1.48}$$

and

$$n'_p(-x') = n'_p\,(0)\left(1 + \frac{x'}{W'_p}\right) \tag{1.49}$$

as shown in Fig. 1.8 for $N_A > N_D$.

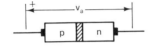

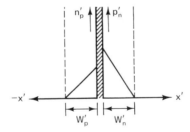

Figure 1.8 Excess carrier distribution in uniformly doped regions.

These are straight line distributions of excess carriers in uniformly doped regions, with the slopes given by

$$\frac{dp'_n\,(x)}{dx'_n} = -\frac{p'_n\,(0)}{W'_n} \tag{1.50}$$

$$\frac{dn'_p\,(x')}{dx'_p} = \frac{n'_p\,(0)}{W'_p} \tag{1.51}$$

The currents throughout the n and p regions are given by those determined at the depletion layer boundaries and are given by Eqs. 1.46 and 1.47, using Eqs. 1.50 and 1.51 along with Eqs. 1.44 and 1.45. These are

$$i_p = \frac{qAD_{pn}p_n'(0)}{W_n'} = \frac{qAD_{pn}n_i^2}{W_n'N_D}[e^{q(v_a/kT)} - 1] \qquad (1.52)$$

$$i_n = \frac{qAD_{np}n_p'(0)}{W_p'} = \frac{qAD_{np}n_i^2}{W_p'N_A}[e^{q(v_a/kT)} - 1] \qquad (1.53)$$

The total current crossing the depletion layer is made up of $i_p + i_n$; hence

$$i = qAn_i^2\left(\frac{D_{pn}}{N_DW_n'} + \frac{D_{np}}{N_AW_p'}\right)[e^{q(v_a/kT)} - 1] \qquad (1.54)$$

The current–voltage relationship is shown in Fig. 1.9, increasing rapidly for positive values of v_a and reaching an asymptotic value

$$i]_{v_a = -\infty} = -qAn_i^2\left(\frac{D_{pn}}{N_DW_n'} + \frac{D_{np}}{N_AW_p'}\right) \qquad (1.55)$$

The magnitude of this value is sometimes referred to as the reverse-bias saturation current I_o.

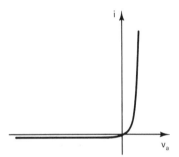

Figure 1.9 Junction voltage–current characteristic.

The development just concluded was based not only on homogeneous impurity distributions in the crystal but on the assumption that no electron–hole recombinations occur as minority carriers diffuse through their respective regions.

Effect of Hole–Electron Recombinations

Injected minority carriers into a p or n region recombine with majority carriers at a somewhat predictable rate. On the average, carriers will remain without recombination for a finite period of time, τ_p for holes and τ_n for electrons. These are referred to as *carrier lifetimes,* which appeared in the continuity equations, Eqs. 1.15 and 1.16. Because of such continuous recombinations, it would be expected that the minority carrier density given by Eq. 1.48 for electrons and Eq. 1.49 for holes would decrease with distance from the junction at a faster rate approximated by an exponential decay factor, and that Eq. 1.48 for excess hole distribution might be modified by this factor and written as

$$p_n'(x') = p_n(0)\left(1 - \frac{x'}{W_n'}\right)e^{-(x'/L_p)} \qquad (1.56)$$

where L_p is referred to as the diffusion length.

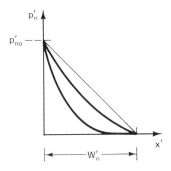

Figure 1.10   Excess carrier distributions resulting from hole–electron recombinations.

Various distributions are shown in Fig. 1.10. For $L_p \ll W_n'$,

$$p_n'(x') = p_n'(0)e^{-(x'/L_p)}$$

This is plotted as the lower curve. For $L_p \gg W_n'$,

$$p_n'(x) = p_n'(0)\left(1 - \frac{x'}{W_n'}\right)$$

which is the straight line distribution previously determined when recombinations were neglected. The middle curve is for L_p and W_n' at roughly comparable values.

The hole current throughout the n region by taking $dp'(x')/dx_n'$ from Eq. 1.56 and using it in Eq. 1.46 yields

$$i_p = qAD_{pn}p_n'(0)\left(\frac{1}{L_p} - \frac{x'}{W_n'L_p} + \frac{1}{W_n'}\right)e^{-(x'/L_p)} \tag{1.57}$$

Similarly, the electron current in the p region is

$$i_n = qAD_{np}n_p'(0)\left(\frac{1}{L_n} + \frac{x'}{W_p'L_n} + \frac{1}{W_p'}\right)e^{x'/L_n} \tag{1.58}$$

These currents decay exponentially with distance from the junction. However, the total current is constant because the majority carriers increase to match the decrease in minority carriers to preserve charge neutrality. Hence the total current can be obtained from the hole and electron currents at the depletion layer boundaries as before and is given by

$$i_p\bigg|_{x'=0} = qAD_{pn}p_n'(0)\left(\frac{1}{L_{pn}} + \frac{1}{W_n'}\right) \tag{1.59}$$

$$i_n\bigg|_{x'=0} = qAD_{np}n_p'(0)\left(\frac{1}{L_{np}} + \frac{1}{W_p'}\right) \tag{1.60}$$

For the case of $W_n' \ll L_p$ and $W_p' \ll L_p$, the resultant current is the same as that given by Eq. 1.54, but for $L_p \ll W_n'$ or $L_n \ll W_p'$,

$$i = qAn_i^2\left(\frac{D_{pn}}{N_DL_p} + \frac{D_{np}}{N_AL_n}\right)[e^{q\,(v_a/kT)} - 1] \tag{1.61}$$

In this case,

$$p'_n(x') \cong p'_n(0)e^{-(x'/L_p)}$$

and

$$\frac{d^2p'_n(x')}{dx_p^2} = \frac{1}{L_p^2} p'_n(x')$$

which may be written as

$$D_{pn} \frac{d^2p'_n(x')}{dx'^2} = \frac{p'_n(x')}{\tau_p} \tag{1.62}$$

where $\tau_p = L_p^2/D_{pn}$ is the carrier lifetime, which is related to diffusion length by $L_p = \sqrt{D_p\tau_p}$.

It is interesting to note that Eq. 1.62 is a special case of the continuity equation, Eq. 1.15, for steady-state conditions (i.e., $\partial p/\partial t = 0$) and that the preceding equations that were deduced could have been derived directly from the continuity equation.

Solution for Nonuniform Impurity Distribution

It was assumed in the previous discussion that Eqs. 1.17 and 1.18 established for zero current flow in Eqs. 1.8 and 1.9 were approximately valid across the depletion layer even when a voltage (with a resultant current) was applied using N_A and N_D as n_o and p_o if the free carriers in the depletion region were a very small fraction of the immobile ions. Furthermore, it was assumed that the currents outside the depletion region were entirely diffusion currents because there was no field in the case of uniformly doped semiconductors. However, if the impurity distribution is not uniform, there is a field throughout the regions, both drift and diffusion currents are present, and the currents include both components of Eqs. 1.8 and 1.9. Nevertheless, the approximation can still be made that the field established in terms of carrier distribution is valid even when current flows using n or n' in Eq. 1.17 and p or p' in Eq. 1.18. Hence Eqs. 1.8 and 1.9 incorporating Eq. 1.10 along with Eqs. 1.17 and 1.18 using $n \cong N_D$ and $p \cong N_A$, with $I = qAJ$, can be written as

$$i_p = -qAD_{pn} \left[\frac{dp'_n(x')}{dx'} + \frac{p'_n \, dN_D(x')/dx'}{N_D(x')} \right] \tag{1.63}$$

$$i_n = qAD_{np} \left[\frac{dn'_p(x')}{dx'} + \frac{n'_p \, dN_A(x')/dx'}{N_A(x')} \right] \tag{1.64}$$

In these equations, x' is the distance measured from the edge of the respective depletion regions; that is, $x' = x - d_n$ or $x' = x + d_p$ with $x = 0$ defined at the junction.

If the p and n regions are sufficiently narrow that recombinations can be neglected, i_n and i_p are constant through their respective regions, and solutions for the preceding equations are

$$p'_n(x') = \frac{I_p}{qAD_{pn}} \frac{\displaystyle\int_x^{W'_n} N_D(x')\,dx'}{N_D(x')} \tag{1.65}$$

$$n'_p(x') = \frac{I_n}{qAD_{np}} \frac{\displaystyle\int_{-x}^{W'_p} N_A(x')\,dx'}{N_A(x')} \tag{1.66}$$

These equations can be rewritten, using $p'_n(0)$ and $n'_p(0)$ at the depletion layer boundaries, as

$$i_p = \frac{qAD_{pn}N_D(0)p'_n(0)}{\displaystyle\int_0^{W'_n} N_D(x')\,dx'} = \frac{qAD_{pn}n_i^2}{\displaystyle\int_0^{W'_p} N_D(x')\,dx'}\,[e^{q(v_a/kT)} - 1] \tag{1.67}$$

$$i_n = \frac{qAD_{np}N_A(0)n'_p(0)}{\displaystyle\int_0^{W'_p} N_A(x')\,dx'} = \frac{qAD_{np}n_i^2}{\displaystyle\int_0^{W'_p} N_A(\mathrm{x}')\,dx'}\,[e^{q(v_a/kT)} - 1] \tag{1.68}$$

since Eqs. 1.44 and 1.45 are valid at the depletion layer boundaries.

Adding these two components, the total current is

$$i = qAN_i^2\left[\frac{D_{pn}}{\displaystyle\int_0^{W'_n} N_D(x')\,dx'} + \frac{D_{np}}{\displaystyle\int_0^{W'_p} N_A(x')\,dx'}\right][e^{q(v_a/kT)} - 1] \tag{1.69}$$

For uniform impurity distributions, this equation reverts to Eq. 1.54 as established previously.

These equations are valid only for narrow p and n regions with $W_n \ll L_p$ and $W_p \ll L_n$, which neglects recombinations. Also, they assume that D_n and D_p are constants, where Fig. 1.1 shows that they are functions of impurity concentrations. However, the variations in D_n and D_p are relatively small by comparison, and average values for $\overline{D}_n$ and $\overline{D}_p$ may be used with relatively small error.

1.4 METAL–SEMICONDUCTOR CONTACTS

The interface of a conducting metal and a semiconductor exhibits varying electrical characteristics depending on the nature of the contact properties of the specific metal, the concentration of impurities in the semiconductor, and whether it is p or n type. The contact may be a rectifying contact that has essentially the properties of a pn junction, with the effective resistance to current flow high in one direction and low in the other, or an ohmic contact where current can flow readily in either direction with very low contact potential at the interface as compared with the built-in voltage of the pn junction. The mechanics of the metal–semiconductor junction are much more complex than those of the simple pn junction, but some approximate relationships can be established rather simply.

Figure 1.11 Depletion layer at a metal –semiconductor boundary.

A metal and lightly *n*-doped semiconductor boundary is shown in Fig. 1.11. Under open circuit conditions, electrons diffuse into the metal, where they simply join the vast supply of electrons already present, leaving an array of positively charged ions in a depletion region as indicated.

The depletion layer and built-in potential lie entirely within the semiconductor; hence, from Eq. 1.26

$$d_n^2 = \frac{\phi_B - v_a}{qN_D/2\varepsilon_o\varepsilon_r} \tag{1.70}$$

with the built-in voltage being dependent on the work function of the metal.

The current is entirely electron current given by an equation of the general form of Eq. 1.9, which, using Eqs. 1.10 and 1.12, may be written as

$$J_x = qD_n\left(-\frac{qn}{kT}\frac{d\phi}{dx} + \frac{dn}{dx}\right) \tag{1.71}$$

From various energy relationiships and boundary conditions, which can be established, several forms for current density can be approximated. One of these is

$$J_x = \frac{q^2 D_n N_D}{kT}\left[\frac{2q(\phi_B - v_a)N_D}{\varepsilon_o\varepsilon_r}\right]^{1/2} e^{-q(\phi_B/kT)}\,[e^{q(v_a/kT)} - 1] \tag{1.72}$$

The coefficient of $[e^{q(v_a/kT)} - 1]$ is in itself a rather complex function involving ϕ_B and v_a, but it is often approximated by a constant J_0 such that for forward-bias conditions

$$J_x = J_0\,[e^{q(v_a/nkT)} - 1] \tag{1.73}$$

when J_0 can be experimentally determined and n is a constant slightly greater than unity ($1.02 < n < 1.2$).

Such a rectifying junction is referred to as a Schottky junction or Schottky barrier and can be seen to have properties similar to those of the *pn* junction, although the carriers are majority rather than minority carriers.

For much heavier impurity concentrations near the contact, the situation changes dramatically. In Fig. 1.12(a), the symbol n^+ indicates very heavy doping with donors and is not an indication of charge polarity. From Eq. 1.70, the depletion layer width decreases very rapidly with increasing N_D, and at very heavy dopings the depletion layer or barrier width becomes very thin (perhaps a few angstroms). Under such conditions, electrons from the semiconductor to the metal or from the metal into the semiconductor pass directly through the potential barrier by a mechanism known as *tunneling* such that the resistance of the contact and hence the contact potential are very small regardless of the direction of current flow. An analogous situation exists for a very heavily doped *p-*

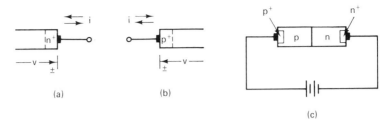

Figure 1.12 Enhanced ohmic contacts using highly doped *p* or *n* regions.

type semiconductor, as indicated in Fig. 1.12(b), where, as the barrier width almost vanishes, electrons can tunnel through the barrier by the same process.

A *pn* junction with enhanced ohmic contacts is shown in Fig. 1.12(c). For the polarity shown, electrons flow readily from the metal contact on the *n* side into the *n* region by tunneling, flow through the *p* region as minority carriers, and tunnel readily into the metal contact on the *p* side. Hence, properly formed external ohmic contacts have negligible effect on the junction *v*–*i* characteristics.

1.5 PN JUNCTION DIODE

The discussion of semiconductor junction properties can best be continued within the framework of specific devices having practical structural and dimensional properties, of which the semiconductor junction diode is an important example. The diode shown symbolically in Fig. 1.13(a) has a mathematical model based on the equation

$$i_D = I_0(e^{v_D/V_T} - 1) \tag{1.74}$$

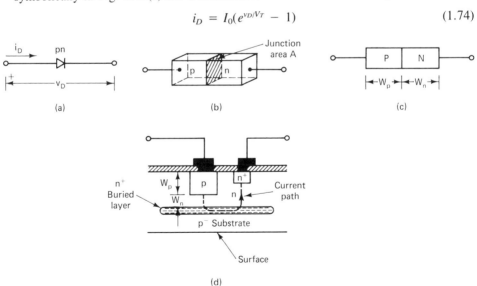

Figure 1.13 *PN* junction diode structures.

where $V_T \cong kT/q \cong 0.026$ V at 300 K, and I_0 is the magnitude of the maximum reverse-bias current given by one of the previously established equations (e.g., Eq. 1.55 for the uniformly doped narrow p and n regions). I_0 is called the reverse-bias saturation current.

The structure may be the symmetrical structure shown in Fig. 1.13(b), usually represented by the cross section of Fig. 1.13(c), but it is more often some variation of the planar structure shown in cross section in Fig. 1.13(d), which permits external connections to be made at a common surface.

For this structure, the critical dimensions are W_p and W_n and the pn junction area. The holes in the n region terminate at the edge of the n^+ buried layer as though it were an external contact, which then provides a high conductivity path along its axis for electrons, which continues through the n region, making an ohmic contact to the outside through the n^+ region at the surface.

Such a heavily doped transition region is not necessary for the p contact because under appropriate conditions the fusion of aluminum and a normal silicon p region itself makes a good ohmic contact.

Assuming that V_T and I_0 are constants, Eq. 1.74 is plotted in normalized form on two different scales in Fig. 1.14. Even for relatively small forward bias, $v_D \gg v_T$ and $i_D/I_0 \cong e^{v_D/V_T}$, and for even relatively small negative bias, $i_D/I_0 \rightarrow -1$.

The two curves of Fig. 1.14 illustrate different modes of diode operation. Scale 1 indicates characteristics for very small values of v_D and i_D, and the nonlinear characteristic

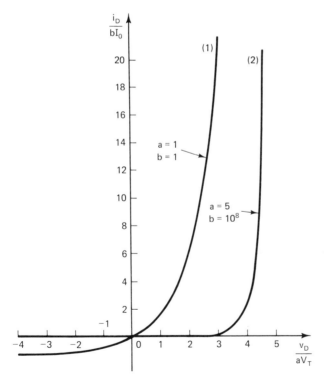

Figure 1.14 Theoretical diode voltage–current characteristic.

Semiconductors, Junctions, and the Junction Diode   Chap. 1

can be used for applications that require such nonlinear characterstics. Scale 2 illustrates a much larger voltage–current range, which suggests a switching-type characteristic where the diode has an extremely large resistance with reverse bias and a very small resistance for forward bias.

Diode Equation Related to Structural Parameters

The application and limitations of the basic relationships that have been established in previous sections when applied to practical devices can be illustrated by the *pn* diode shown in Fig. 1.15, where it is assumed that impurity concentrations are uniform and that W_n and W_p are small compared to their respective diffusion lengths.

The following calculations are made at 300 K ($V_T = 0.026$ V).

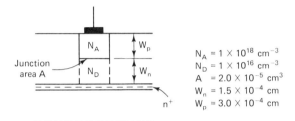

$$N_A = 1 \times 10^{18} \text{ cm}^{-3}$$
$$N_D = 1 \times 10^{16} \text{ cm}^{-3}$$
$$A = 2.0 \times 10^{-5} \text{ cm}^3$$
$$W_n = 1.5 \times 10^{-4} \text{ cm}$$
$$W_p = 3.0 \times 10^{-4} \text{ cm}$$

Figure 1.15 *P-N* diode geometry.

(1) Reverse-bias saturation current from Eq. 1.55 (depletion layer width neglected), with μ_n and μ_p taken from Fig. 1.1 and D_{np} and D_{pn} calculated.

$$I_0 = 1.602 \times 10^{-19} \times 2 \times 10^{-5} \times (1.45 \times 10^{10})^2 \times$$

$$\left(\frac{9.3}{1 \times 10^{18} \times 3.0 \times 10^{-4}} + \frac{10.4}{1 \times 10^{16} \times 1.5 \times 10^{-4}} \right)$$

$$= 6.74 \times 10^{-4} (3.1 \times 10^{-14} + 6.9 \times 10^{-12})$$

$$\cong 6.74 \times 10^{-4} (6.9 \times 10^{-12}) \text{ amperes (A)}$$

$$I_0 \cong 4.65 \times 10^{-3} \text{ pA}$$

(2) Built-in voltage from Eq. 1.25:

$$\phi_B = 0.026 \left(\ln \frac{1 \times 10^{18}}{1.45 \times 10^{10}} + \ln \frac{1 \times 10^{16}}{1.45 \times 10^{10}} \right)$$

$$= 0.026(18.05 + 13.44)$$

$$\phi_B = 0.82 \text{ V}$$

(3) Depletion layer widths in *n* and *p* regions from Eqs. 1.28 and 1.29 at (a) $v_D = 0$, (b) $v_D = 0.72$ V, and (c) $v_d = -10.0$ V:

(a) $d_n = \left[\dfrac{1 \times 10^{18}}{1 \times 10^{18} \times 10^{16} + (1 \times 10^{16})^2} \right.$

$\left. \times \dfrac{2 \times 8.85 \times 10^{-14} \times 11.7}{1.602 \times 10^{-19}} \right]^{1/2} (0.82)^{1/2}$

$= (1 \times 10^{-16} + 1.29 \times 10^7)^{1/2} (0.82)^{1/2}$

$\cong 3.60 \times 10^{-5} (0.82)^{1/2}$

$= 3.26 \times 10^{-5} \text{ cm}$

$d_p = \left[\dfrac{1 \times 10^{16}}{1 \times 10^{18} \times 10^{16} + (1 \times 10^{18})^2} \times 1.29 \times 10^7 \right]^{1/2} (0.82)^{1/2}$

$= (1 \times 10^{-20} \times 1.29 \times 10^7)^{1/2} (0.82)^{1/2}$

$= 3.6 \times 10^{-7} (0.82)^{1/2}$

$= 3.26 \times 10^{-7} \text{ cm}$

(b) $d_n + d_p \cong d_n = (3.60 \times 10^{-5})(0.82 - 0.72)^{1/2}$

$d_n + d_p = d_n = 1.4 \times 10^{-5} \text{ cm}$

(c) $d_n + d_p \cong d_n - (3.6 \times 10^{-5})(0.82 + 10)^{1/2}$

$d_n + d_p \cong d_n = 1.2 \times 10^{-4} \text{ cm}$

(4) Current at large negative voltage; for example, -10 V as in part (c):

$i \cong -qAN_i^2 \dfrac{D_{pn}}{N_D W_n'}, \qquad$ where $W_n' \qquad = W_n - d_n$ at -10 V

$\qquad\qquad\qquad\qquad\qquad\qquad\qquad\qquad = (1.5 - 1.2)10^{-4}$

$= -6.74 \times 10^{-4} \dfrac{10.4}{1 \times 10^{16} \times 0.3 \times 10^{-4}} \qquad = 0.3 \times 10^{-4} \text{ cm}$

$= -2.33 \times 10^{-2} \text{ pA}$

Note: This (at -10 V) is approximately five times the value for I_0 calculated from part (1) assuming no variation in depletion layer width.

(5) Assuming minority carrier lifetimes $\tau_n = \tau_p \cong 1 \times 10^{-6}$ s, verify the validity of using the short-base approximation. For holes in the n region,

$$L_p = \sqrt{\tau_p D_p} = \sqrt{1 \times 10^{-6} \times 10.4} = 32 \times 10^{-4} \text{ cm}$$

For electrons in the p region,

$$L_n = \sqrt{\tau_n D_n} = \sqrt{1 \times 10^{-6} \times 7.3} = 27 \times 10^{-4} \text{ cm}$$

The diffusion lengths L_p or $L_n \gg W_n$ or W_p by a sufficiently large factor that short-base equations represent reasonable approximation but have some error.

Actual Diode Compared with Ideal Diode

The relationship expressed by Eq. 1.74 is generally referred to as the diode equation, and for purposes of discussion it might be referred to as representing the ideal semiconductor diode because it is the result of the more elementary diffusion processes. However, in the example just concluded there was shown to be a large increase in reverse-bias current as a result of depletion layer widening with reverse junction voltage. There are further differences between the ideal diode and most actual diodes, which can be discussed in conjunction with Fig. 1.16 using the dashed curve as the ideal diode.

In the reverse-bias region, the current increases even more rapidly than that predicted by depletion layer widening alone. For example, there are hole–electron pairs generated in a depletion region that would be proportional to the volume of the region and the intrinsic carrier concentration and inversely proportional to a lifetime τ_0 analogous to, but usually greater than, τ_n or τ_p for the neutral regions.

The magnitude of this excess generated current can be shown to be approximately

$$I = \frac{q n_i d A}{2 \tau_o} \tag{1.75}$$

where d is the width of the depletion region.

This excess reverse current can range from a negligible to a dominant factor, depending on the dimensions of a particular diode. For the example just discussed in the preceding section for $v_D = -10$ V, and $d_n = 1.2 \times 10^{-4}$ cm, and τ_o assumed to be 10 μs,

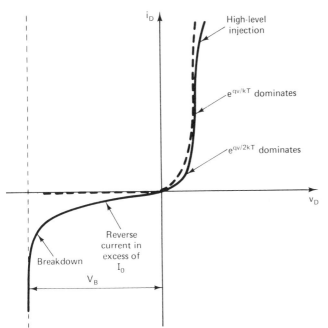

Figure 1.16 Nonideal diode characteristics.

$$I_{\text{gen}} = \frac{1.602 \times 10^{-19} \times 1.45 \times 10^{10} \times 1.2 \times 10^{-4} \times 2 \times 10^{-5}}{2 \times 1.0 \times 1.0^{-5}}$$

$$= 27.8 \times 10^{-14} \text{ A}$$

$$= 27.8 \times 10^{-2} \text{ pA}$$

This depletion layer current for this particular example is approximately an order of magnitude greater than that accounted for by the narrowing of the n region alone.

This generation–recombination current also modifies the forward characteristic. It can be shown that the forward component of the current arising from carrier recombination in the depletion region is approximately

$$i = \frac{qn_idA}{2\tau_o}(e^{qV_a/2kT} - 1) \tag{1.76}$$

At very low voltages, this term might dominate the forward characteristic, so the voltage dependency tends toward the $e^{qv_a/2kT}$ value, whereas at higher voltages the narrowing of the depletion layer reduces its significance, and the $e^{(qv_a/kT)}$ dependency dominates.

To approximate the depletion recombination effect at low voltages in the forward region, the equation is sometimes written as

$$i = I_0(e^{qv_a/nkT} - 1) \tag{1.77}$$

where n is a function of forward voltage; that is, $n \to 2$ at extremely low voltages and $n \to 1$ as the voltage is increased.

Looking again at the large reverse-bias situation, as might be predicted by the previous example, as the reverse bias increases a value will finally be reached where the depletion region widens to encompass the entire more lightly doped region. This voltage is referred to as *punch-through* and causes a dramatic increase in conductivity. However, usually a rapid increase in current occurs at a voltage less than the punch-through value due to a rapid multiplication of free carriers far beyond that predicted by Eq. 1.75, which is so rapid that it is referred to as *breakdown*. There are at least two possible breakdown mechanisms. One is referred to as *zener* breakdown, where for large electric fields in the depletion region covalent bonds will be broken and carriers transported from the valence band on one side of the junction to the other side by the tunneling mechanism. The other is *avalanche* breakdown, which more often occurs before the zener breakdown voltage is reached. In this case, carriers at high velocity in the high field region may have sufficient energy to create additional hole–electron pairs by *impact ionization*, which in turn has a multiplying effect as additional pairs themselves acquire high velocities. The breakdown, by whatever mechanism, is often modeled empirically by a multiplying factor M applied to the reverse current and given by

$$M = \frac{1}{1 - (v/V_B)^n} \tag{1.78}$$

where V_B is the asymptotic value referred to as the breakdown voltage and n is a factor (usually $3 < n < 6$) that defines the rapidity of the onset of breakdown. Although it is not accurate and does not model actual physical mechanisms, it is nevertheless a useful representation of the breakdown phenomenon.

Again, at very large forward currents, low-level injection theory no longer applies (i.e., the minority concentration is no longer small compared to the impurity concentration). As a result, to preserve charge neutrality, the majority carrier concentration rises, which has the same effect as increasing the impurity concentration; this slows down the rate of increase of current with voltage, which can be approximated by using the $e^{qv/nkT}$ factor, where n slowly approaches 2. The contact ohmic resistance and the bulk resistance slow down the rate of increase still more.

The overall characteristic of a semiconductor diode may be modeled by an equation that includes most of the nonideal effects that have been described:

$$i_D = [I_0 f(v_D)][e^{(qv_D/nkT)} - 1] \tag{1.79}$$

where I_0 is the value of reverse-bias saturation current obtained by pure diffusion currents with depletion layer widths neglected, $f(v_D)$ is a nonlinear factor, which includes the base narrowing effect on diffusion current as well as depletion layer carrier generation as a function of depletion layer widening, and n is a forward voltage-dependent factor that models the low voltage depletion layer recombination effect and high-level injection effects ranging with increasing voltage; that is, $n \to 2$ at low currents, $n \cong 1$ at medium currents, and $n \to 2$ again at high injection currents.

1.6 PN JUNCTION DIODE DYNAMICS

There are two separate components involved in the overall time-varying voltage–current relationship in a *pn* junction device: (1) the charge–voltage relationship across the depletion region, and (2) the charge–voltage relationship in the *n* and *p* regions outside the depletion layer. These will be described next.

Depletion Layer

For the depletion layer the effective charge separation as a function of applied voltage as a consequence of the variation of depletion layer width with applied voltage can be shown to be of the general form

$$q_v = K[\phi_B^n - (\phi_B - v_D)^n] \tag{1.80}$$

where K and n are constants that are dependent on the junction area and impurity profiles adjacent to the junction in the *n* and *p* regions, while ϕ_B is the built-in voltage.

At a specified applied voltage, it is common to define junction capacitance, given by the usual definition of capacitance, as

$$C_j = \frac{dq_v}{dv} = nK(\phi_B - v_D)^{n-1} \tag{1.81}$$

For uniform impurity distributions or the abrupt junction as discussed in Sec. 1.3, K and n are readily obtained analytically as follows:

The magnitude of the depletion layer charge on either side of the *pn* boundary, assuming the validity of the depletion approximation, is

$$|Q_j| = qAN_Dd_n = qAN_Ad_p \qquad (1.82)$$

Use of this relationship in Eq. 1.28 or 1.29 yields

$$|Q_J| = A(2q\varepsilon_o\varepsilon_r)^{1/2} \left(\frac{N_AN_D}{N_A + N_D}\right)^{1/2} (\phi_B - v_D)^{1/2} \qquad (1.83)$$

Then defining $q_v(v_0) = Q_j(0) - Q_j$,

$$q_v(v_D) = A(2q\varepsilon_o\varepsilon_r)^{1/2} \left(\frac{N_AN_D}{N_A + N_D}\right)^{1/2} [\phi_B^{1/2} - (\phi_B - v_D)^{1/2}] \qquad (1.84)$$

Hence

$$K = A(2q\varepsilon_o\varepsilon_r)^{1/2} \left(\frac{N_DN_A}{N_A + N_D}\right)^{1/2}$$

and $n = \frac{1}{2}$. Then, from Eq. 1.81,

$$C_j = \frac{K}{2}(\phi_B - v_D)^{-1/2} \qquad (1.85)$$

or, by substituting $(\phi_B - v_D)^{-1/2}$ from Eq. 1.30,

$$C_j = \frac{\varepsilon_o\varepsilon_rA}{d} \qquad (1.86)$$

which is equivalent to the capacitance of a parallel plate capacitance at variable spacing *d* corresponding to the depletion layer width. Hence varying the voltage applied to a *pn* junction creates a voltage-variable capacitance; such a device is referred to as a varactor or varactor diode.

For the linearly graded junction shown in Fig. 1.6, the charge on either side of the junction can be expressed in terms of the integral of the charge distribution; for example, on the *n* side,

$$Q_{xn} = \int_0^{d_n} qAmx_n \, dx_n$$

or $\qquad\qquad\qquad\qquad\qquad\qquad\qquad\qquad\qquad\qquad\qquad\qquad (1.87)$

$$Q = qAm\frac{d^2}{8}$$

where $d = d_n + d_p = 2d_n$.

Substitituting this result into Eq. (1.34) for *d* yields a charge–voltage relationship:

$$Q = K(\phi_B - v_D)^{2/3} \qquad (1.88)$$

where $n = 2/3$ and

$$K = 0.655A(qm)^{1/3}(\varepsilon_o\varepsilon_r)^{2/3}$$

$$= 1.932 \times 10^{-13} Am^{1/3} \quad \text{for silicon}$$

Then, as before

$$q_v = K[\phi_B^{2/3} - (\phi_B - v_a)^{2/3}] \tag{1.89}$$

with the junction capacitance

$$C_j = \frac{dq_v}{dv} = \frac{2}{3}K(\phi_B - v_D)^{-1/3} \tag{1.90}$$

Since capacitance is directly proportional to area it is sometimes useful to work with C_j/A or capacitance per unit area, defining a new constant per unit area as $K_j = K/A$.

Junction capacitance relative to its value at $v_D = 0$ has the general form plotted in Fig. 1.17(a) from

$$\frac{C_j}{C_{j0}} = \left(1 - \frac{v_D}{\phi_B}\right)^{n-1} \tag{1.91}$$

In the limit as $v_D \to \phi_B$ the equation is not valid, and more complex relationships are required.

Another useful form for the abrupt junction is

$$\frac{1}{C_j^2} = \frac{4}{K}(\phi_B - v_D) \tag{1.92}$$

plotted in Fig. 1.17(b).

If points plotted from capacitance measurements at a succession of applied voltages define a straight line, the existence of an abrupt junction is predicted and the extrapolated value permits an estimate of ϕ_B to be made. The slope determines the value of K_j, which allows N_D or N_A to be determined if it is known that $N_D \gg N_A$ or $N_A \gg N_D$.

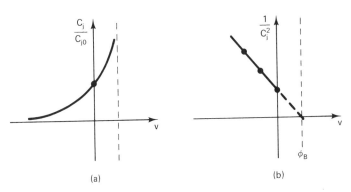

(a)   (b)

Figure 1.17 Capacitance–voltage characteristic of a *pn* junction. (a) General characteristic. (b) Abrupt junction.

The estimated value for ϕ_B is useful for either the *pn* junction or the metal–semiconductor (Schottky) junction.

For the general case, using Eq. 1.81 capacitance measurements at three values of v_D yields three nonlinear simultaneous equations from which n, K, and ϕ_B can be determined by a variety of techniques. By one suggested method, capacitance measurements C_0, C_1, and C_2 can be made at zero voltage and two successively increasing negative voltages V_1 and V_2. After first forming C_1/C_0 and C_2/C_0, the three equations can be written as

$$\frac{(C_1/C_0)^{1/(n-1)} - 1}{(C_2/C_0)^{1/(n-1)} - 1} = \frac{V_1}{V_2} \qquad (1.93)$$

$$\phi_B = \frac{-V_1}{(C_1/C_0)^{1/(n-1)} - 1} \qquad (1.94)$$

$$K = \frac{C_0}{n\phi_B^{n-1}} \qquad (1.95)$$

Equation 1.93 may first be solved for n by using successive approximation methods, and then ϕ_B and K may be determined.

Minority Carrier Diffusion Dynamics

Dynamically, as a junction is abruptly switched from zero or reverse to forward bias, a finite time is required for minority carriers to reach a final or steady-state value. Similarly, when the junction is restored to zero or reverse-bias conditions, a finite time is required to remove the stored charge and reach steady-state conditions for the zero or reverse bias. Even for incremental changes in junction voltage about a specific forward-bias condition, finite times are required for the stored charges to respond, and at a particular forward-biased current a *diffusion capacitance* may be defined as

$$C_d = \frac{dq_F}{dv} \bigg]_{v=V} \qquad (1.96)$$

where q_F is the total minority carrier charge stored in the n and p regions at a forward-bias voltage, V.

The equation for total charge due to minority carriers (holes) in the n region is

$$q_{pn} = qA \int_0^{W'_n} p'_n(x'_n) \, dx'_n \qquad (1.97)$$

Also, for electrons in the p region

$$q_{np} = qA \int_0^{W'_p} n'_p(x'_p) \, dx'_p \qquad (1.98)$$

The hole current can be written as

$$I_p = \frac{q_{pn}}{\tau_{Fn}} \qquad (1.99)$$

where τ_{Fn} is defined as the transit time and is the time required for holes to progress from the edge of the depletion layer boundary through the n region. Conversely,

$$\tau_{Fn} = \frac{q_{pn}}{I_p} \tag{1.100}$$

If the n region is sufficiently narrow that $W_n \ll L_p$, Eqs. 1.65 and 1.67 are valid, and τ_{Fn} can be written as

$$\tau_{Fn} = \frac{1}{D_{pn}} \int_0^{W_n'} \frac{\int_{x_n'}^{W_n'} N_D \ (x_n') \ dx_n'}{N_D(x_n')} \ dx_n' \tag{1.101}$$

Similarly, for electrons in the p region, using Eqs. 1.66 and 1.68 in Eq. 1.98

$$\tau_{Fp} = \frac{1}{D_{np}} \int_0^{W_p'} \frac{\int_{x_p'}^{W_p'} N_A \ (x_p') \ dx_p'}{N_A \ (x_p')} \ dx_p' \tag{1.102}$$

For the special cases of uniform doping in the n and p regions, Eq. 1.101 reduces to

$$\tau_{Fn} = \frac{W_n'^2}{2D_{pn}} \tag{1.103}$$

and Eq. 1.102 reduces to

$$\tau_{Fp} = \frac{W_p'^2}{2D_{np}} \tag{1.104}$$

These specific values of transit time can be obtained alternatively and more simply directly from Eqs. 1.48 and 1.49 applied to Eqs. 1.97 and 1.98, along with Eqs. 1.52 and 1.53.

The diffusion capacitance as defined by Eq. 1.96 can be obtained in general terms using Eqs. 1.97 and 1.98, with p_n' and n_p' expressed in terms of voltage using Eq. 1.65 or 1.68 in Eq. 1.66 or 1.65. The result for holes only, assuming $N_A \gg N_D$ (i.e., $i_n \ll i_p$), is

$$C_d = qAn_i^2 \frac{q}{kT} (e^{qv/kT} - 1) \int_0^{W_n'} \left\{ \frac{\int_{x'}^{W_p'} N_D \ (x') \ dx'}{N_D(x) \int_0^{W_n'} N_D \ (x') \ dx'} \right\} dx' \tag{1.105}$$

Alternatively, C_d can be expressed in terms of current using Eq. 1.67 as

$$C_d = \frac{q}{kT} \frac{I_p}{D_{pn}} \int_0^{W_n'} N_D \ (x') \ dx' \int_0^{W_n'} \frac{\int_{x'}^{W_p'} N_D \ (x') \ dx'}{N_D \ (x') \int_0^{W_n'} N_D \ (x') \ dx'} \tag{1.106}$$

For uniform doping [i.e., $N_D(x') = N_D$], Eq. 1.105 reduces to

$$C_d = qAn_i^2 \frac{q}{kT} \frac{W_n'}{2N_D} (e^{qv/kT} - 1) \tag{1.107}$$

and Eq. 1.106 reduces to

$$C_d = \frac{q}{kT} \frac{W_n'^2}{2D_p} I_p \tag{1.108}$$

or

$$C_d = \frac{q}{kT} \tau_{Fn} I_p \tag{1.109}$$

Similar equations can be derived for electrons in the p region involving N_A, W_p', D_{np}, and I_n for $N_D >> N_A$ (i.e., $I_p << I_n$).

Where hole and electron currents must both be considered, the diffusion capacitance is the sum of that calculated separately for hole and electron flow.

The junction capacitance and diffusion capacitance may be considered as acting in parallel such that

$$C_{pn} = C_j + C_d \tag{1.110}$$

This total capacitance may be considered as acting in parallel with a forward-bias resistance at a particular value of forward voltage V_D given by

$$r_f = \frac{dv}{di} \Big|_{v=V} \tag{1.111}$$

as indicated in the small-signal equivalent circuit of Fig. 1.18(a).

For reverse bias, there is no stored charge, and the capacitance is the depletion layer capacitance only, which as indicated in Fig. 1.18(b) acts in parallel with a reverse or back resistance.

$$r_b = \frac{dv}{di} \Big|_{v=V(-)} \tag{1.112}$$

These are the incremental or small-signal circuit models with small voltage and current deviation about a specific operating point.

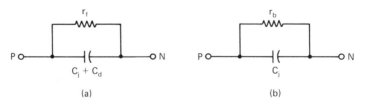

Figure 1.18 Small-signal equivalent circuit of a diode. (a) Forward characteristic. (b) Reverse characteristic.

1.7 DIODE SWITCHING CHARACTERISTICS

Inasmuch as junction and diffusion capacitances are both nonlinear (voltage or current dependent), the response to large input signal currents or voltages having rapid time variations is much more complicated to determine using the capacitance concept directly, although it can be done with piecewise linear models and iterative techniques with appropriate computer programs. However, a more direct solution for the transient or switching response where operating points are switched between forward and reverse bias is indicated in Fig. 1.19(a).

Such solutions are quite complicated unless some rather major assumptions are made. Very often the analysis is somewhat simplified by assuming a piecewise linear model for the steady-state diode characteristic, as indicated in Fig. 1.19(b) or even by the more idealized characteristic of Fig. 1.19(c).

In any case, it is important to explain the mechanisms by which charges are stored or removed as the diode is switched between forward and zero or reverse bias. Holes stored in the n region and electrons in the p region are both involved. However, if one component (hole storage) is explained, the other follows by analogy. The continuity equation (Eq. 1.15) is a useful starting point. This equation, written for holes stored in the n region under forward-bias conditions, is

$$q \frac{\partial p_n'}{\partial t} = -q \frac{p_n'}{\tau_p} - \frac{\partial J_p}{\partial x'} \qquad (1.113)$$

Under dynamic (switching) conditions, p_n' and J_P are functions of both x' and t, and the buildup of stored charge as the forward bias is increased or the decay of charge as the bias is decreased are both complex solutions of the continuity equation for which manageable approximations can be made.

At any time t, Eq. 1.113 may be integrated with respect to x' and written in terms of total current, $I = JA$. Using instantaneous values, the result for any x' is

$$I_P(0,t) - i_p(x',t) = \frac{qA}{\tau_p} \int_0^{x'} p_n'(x',t)\, dx' + qA \frac{\partial}{\partial t} \int_0^{x'} p_n'(x',t)\, dx' \quad (1.114)$$

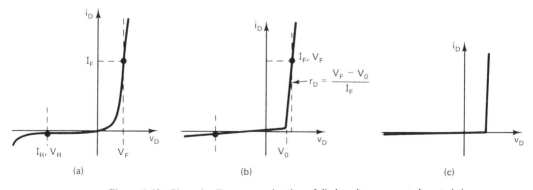

Figure 1.19 Piecewise linear approximation of diode voltage–current characteristic.

and specifically for $x' = W'_n$,

$$i_p(0, t) - i_p(W'_n, t) = \frac{qA}{\tau_p} \int_0^{W'_n} p'_n(x', t)\, dx' + qA \frac{\partial}{\partial t} \int_0^{W'_n} p'_n(x', t)\, dx' \quad (1.115)$$

Using Eq. 1.97 for total charge stored in the n region and Eq. 1.100 for i_p and W'_n, with τ_T used instead of τ_F to generalize it for either forward or reverse transit time, Eq. 1.115 can be written as

$$i_p(0, t) = q_{pn}(t) \left(\frac{1}{\tau_p} + \frac{1}{\tau_T} \right) + \frac{dq_{pn}(t)}{dt} \quad (1.116)$$

where $\tau_T = \tau_F$ for increasing currents and $\tau_T = \tau_R$ for decreasing current. For uniform doping in the region, $\tau_F = \tau_R = \tau_T$; otherwise, the values are different.

In the steady state, $q_{pn} = \bar{q}_{pn}$ is a function of $p'_n(x')$, which in turn is a function of $N_D(x')$ and the relative values of L_p and W'_n, as discussed in Sec. 1.3. In any case, the voltage was related to the charge at the depletion layer boundary by $p'_n(0) = p_{no}$ $(e^{qv/kT} - 1)$. Also, the form of the charge distribution did not change as $p'_n(0)$ increased; hence the voltage dependency of total charge in Eq. 1.116 retained the same exponential form. However, under switching conditions, the charge distribution is a function of both time and distance, and charge–voltage relationship becomes more complicated, which makes exact solutions for Eq. 1.116 with the voltage variable included extremely complex. Approximate solutions are possible assuming that the shape of the distribution does not change. However, even a simpler approach to dynamic switching solutions yields useful results, which simply makes use of the piecewise linear voltage–current models suggested in Fig. 1.19(b) or (c).

An approximate solution for a turn-off transient where hole current dominates is suggested as follows:

It is first assumed that the diode is forward biased under steady-state conditions with $v_D = V_F$ and $i_D = I_F \cong I_p$ as supplied from an external circuit, as indicated in Fig. 1.20(a) with

$$v_D = V_F = V_{SF} - I_F R_F$$

Also, from Eq. 1.116 for $dq_{pn}/dt = 0$,

$$q_{PN} = I_F \tau_{SF} \quad (1.117)$$

where q_{PN} is the maximum value of stored charge in the n region and

$$\tau_{SF} = \frac{\tau_p \tau_F}{\tau_p + \tau_F}$$

which is defined as the forward storage time constant.

Then, as indicated in Fig. 1.20(b) at $t = 0$, the input is abruptly switched to a negative value acting through a series resistance R_R, which may or may not be the same as R_F, (i.e., resistance may be switched at the same time).

The diode charge is represented by the stored charge indicated by a nonlinear capacitance. The initial value of the reverse current that flows is

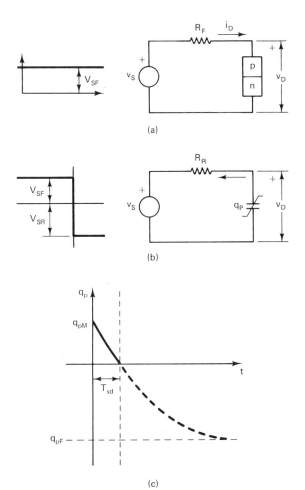

Figure 1.20 Decay of stored charge in a diode. (a) Structure. (b) Circuit. (c) Charge decay.

$$I_R = \frac{V_{SR} - V_F}{R_R}$$

which is negative (i.e., $I_R = -|I_R|$).

The important simplifying approximation now to be made is based on the assumption that as long as stored charge remains the forward diode voltage does not change substantially, as suggested by the model of Fig. 1.19(b), and hence i_R to be used in Eq. 1.116 remains approximately constant during the removal of stored charge. Then, using the initial value of I_R, Eq. 1.116 may be written as

$$I_R = \frac{q_{pn}(t)}{\tau_{SR}} + \frac{d(q_{pn}\,t)}{dt} \tag{1.118}$$

where

$$\tau_{SR} = \frac{\tau_p \tau_R}{\tau_p \tau_R}$$

Sec. 1.7   Diode Switching Characteristics   **35**

The solution for this equation has the form

$$q_{pn}(t) = q_{PM} - (q_{PM} - q_{PF})[1 - e^{-(t/\tau_{SF})}] \qquad (1.119)$$

where q_{PM} is the initial value of stored charge from Eq. 1.117 and q_{pF} the final value of stored charge, $q_{pF} = I_R\tau_{SR}$, from Eq. 1.118 for $d_{qpn}(t)/dt = 0$.

This equation is plotted in Fig. 1.20(c). At the time q_{pn} reaches zero, stored charge is removed and Eq. 1.119 no longer applies, and the time taken for q_{pn} to reach zero is called the storage delay time, T_{sd}.

As determined from Eq. 1.119 for $q_{pn} = 0$ and using the currents for q_{sM} and q_{sF},

$$T_{sd} = \tau_S \ln\left(1 + \frac{I_F\tau_{SF}}{|I_R|\tau_{SR}}\right) \qquad (1.120)$$

In addition to the approximation of constant reverse current, the depletion layer charge was not included in the above analysis. A few words of explanation are in order. The depletion layer charge is voltage related according to Eq. 1.80. Hence the total charge that must be reckoned with includes both minority carrier storage and depletion layer charge. However, under the assumption of almost constant voltage, as minority carrier charge is removed, there will be negligible dq_v/dv. Hence *most* of the depletion layer charge takes place after the minority carrier charge has been removed. In Fig. 1.21(a), the decay up to $t = T_{sd}$ is repeated, while the majority of the change in depletion layer charge is indicated in Fig. 1.21(b). The dashed line for $0 < t < T_{sd}$ indicates some charge removal during this interval.

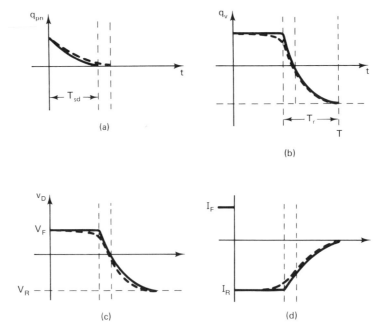

Figure 1.21 Diode transient recovery characteristics.

The corresponding voltage and current curves are shown in Figs. 1.21(c) and (d). The solid curves for $0 < t < T_{sd}$ indicate the approximations made during the removal of depletion layer charge, while the dashed curves indicate the changes that actually take place. (In reality, the stored charge is not totally removed until $v_D \to 0$).

The continuation of the solid curves for $t > T_{sd}$ anticipates the approximate final recovery as depletion layer charge is changed to its final value, assuming such change starts as $t = T_{sd}$. This time is referred to as the *recovery time*.

An exact solution for the recovery time is complicated because of the nonlinearity of the charge–voltage relationship as indicated in Eq. 1.80. However, a reasonable estimate of recovery time can be made starting with

$$\int_{T_{sd}}^{T_{sd}+T_r} i_R (t) \, dt = q_v (T_{sd} + T_r) - q_v (T_{sd}) \tag{1.121}$$

with reference to Fig. 1.22. The actual value of V_R is zero or negative, and the current actually flows away from the depletion layer as indicated. The form of $i_R(t)$ is not known, but its integral may be estimated by using its average value, given approximately by

$$\bar{I}_R = \frac{V_{SR} - V_D}{2R_R} \tag{1.122}$$

where V_D is the diode forward voltage previously determined and V_{SR} is the reverse-bias voltage indicated in Fig. 1.20. Thus, using the average value in Eq. 1.121, the final recovery time is given approximately by

$$T_r = \frac{q_v (V_D) - q_v(V_R)}{|\bar{I}_R|} \tag{1.123}$$

where the q_v's are determined from Eq. 1.80 using appropriate values of K and n determined by the junction grading.

The relative and absolute values of the components of total switching time ($T_{sd} + T_r$) are determined by V_{SR} and R_R. If, for some reason, R_R must be large, then V_R should be a highly negative value if stored charge is to be removed rapidly by having a large I_R. This condition, however, causes the recovery of depletion layer charge to be slower. Thus, ideally, R_R should be small and $V_R \to 0$ to minimize *both* T_{sd} and T_r.

When a diode is switched from zero or reverse bias to forward bias, the preceding sequence is reversed. When starting from a negative bias, forward current flows through the forward resistance R_F to charge the depletion layer capacitance alone, which continues until almost normal forward bias is reached. Then charge begins to accumulate in the neutral regions until steady-state values are reached. Sequential charging of the depletion layer and the neutral regions similar to those suggested for turn-off may be used to approximate the forward conditions.

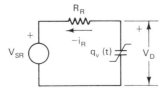

Figure 1.22 Diode circuit model for depletion layer charge recovery.

1.8 APPROXIMATE TEMPERATURE CHARACTERISTICS OF PN JUNCTIONS

The voltage–current characteristic of a *pn* junction is temperature dependent primarily through the $e^{qv/kT}$ relationship and the temperature dependence of intrinsic carrier concentration and diffusion constants in the various relevant equations that have been developed in foregoing sections.

The intrinsic carrier concentration of a semiconductor is temperature dependent according to

$$n_i^2 (T) = K_i^2 T^3 e^{-qV_g/kT} \qquad (1.124)$$

where K_i involves Boltzmann's constant, Planck's constant, and the effective masses of holes and electrons. V_g is the band gap (approximately 1.12 for silicon at 300 K), which itself is slightly temperature dependent but as an approximation can be assumed to be constant over a relatively wide range.

Electron and hole mobilities are temperature dependent as a result of two mechanisms known as *impurity scattering* and *lattice scattering*. Impurity scattering is a result of lattice defects that increase the density of ionized impurities and hence the random scattering of carriers as a result of interaction with such ionized impurities. These interactions become greater at lower temperatures because carriers move more slowly at low temperatures and will interact more strongly with charged ions. The overall result is a mobility that is observed to be approximately proportional to $T^{3/2}$.

Lattice scattering is a result of carrier interaction with atoms that are displaced from their normal lattice position in the crystal as a result of vibrations due to thermal agitation at higher temperatures. These interactions result in scattering, which is an increasing function of temperature with a corresponding decrease in mobility, which is approximately proportional to $T^{-3/2}$. In a semiconductor, both mechanisms are present, but the relative dominance of each is illustrated in Fig. 1.23. Lattice scattering tends to dominate at temperatures above 300 K.

At any given temperature, scattering probabilities increase with impurity concentration; hence mobility is a decreasing function of impurity concentration in a manner illustrated earlier by Fig. 1.1. Increasing the impurity concentration raises the temperature at which the transistion from impurity scattering dominance to lattice scattering dominance occurs.

For temperatures at which diodes are normally used (in the vicinity of 300 K), lattice scattering is dominant, but impurities increase the scattering probability, which makes the decrease in mobility as a function of temperature even faster. In a normally doped silicon semiconductor, a variation of μ according approximately to $T^{-5/2}$ is most often observed as indicated in Fig. 1.23.

The diffusion constants appearing in the diode voltage–current relationships are related to mobility through $D/\mu = kT/q$; hence for impurity scattering dominance

$$D = K_D T^{5/2}$$

and for lattice scattering

$$D = K_D T^{-1/2}$$

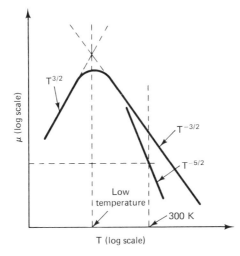

Figure 1.23 Semiconductor mobility versus temperature characteristics.

T (log scale)

However, for a doped silicon semiconductor, it has been observed that $D \cong K_D T^{-3/2}$, using the $\mu \propto T^{-5/2}$ relationship.

This temperature dependence affects the hole and electron diffusion currents and recombination currents in slightly different ways because of the temperature-dependent carrier lifetimes. However, an approximate temperature characteristic of the *pn* junction can be obtained considering hole and electron diffusion currents only, assuming negligible recombinations. Combining hole and electron diffusion components, the theoretical reverse-bias saturation current may be written as

$$I_O = A_o T^n e^{-qV_g/kT} \qquad (1.125)$$

where A_o is determined by segregating the non-temperature-dependent terms and n is a constant combining the temperature dependent n_i and D, whose value depends on the combination of scattering mechanisms. (For the assumption of lattice scattering dominance, $n \cong 2.5$ or $n \cong 1.5$ for a doped silicon semiconductor.) Then, for reasonably large forward bias, the diode equation may be expressed as

$$i_D = A_o T^n e^{-qV_g/kT} \, e^{qV_D/kT} \qquad (1.126)$$

Equation 1.126 may be modified slightly to take into account the slight temperature dependence of V_g, which over normal operating temperatures is a slightly decreasing function. This variation is quite linear over a rather wide temperature range and can be approximated by the equation

$$V_g = V_{g0} \left[1 - \alpha T \right]$$

where

$$\alpha = -\frac{\Delta V_{g0}}{\Delta T}$$

and V_{g0} is the projection to $T = 0$, which is referred to as the extrapolated value. $V_{g0} \cong 1.21$ for silicon, while the actual value at 300 K is approximately 1.12. The actual value at 0 K is somewhat less as the slope of the curve inverts at very low temperatures.

Using the extrapolated value, Eq. 1.126 may be written as

$$i_D = A_o' T^n e^{-qV_{g0}/kT} e^{qV_D/kT} \tag{1.127}$$

where $A_o' = A_o e^{\alpha q/k}$.

If junction diode voltage and current can be computed or measured at a reference temperature, T_r, the need for knowing A_o' can be avoided by using the ratio of i_D/i_{Dr}. In these terms, the diode voltage at any temperature can be obtained from Eq. 1.127 as

$$v_D = V_{Dr} \frac{T}{T_r} + V_{g0} \left(1 - \frac{T'}{T_r}\right) - \frac{nkT}{q} \ln \frac{T}{T_r} + \frac{kT}{q} \ln \frac{i_D}{I_{Dr}} \tag{1.128}$$

where V_{Dr} and I_{Dr} are the diode voltage and current at the reference temperature, T_r.

The variation of diode voltage with temperature at a constant $i_D = I_D$ is

$$\left. \frac{dv_D}{dT} \right]_{i-I_D} = \frac{1}{T_r} (V_{Dr} - V_{g0}) - \frac{nk}{q} - \frac{nk}{q} \ln \frac{T}{T_r} + \frac{k}{q} \ln \frac{I_D}{I_{Dr}} \tag{1.129}$$

which at $T = T_r$ is

$$\left. \frac{dv_D}{dT} \right]_{T=T_r} = -\frac{1}{T_r} \left(V_{g0} - V_{Dr} + \frac{nkT_r}{q}\right)$$

For a typical silicon junction with $V_{Dr} = 0.60$ and $n = 1.5$ for $T_r = 300$ K,

$$\frac{dv_D}{dT} = -\frac{1}{300} [1.21 - 0.60 + 1.5 (0.026)]$$

$$\cong -2.2 \text{ mV/°C}$$

Examination of Eqs. 1.128 and 1.129 shows that v_D and dv_D/dT are both relatively insensitive functions of i_D; hence dv_D/dT remains reasonably constant over a wide range of currents. Also, the term $(nk/q) \ln (T/T_r)$ in Eq. 1.129 is relatively small and does not vary rapidly with temperature; therefore, dv_D/dT is also relatively constant over wide temperature changes. For these reasons, a number in the neighborhood of -2.2 mV/°C is often quoted as the change in diode voltage required to maintain a constant current without reference to a specific value of current level.

As an alternative to using the current ratio as expressed in Eqs. 1.128 and 1.129, expressions for v_D and dv_D/dT can be obtained directly from Eq. 1.127, with the results

$$v_D = \frac{kT}{q} \left(\ln \frac{i_D}{A_o'} - n \ln T\right) + V_{g0}$$

and

$$\frac{dv_D}{dT} = -\frac{nk}{q} \left(1 + \ln T - \frac{1}{n} \ln \frac{I_D}{A_o'}\right)$$

These forms might be preferable to use where A_o' can be determined from known process and structural parameters.

1.9 NOISE IN SEMICONDUCTOR MATERIALS AND JUNCTIONS

There are always rapid current or voltage fluctuations of an apparently randomly varying nature present in circuit elements or electronic devices to which external sources of voltage or currents are applied. These fluctuations are due to statistical variations in the various thermal and electronic processes, such as in charge carrier production, diffusion, recombination, and drift.

Such fluctuations may be uncorrelated with any applied signals and are commonly termed *random noise* or simply *noise*. If the levels of applied voltages or current are sufficiently high, these fluctuations are unobtrusive in the output; but at very low signal levels, the internal random variations tend to mask the desired signal and in the limit render the device useless, which therefore places a lower limit on the voltage or current level at which a device can be successfully used. In this section, specific noise sources associated with semiconductor junctions will be considered, but first some general characteristics of noise irrespective of the devices that produce them will be reviewed.

Characteristic of Noise and Some Basic Concepts

Being random or quasi-random in nature, no specific frequency or group of frequencies or relative phases can be assigned to noise. Furthermore, the average or integrated value of random-noise voltage or current over a sufficiently long time interval would be zero. In other words, there is no dc component associated with noise. However, noise will be found to be characterized by an effective or root-mean-square (rms) value, and power-measuring instruments can record such values. The mean-square value of a noise component of current is symbolized by $\overline{i_N^2}$ and its rms value by $\sqrt{\overline{i_N^2}}$, and for voltages by $\overline{v_N^2}$ and $\sqrt{\overline{v_N^2}}$, respectively. Ordinarily, a measurement of noise produced at the terminals of an electrical circuit or device can be made using power-measuring instruments limited to a specific bandwidth. If such a measurement is made over a specific incremental bandwidth Δf at a specific center frequency f_k, the output noise voltage can be expressed in terms of $\overline{v_k^2}/\Delta f$, which may be referred to as mean-square noise voltage per unit bandwidth or more commonly power spectral density. This value may or may not be the same at other center frequencies.

The *frequency spectrum* of a source of noise can be approximated experimentally from successive measurements at adjacent intervals of small values of Δf by plotting a curve through the points of $\overline{i_k^2}/\Delta f$ or $\overline{v_k^2}/\Delta f$ as a function of frequency, as indicated in Fig. 1.24.

Three different spectral distributions of mean-square current over a bandwidth BW $= f_n - f_1$ made up of n intervals of Δf spacing are shown in Fig. 1.24, along with

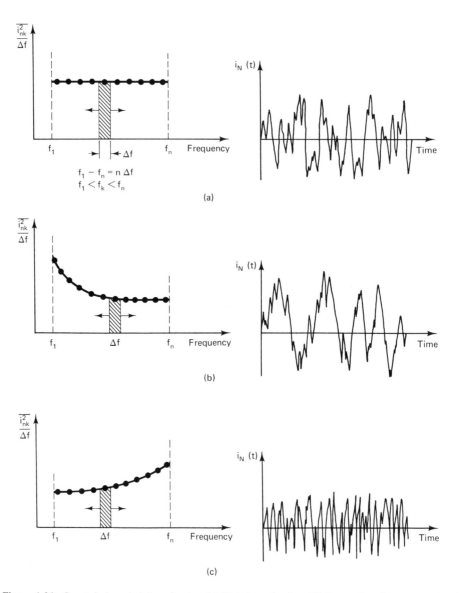

Figure 1.24 Spectral characteristics of noise. (a) Flat channel noise. (b) Excess, low-frequency noise. (c) Excess high-frequency noise.

corresponding time variations of noise presented over a specific time interval of a continuous recording.

The total noise current over the band may be written as

$$\overline{i_N^2} = \sum_{k=1}^{n} \overline{i_k^2} \tag{1.130}$$

In the limit as $\Delta f \to 0$ and $n \to \infty$, the current may be written in integral form as

$$\overline{i_N^2} = \int_{f_l}^{f_n} (\overline{i_n^2}) \, df \qquad (1.131)$$

where

$$(\overline{i_N^2}) = \lim_{\Delta f \to 0} \frac{\overline{i_n^2}}{\Delta f} \qquad (1.132)$$

defined as the current squared spectral density or, since power is proportional to the square of current, as power spectral density.

Each frequency spectrum curve shown in Fig. 1.24 *may* have the same value of $\overline{i_n^2}$ over the entire bandwidth, BW. However, the structure of the corresponding time plots may be different. Noise having the flat spectrum shown in Fig. 1.24(a) is termed *white noise* in general, and if limited to a specific frequency range as indicated, is termed band-limited white noise. Sometimes the term flat channel noise is used.

The coarser-grained structure shown in Fig. 1.24(b) indicates a proportionately larger low-frequency content and the finer structure of Fig. 1.24(c) indicates a proportionately higher frequency content. The spectral distribution of noise, as well as its overall mean-square value, is of importance when determining how much noise is tolerable in a particular system.

Thermal Noise

Independent of any signal applied to the terminals of a resistive circuit element, there is a fluctuation voltage or noise that can theoretically be measured across its open-circuit terminals. This noise is generally attributed to thermal interactions between free electrons in random motion and ions that are in vibration about their normal positions. This noise is temperature dependent and is referred to as thermal noise or Johnson noise. The mean-square value of thermal noise of any material having a resistance of value R is given by

$$\overline{v_R^2} = 4kTR \, \Delta f \qquad (1.133)$$

where T is the temperature in degrees Kelvin, Δf is the bandwidth over which the noise is determined, and k is Boltzmann's constant. Theoretically, $\overline{v_R^2}$ is independent of frequency and its spectrum is consequently flat. However, such thermal noise is not limited to elements specifically designated as resistors but is present in other passive devices where the spectrum may not be flat, even in antennas, which may be characterized by their radiation resistance. In more general form,

$$\overline{v_N^2} = 4kT \int_{f_a}^{f_b} |R| df \qquad (1.134)$$

where R may itself be frequency dependent.

The thermal noise of a resistor with a uniform spectral density may be modeled as a voltage source in series with the open-circuit terminals of a noiseless resistance, as indicated in Fig. 1.25(a).

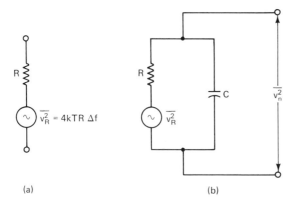

(a)

(b)

Figure 1.25 Circuit model for thermal noise. (a) Noise generator model. (b) Capacitance effects.

As a practical matter, to illustrate how frequency dependence enters into noise calculations, even though the noise source itself is not frequency dependent, let us assume that there is a capacitance C associated with the circuit shown in Fig. 1.25(b), which may be external or may be distributed capacitance associated with the circuit element itself.

At a single frequency, the magnitude of the mean-square noise voltage per unit bandwidth at the terminals is

$$(\overline{v_n^2}) = \frac{1}{1 + \omega^2 R^2 C^2} (\overline{v_R^2}) \tag{1.135}$$

where $[\overline{v_R^2}]$ is the generated noise per unit bandwidth assumed to be uniform, that is, white noise.

The total mean-square noise integrated over a frequency spectrum $f_b - f_a$ is given by

$$\overline{v_N^2} = 4kT \int_{f_a}^{f_b} \frac{R}{1 + 4\pi^2 R^2 C^2} \, df \tag{1.136}$$

where R may or may not be itself frequency dependent.

Shot Noise

There are random variations in all current-producing mechanisms, such as electrons produced at an emitting surface or charge carries generated in the p and n regions of semiconductors or produced in the depletion region of a semiconductor junction. Shot noise is proportional to the generated current itself and is usually independent of the specific current-generating mechanism. Its mean-square value is given by

$$\overline{i_N^2} = 2qI \, \Delta f \tag{1.137}$$

where I is the average or dc value of the current involved and q is the magnitude of electronic charge.

Excess Low-Frequency Noise

In addition to thermal noise and shot noise, which have a firm basis in purely theoretical considerations, there are components of noise in most devices that are found to be inversely related to frequency at quite low frequencies (below ~ 1000 Hz). Such noise, if the spectral variation is quite orderly, varying smoothly as K/f^n, is referred to as *flicker noise* or $1/f$ noise and has the spectral distribution of the general form illustrated in Fig. 1.24(b). If the corresponding time variations have somewhat irregular and long-time spacings between zero crossings with high-frequency components superimposed on positive and negative limits, as indicated in the time scale of Fig. 1.24(b) if it were somewhat exaggerated, the noise that it illustrates is referred to as *burst noise*.

Excess low-frequency noise may be due to various processing imperfections, such as contamination and crystal imperfections, and is much less amenable to analytical treatment than pure shot noise or thermal noise. A general equation that crudely models the various components of excess low-frequency noise is suggested as

$$\overline{i_{ex}^2} = K_1 \frac{I^x}{K_2 + (f/f_1)^y} \Delta f \tag{1.138}$$

where K_1 is a device parameter, K_2 is a small constant, f_1 is a reference frequency below which excess noise rapidly increases, and x and y are positive quantities ranging roughly between 0.5 and 2.0.

For pure orderly flicker noise, $K_2 \to 0$, and x and y are close to unity. For a large component of burst noise, $K_2 \to 1$.

Components of burst and flicker noise may result from related but separate phenomena, and more than one term of the general form of Eq. 1.38 may be necessary for a complete excess low-frequency noise model. As a result, where large proportions of burst noise are present, the spectral distribution may have irregularities or "humps" rather than the orderly $1/f$ distribution illustrated in Fig. 1.24(b). Excess noise is also generated in avalanche and zener breakdown processes in the mechanism of the cumulation generation of increased hole–electron pairs, resulting in large spikes of excess noise that have a substantial excess low-frequency constant.

Excess High-Frequency Noise

At extremely high frequencies, it is possible to find a noise spectrum in a device that is an increasing function of frequency, as shown in Fig. 1.24(c). Such effects are present where operating frequencies begin to be transit time limited and involve initial random velocities of charge carriers.

Noise Models for the Junction Diode

Components of both thermal noise and current-dependent noise are generated within the junction diode. Thermal noise is associated with the bulk resistance of the n and p semiconductor materials, while current dependent noise is primarily shot noise associated with the various current-producing mechanisms, but may include excess low-frequency

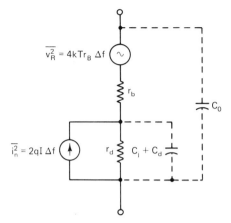

Figure 1.26 Noise circuit model for junction diode.

components. Such noise sources are added to the small-signal diode model as indicated in Fig. 1.26.

The resistance r_d arises from the current–voltage diode equation and is given approximately by $r_d \cong kT/qI_D$. The noise associated with this process is shot noise as indicated.

The resistance r_b is the incremental value of the bulk resistance at the operating point, while r_B from which thermal noise is determined is its total dc value.

These two noise sources are uncorrelated and calculations of the noise of the terminals of the diode when connected to an external circuit are made from considering each source separately. At low frequencies, the capacitances can be ignored, but they, as well as capacitance in the external circuit, must be included as the frequency is increased.

PROBLEMS

1.1 Determine the resistance in ohms of a bar of doped silicon with dimensions of length $= 50$ μm, width $= 5$ μm, and thickness $= 0.5$ μm. Impurity concentration is $N_D = 1 \times 10^{17}$ cm^{-3}.

(a) What is the sheet resistance in $\Omega/\square$?

(b) What is the resistance of a pure (intrinsic) silicon bar having the same dimensions?

1.2 A resistor is formed by diffusing n-type silicon into an almost intrinsic p-type substrate in the pattern shown with an approximate thickness of 1.8 μm.

(a) Assuming that the impurity concentration can be maintained uniform through the entire thickness, what impurity concentration N_D would produce a sheet resistance of 500 $\Omega/\square$?

(b) What is the total linear resistance from contact to contact assuming that the resistance through each corner square is the same as through all others?

(c) Because the current density is not uniform around a corner square, the linear resistance of a corner square will be less than that in a linear region. It has been found that the

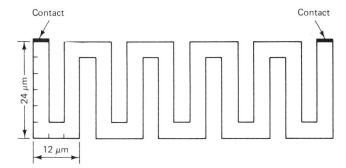

24 μm

12 μm

Prob. 1.2

resistance factor is approximately 70%. Devise an approximate distribution that will lead to this approximate value of resistance reduction.

1.3 An abrupt junction is formed by joining uniformly doped n- and p-type silicon. The respective impurity concentrations are $N_A = 10^{17}$ cm^{-3} and $N_D = 10^{16}$ cm^{-3}, respectively.
  (a) Determine the diffusion constants D_n and D_p for minority barriers at a temperature of 300 K (i.e., $kT/q \cong 0.026$ V).
  (b) Determine the built-in potential at the junction.
  (c) What are the depletion layer widths, d_p and d_n, with zero external applied voltage.

1.4 A silicon pn junction has uniformly doped p and n regions with $N_D = 10^{18}$ and $N_A = 10^{16}$ atoms/cm^3 and a junction area of 3×10^{-5} cm^2.
  (a) Plot the built-in potential ϕ_o over a temperature range of $-25°$ to $+125°$C.
  (b) Now suppose that the junction is reverse biased at a voltage $V_D = -20$ V. What is the width of the depletion region? Does it extend mostly into the p region or the n region?
  (c) What is the theoretically limiting forward-bias voltage? What practical factors prevent this limit from being approached?

1.5 The diode shown has $N_D \gg N_A$ with the following parameters.

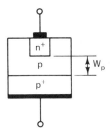

Prob. 1.5

$$N_A = 10^{16} \text{ atoms/cm}^3, \quad W_p = 2 \times 10^{-4} \text{ cm}, \quad A_j = 2 \times 10^{-5} \text{ cm}^2$$

  (a) For the junction reverse biased, determine I_0 and $f(v_D)$ in the expression $i_D = I_0 f(v_D)$, where $f(v_D)$ results from variation in depletion layer width with reverse voltage.
  (b) At a reverse-bias voltage $v_D = -10$ V, compare the actual value of reverse-bias current with I_0.

1.6 For the linearly graded junction shows in Fig. 1.6:
  (a) Verify Eq. 1.33, noting that $\xi_x = 0$ when the region is fully depleted at $x = \pm d/2$, which allows a constant of integration in the first integration to be determined.
  (b) Using Eqs. 1.33 and 1.36 to eliminate d, write an equation for the built-in potential.
  (c) At 300 K, plot ϕ_B as a function of m for m ranging from 10^{18} cm^{-4} to 10^{23} cm^{-4}.

Chap. 1   Problems

1.7 Consider the uniform geometry pn junction diode with uniform dopings in the p and n regions of $N_A = 10^{15}$, $N_D = 10^{17}$. Also, $W_p = 10 \ \mu m$, $W_n = 3 \ \mu m$, and $A_j = 5 \times 10^{-5} \ cm^2$. Also assume minority carrier lifetimes of $\tau_n = \tau_p = 1 \times 10^{-7} \ s$.

(a) Determine approximately the built-in voltage and the depletion layer width with zero applied voltage. Repeat the calculation of depletion layer width for a forward applied voltage $V_D = 0.6 \ V$.

(b) Determine the diffusion length for minority carriers in both the n and p regions.

(c) At a forward applied voltage $v_D = 0.68 \ V$, calculate the excess minority carrier concentrations at the boundary.

(d) Calculate the total diode terminal current, and compare it with what it would be for $L_p \gg W_n$ and $L_n \gg W_p$.

1.8 Junction capacitances for a junction diode are measured at zero and two negative voltages and found to be $C_0 = 2 \ pF$ at $V_D = 0$, $C_1 = 1.0415 \ pF$ at $V_D = -5.0 \ V$, and $C_2 = 0.847$ pF at $V_D = -10 \ V$.

(a) Determine the built-in potential, ϕ_B, and also n and k.

(b) What would you expect the impurity profile in the vicinity of the junction to be?

1.9 A silicon diode has a junction area $A_j = 5 \times 10^{-5} \ cm^2$ and $W_p = W_n = 1 \times 10^{-4} \ cm$, with dopings $N_A = 10^{16}$ and $N_D = 10^{18}$.

(a) What is the theoretical reverse-bias saturation current?

(b) Determine the diode voltage for $I_D = 5 \ mA$. What is the junction capacitance at this current and voltage level?

(c) At the operating point of part (b), determine the diffusion capacitance, the electron transit time through the p region, and the hole transit time through the n region.

1.10 A pn junction diode has $\overline{N_A} \gg \overline{N_D}$ with the impurity distribution in the n region given by $N_D(x) = N_D(0)e^{-x/W_n}$.

(a) Derive an expression for forward transmit time τ_{Fn} by first obtaining an expression for $p'_n(x)$, then the total minority excess charge in the n region, and finally the hole current i_p.

(b) Check the result of part (a) by using Eq. 1.101 directly.

1.11 A pn junction diode with uniform p and n regions has the following parameters:

$$N_A = 10^{18} \ cm^{-3}, \qquad N_D = 10^{16} \ cm^{-3}, \qquad A_j = 4 \times 10^{-5} \ cm^2$$
$$W_p = 2 \times 10^{-4} \ cm, \qquad W_n = 3 \times 10^{-4} \ cm$$

In the calculations to follow, assume that $L_P \gg W_n$ and $L_n \gg W_p$, and neglect the variation in width of the depletion layer with voltage.

(a) Determine the minority carrier forward transit times, τ_{Fp} and τ_{Fn}.

(b) This diode is forward biased with a forward current of $I_D = 5 \ mA$. Determine the hole and electron currents.

(c) Calculate the maximum stored charge in the n and p regions, respectively.

(d) The input terminals of the diode are now short-circuited through a 1000-Ω series resistor. What is the approximate time required to remove stored charge in both the n and p regions? (Assume that the change in forward voltage is negligible during this interval.)

(e) Compare the result of part (d) with that obtained with a negative reverse voltage such that $I_R = -I_F = -5 \ mA$.

1.12 Having dealt with Prob. 1.11 and with excess charges stored in the bulk regions removed:

(a) Calculate the charge stored in the depletion layer and determine the junction capacitance.

(b) With the diode switched to a reverse bias of $-5 \ V$ through a 5.2-kΩ resistor, determine the final value of stored charge and make an estimate of the recovery time (i.e., for $v_D = 5$ V and $i_D = 0$).

1.13 Using Eq. 1.129 and the diode in the example following it, and assuming that n and I remain constant.

    (a) Calculate dv_D/dT at $-55°$, $0°$, $+75°$, and $+125°$C, and carefully note the relative contributions of the various terms.

    (b) Suppose that at the lowest temperature, $-55°$C, the temperature-dependent mobility becomes proportional to T; determine the corresponding value of n and recalculate dv_D/dT for this value of n at $-55°$C.

1.14 It has been suggested that A_0' and n might be computed from measurements on an actual diode at various voltages, currents, and temperatures. Using Eq. 1.127 and various derivatives from it, consider for example the following measurements: at 300K, $I_D = 2.5$ mA and $V_D = 0.7310$ V. Then with I_D maintained at 2.5 mA as the temperature is raised, it is found that the diode voltage decreases to $V_D = 0.712$ when T is raised by 10°C.

    (a) Determine a value for A_0' and n in Eq. 1.127.

    (b) If the measured V_D is greater than the theoretical because of ohmic resistance or other reasons, what is the effect on your calculated value of n?

    (c) Discuss other reasons why the values obtained experimentally might differ from those obtained from structural and process parameters.

    (d) Unless measurements are measured very quickly after application of voltage, results will be in error. Why?

1.15. A single silicon diode is used in the circuit shown.

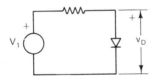

Prob. 1.15

    (a) Using Eq. 1.128 as a starting point, write an equation for v_D in terms of V_1 and R, and find the simplest expression you can for dv_D/dt for fixed values of V_1 and R.

    (b) For a diode at $I_D = 5 \times 10^{-3}$ A with $V_D = 0.7$ V and $V_1 = 5$ V, at 300 K determine R and obtain dv_D/dT using $n = 1.5$.

    (c) Repeat part (b) for two identical series diodes for the voltage $2v_D$.

REFERENCES

GROVE, A. S. *Physics and Technology of Semiconductor Devices*, Wiley, New York, 1967.

MULLER, R. S., AND T. I. KAMINS. *Device Electronics for Integrated Circuits*, Wiley, New York, 1977.

STREETMAN, B. G., *Solid State Electronic Devices*, 2nd ed., Prentice-Hall, Englewood Cliffs, N.J., 1980.

SZE, S. M. *Physics of Semiconductor Devices*, 2nd ed., Wiley, New York, 1980.

Van Der Ziel, A. *Noise, Sources, Characterization, Measurement*, Prentice-Hall, Englewood Cliffs, N.J., 1970.

————. *Solid State Physical Electronics*, Prentice-Hall, Englewood Cliffs, N.J., 1976.

Chapter 2

Bipolar Junction Transistors: Mathematical and Circuit Models

INTRODUCTION

The bipolar junction transistor is a two-junction device that makes use of the interactive properties of two *pn* junctions in proximity, where the two junctions have a common *p* or *n* region. Used correctly, such a device functions either as a current amplifier or as a switch.

It is the purpose of this chapter, first, to show how a reasonably good electrical model for the device can be evolved from single *pn* junction considerations as discussed in Chapter 1, and then how more accurate models can be obtained by taking additional factors into account, and finally how complete models can be derived and compared when frequency- and time-dependent parameters are considered.

The emphasis is on the development of the several commonly used incremental models and the relationships among their model parameters through their structural and process parameters. Factors affecting linearity in both high- and low-current regions are evaluated, empirical nonlinear models are suggested, and it is shown how they relate to the structural and process parameters.

The temperature dependency of the various parameters is discussed and the effect of temperature on overall performance is described. Sources of noise are considered and device noise models are developed.

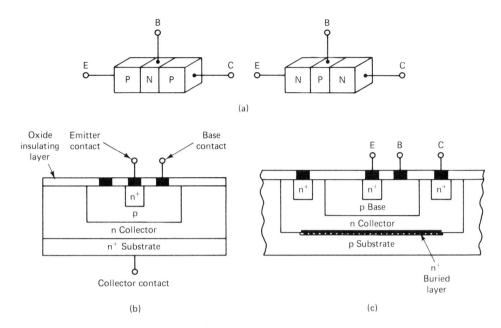

Figure 2.1 Bipolar junction transistor structures.

2.1 STRUCTURES, TERMINOLOGY, AND GENERAL CHARACTERISTICS

The two structures shown in Fig 2.1(a) represent the simplest concept of the bipolar junction transistor. The region separating the two junctions is referred to as the *base,* and the other regions as the *emitter* and *collector*. When the base is an *n*-type semiconductor, the transistor is referred to as a *pnp* transistor, and if it is *p* type, the structure is an *npn* transistor. The original "grown-junction" transistors were representative of this configuration.

Modern transistors are more commonly formed by a succession of diffusion or ion implantation processes resulting in structures that are some variation of those shown in cross section in Fig. 2.1(b) or (c) for *npn* transistors. In Fig. 2.1(b), the collector terminal is an ohmic contact with the *n*-collector region on the side opposite the emitter and base contacts, whereas in Fig. 2.1(c) the structure is diffused into a lightly doped *p*-substrate that acts as an isolating barrier between adjacent transistors, which might be formed at the same time. In this case, the ohmic collector contact is through a highly doped (n^+) region, with terminals in the same plane as the base and emitter contacts.

Such structures are useful in integrated-circuit design where many such transistors can be processed simultaneously, including appropriate interconnections, which also may include resistors and *pn* junctions used as capacitors.

The most useful configuration for the transistor is the common-emitter connection shown schematically for the *npn* transistor in Fig. 2.2(a), with the device functioning as a two-port network element with the base–emitter terminals as the input port and the collector–emitter terminals as the output port.

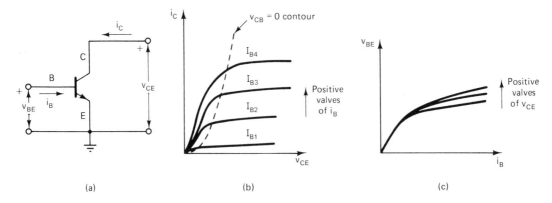

Figure 2.2 Suggested transistor voltage–current characteristics. (a) Transistor symbol. (b) Collector current–voltage characteristics. (c) Base voltage–current characteristics.

Graphically, the common-emitter characteristics for the common-emitter *npn* transistor are usually represented as shown in Fig. 2.2(b) and (c). The region to the right of the $v_{CB} = 0$ contour is referred to as the normal active region with the base–emitter junction forward biased (as for the forward-biased diode) and the collector–base junction reverse biased.

In such connections, the transistor functions as a current amplifier, with the current gain given by

$$\beta \equiv \frac{\partial i_C}{\partial i_B}\bigg|_{v_{CE} = \text{constant}}$$

2.2 BASIC MATHEMATICAL MODEL FOR THE NPN TRANSISTOR

A simple model for transistor action may be evolved by combining the action at the two junctions illustrated by the simple geometrical structure shown in Fig. 2.1(a). In this model, only diffusion currents will be considered, and variation of depletion layer widths with voltage will be neglected. These and other secondary effects, which were discussed in Chapter 1 for the single *pn* junction, can be accounted for by subsequent modifications of and additions to the basic model.

Consider first the base–collector junction of the *npn* transistor with the base–emitter junction open circuit as shown in Fig. 2.3(a). The presence of the unused emitter–base junction has negligible effect, and, except for the dissymmetry introduced by the base contact, the diode equation, Eq. 1.74, should be approximately valid for the collector–base junction. This equation, rewritten in terms of the terminal reference conditions shown, is

$$i_C = I_{CO}(1 - e^{-av_{CB}}) \tag{2.1}$$

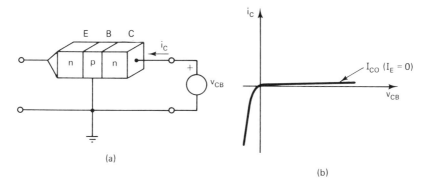

(a)

(b)

Figure 2.3 Transistor collector characteristic with open emitter.

and plotted in Fig. 2.3(b), where

$$a = \frac{q}{kT} = \frac{1}{V_T}$$

and I_{CO} is the open-circuit reverse-bias collector saturation current.

Now consider simultaneous forward bias of the emitter–base junction and reverse bias of the collector–base junction, as shown in Fig. 2.4(a). Normally, although it is not necessary to make most of the following discussion valid, the emitter is much more heavily doped than the base, and *most* of the charge carriers crossing the base–emitter junction are electrons moving by diffusion into the base, where they are referred to as minority carriers in the base region.

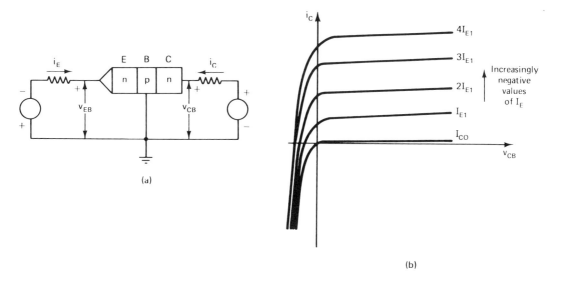

(b)

Figure 2.4 Common-base *npn* collector characteristic.

If the collector–base junction is zero or reverse biased (collector positive), most of the carriers will flow to the collector. This fraction of injected current α_N augments the collector current given by Eq. 2.1, as shown in Fig. 2.4(b), permitting the collector current to be written as

$$i_C = I_{CO}(1 - e^{-a v_{CB}}) - \alpha_N i_E \tag{2.2}$$

For $v_{CB} = 0$, $i_C = -\alpha_N i_E$; so α_N is defined as

$$\alpha_N = -\frac{i_C}{i_E}\bigg|_{v_{CB}=0} \tag{2.3}$$

Defined as the theoretical short-circuit current gain, α_N is a positive constant of $\alpha_N \lessgtr 1$. The negative sign for $\alpha_N i_E$ in Eq. 2.2 arises because the actual direction of i_E is away from the base when the base–emitter junction is forward biased. A set of equations like Eqs. 2.1 through 2.3 may be obtained by reversing the bias conditions for the base–emitter and collector–base junctions. Then the emitter–base and collector–base voltage–current characteristic for open collector–base junction is

$$i_E = I_{EO}(1 - e^{-a v_{EB}}) \tag{2.4}$$

as shown in Fig. 2.5(a).

Then with carriers (electrons) injected from a forward-biased base–collector junction,

$$i_E = I_{EO}(1 - e^{-a v_{EB}}) - \alpha_I i_C \tag{2.5}$$

as shown in Fig. 2.5(b), where

$$\alpha_I = -\frac{i_E}{i_C}\bigg|_{v_{EB}=0} \tag{2.6}$$

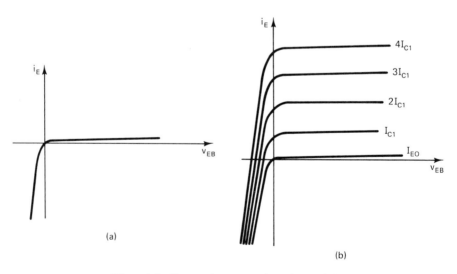

Figure 2.5 Common-base *npn* emitter characteristic.

is defined as the theoretical short-circuit common-base (or inverted) short-circuit current gain and I_{EO} is the open-circuit reverse bias saturation current.

If the emitter and collector regions are identical with respect to dimensions and impurity concentrations and each region is uniformly doped, the transistor is symmetrical, $\alpha_N = \alpha_I$, and the curves of Figs. 2.4(b) and 2.5(b) are identical. If the collector region is much more lightly doped than the emitter region, the currents are still essentially minority carriers in the base (electrons), but $\alpha_N >> \alpha_I$ and the scales of the two sets of curves are vastly different.

It can be shown from theoretical considerations that to a reasonable approximation Eqs. 2.2 and 2.5 simultaneously describe emitter and collector voltage–current relationships in all possible combinations of forward and reverse bias. In other words, a linear superposition of the two equations is a reasonably valid representation of the device characteristics.

Using Eqs. 2.2 and 2.5 together, the following symmetrical relationships may be written:

$$i_C = \frac{I_{CO}}{1 - \alpha_N\alpha_I} (1 - e^{-av_{CB}}) - \alpha_N \frac{I_{EO}}{1 - \alpha_N\alpha_I} (1 - e^{-av_{EB}}) \tag{2.7}$$

$$i_E = \frac{I_{EO}}{1 - \alpha_N\alpha_I} (1 - e^{-av_{EB}}) - \alpha_I \frac{I_{CO}}{1 - \alpha_N\alpha_I} (1 - e^{-av_{CB}}) \tag{2.8}$$

In Eq. 2.7 for $v_{EB} = 0$,

$$i_C = \frac{I_{CO}}{1 - \alpha_N\alpha_I} (1 - e^{-av_{CB}}) \tag{2.9}$$

Then, for v_{CB} large (highly reverse biased),

$$i_C = \frac{I_{CO}}{1 - \alpha_N\alpha_I} \equiv I_{CS} \tag{2.10}$$

which is the maximum value of reverse-bias collector current when the emitter–base junction is short circuited. Thus

$$I_{CS} \equiv \text{short-circuit reverse-bias collector saturation current}$$

Similarly, in Eq. 2.8 for $v_{CB} = 0$,

$$i_E = \frac{I_{EO}}{1 - \alpha_N\alpha_I} (1 - e^{-av_{EB}}) \tag{2.11}$$

and for large v_{EB},

$$i_E = \frac{I_{EO}}{1 - \alpha_N\alpha_I} \equiv I_{ES} \tag{2.12}$$

and

$$I_{ES} \equiv \text{short circuit reverse bias emitter saturation current}$$

Using these definitions, Eqs. 2.7 and 2.8 may be written as

$$i_C = I_{CS}(1 - e^{-av_{CB}}) - \alpha_N I_{ES}(1 - e^{-av_{EB}}) \tag{2.13}$$

$$i_E = I_{ES}(1 - e^{-av_{EB}}) - \alpha_I I_{CS}(1 - e^{-av_{CB}}) \tag{2.14}$$

These two equations are the standard forms of what are known as the Ebers–Moll equations for the *npn* transistor. They can also be derived directly from the basic equations of semiconductor physics, where it can be shown that they are valid under a variety of conditions and that the validity does not depend on either uniform regional doping or geometrical symmetry. It can be demonstrated that even when $\alpha_N \neq \alpha_I$

$$\alpha_N I_{ES} = \alpha_I I_{CS} \equiv I_S \tag{2.15}$$

Therefore, the Ebers–Moll equations are often written as

$$i_C = I_{CS}(1 - e^{-av_{CB}}) - \alpha_N I_{ES}(1 - e^{-av_{EB}}) \tag{2.16}$$

$$i_E = -\alpha_N I_{ES}(1 - e^{-av_{CB}}) + I_{ES}(1 - e^{-av_{EB}}) \tag{2.17}$$

A slightly modified form of these equations can be obtained as follows: Let $I_S = \alpha_N I_{ES} = \alpha_I I_{CS}$; add to and subtract from Eq. 2.16 the term $I_S e^{-av_{CB}}$; add to and subtract from Eq. 2.17 the term $I_S e^{-av_{EB}}$. Then defining

$$\beta_N = \frac{\alpha_N}{1 - \alpha_N} \tag{2.18}$$

$$\beta_I = \frac{\alpha_I}{1 - \alpha_I} \tag{2.19}$$

and using these substitutions, Eqs. 2.16 and 2.17 may be rewritten as

$$i_C = I_S(e^{-av_{EB}} - e^{-av_{CB}}) + \frac{I_S}{\beta_I}(1 - e^{-av_{CB}}) \tag{2.20}$$

$$i_E = -I_S(e^{-av_{EB}} - e^{-av_{CB}}) + \frac{I_S}{\beta_N}(1 - e^{-av_{EB}}) \tag{2.21}$$

Also,

$$i_B = -\frac{I_S}{\beta_I}(1 - e^{-av_{CB}}) - \frac{I_S}{\beta_N}(1 - e^{-av_{EB}}) \tag{2.22}$$

These equations show a component of current.

$$i_{CC} = I_S(e^{-av_{EB}} - e^{-av_{CB}}) \tag{2.23}$$

that is common to the emitter and collector; it is sometimes called the dominant component and would be the only component if $\alpha_N = 1$ and $\alpha_I = 1$. It is that component of current reaching the collector if the collector is zero or reverse biased and that component reaching the emitter if the emitter is reverse biased. It is the minority carrier current in the base in the absence of losses due to recombination.

The base current (Eq. 2.22) is the current that represents base and emitter recombination currents as well as hole current from base to emitter for the forward-bias emitter–base junction, or hole current from base to collector when the collector–base junction is forward biased.

The preceding equations are the form of Ebers–Moll equations that are often modified to take into account factors not included in the basic Ebers–Moll formulation, such as depletion layer recombinations, and high current effects, such as high-level injection, and geometrical effects, such as emitter crowding.

2.3 PNP TRANSISTOR

Essentially everything that has been discussed with respect to the *npn* transistor relative to development of the Ebers–Moll model is valid for the *pnp* transistor except that the minority carriers in the base region are holes rather than electrons. If the same voltage–current reference conditions are maintained at the terminals as shown in Fig. 2.6(a), the Ebers–Moll equations may be obtained from the following equations, analogous to Eqs. 2.2 and 2.5.

$$i_C = I_{CO}(e^{av_{CB}} - 1) - \alpha_N i_E \qquad (2.24)$$

$$i_E = I_{EO}(e^{av_{EB}} - 1) - \alpha_I i_C \qquad (2.25)$$

The collector current–voltage characteristics for Eq. 2.24 are plotted in Fig. 2.6(b). The emitter characteristics (not plotted) are of similar form.

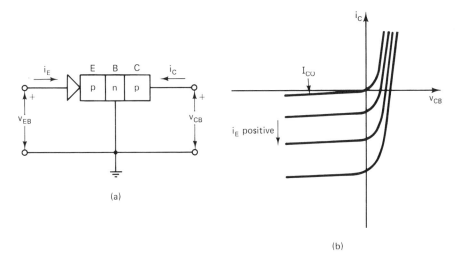

(a)

(b)

Figure 2.6 Common-base *pnp* transistor characteristic.

The resulting equations in standard Ebers–Moll form are

$$i_C = I_{CS}(e^{av_{CB}} - 1) - \alpha_N I_{ES}(e^{av_{CB}} - 1) \qquad (2.26)$$

$$i_E = I_{ES}(e^{av_{EB}} - 1) - \alpha_I I_{CS}(e^{av_{EB}} - 1) \qquad (2.27)$$

The modified form analogous to Eqs. 2.20, 2.21, and 2.22 are

$$i_C = I_S(e^{av_{CB}} - e^{av_{EB}}) + \frac{I_S}{\beta_I}(e^{av_{CB}} - 1) \qquad (2.28)$$

$$i_E = - I_S(e^{av_{CB}} - e^{av_{EB}}) + \frac{I_S}{\beta_N}(e^{av_{EB}} - 1) \qquad (2.29)$$

$$i_B = - \frac{I_S}{\beta_I}(e^{av_{CB}} - 1) - \frac{I_S}{\beta_N}(e^{av_{EB}} - 1) \qquad (2.30)$$

2.4 COMMON-EMITTER CONNECTION

Because of its less-than-unity current gain in the common-base connection, the transistor finds its most widespread use in the common-emitter form, as suggested in Sec 2.1 and as indicated in Fig. 2.7, using two different symbolic representations for the *npn* structure.

Because of the normally forward biased base–emitter junction, it is most useful to consider the transistor as a current-driven device. Then the most useful two-port relationships are

$$i_C = h_2(v_{CE}, i_B) \qquad (2.31)$$

$$v_{BE} = h_1(v_{CE}, i_B) \qquad (2.32)$$

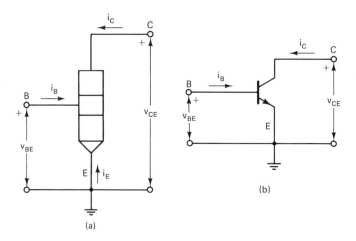

Figure 2.7 Common-emitter transistor configuration.

Also

$$v_{BE} = g_1(i_B, i_C) \tag{2.33}$$

Using the functional relationships of these equations along with $i_B = -i_C - i_E$, $v_{BE} = -v_{EB}$, and $v_{CE} = v_{CB} + v_{BE}$, the Ebers–Moll equations may be manipulated into the forms

$$i_C = \frac{I_{CEO}(1 - e^{-av_{CE}}) + \beta_N \left[1 - e^{-av_{CE}}\left(\dfrac{\beta_N}{1 + \beta_N} + \dfrac{I_{CEO}}{\beta_N I_{ES}}\right)\right] i_B}{1 + e^{-av_{CE}}\left(\dfrac{I_{CEO}}{I_{ES}} - \dfrac{\beta_N}{1 + \beta_N}\right)} \tag{2.34}$$

$$e^{av_{BE}} = \frac{I_{CEO} + \dfrac{1}{1 + \beta_N} I_{ES} + (1 + \beta_N) i_B}{I_{ES}\left[1 + e^{-av_{CE}}\left(\dfrac{I_{CEO}}{I_{ES}} - \dfrac{\beta_N}{1 + \beta_N}\right)\right]} \tag{2.35}$$

$$e^{av_{BE}} = 1 + \frac{1}{I_{ES}}\left(\frac{\beta_N^2}{1 + \beta_N}\frac{I_{ES}}{I_{CEO}} + 1\right) i_B + \frac{1}{I_{ES}}\left(1 - \frac{I_{ES}}{I_{CEO}}\frac{\beta_N}{1 + \beta_N}\right) i_C \tag{2.36}$$

where

$$\beta_N = \frac{\alpha_N}{1 - \alpha_N} \qquad \text{and} \qquad I_{CEO} = \frac{I_{CO}}{1 - \alpha_N}$$

The correct interpretation of β_N and I_{CEO} can be made with reference to Fig. 2.8 along with the common-base Ebers–Moll equations for $v_{CB} = 0$. These are

$$i_C = -\alpha_N I_{ES}(1 - e^{-av_{EB}})$$

$$i_E = I_{ES}(1 - e^{-av_{EB}})$$

Also, $i_B = -i_C - i_E$.

Then

$$\left.\frac{i_C}{i_B}\right]_{v_{CB}=0} = \frac{\alpha_N}{1 - \alpha_N} \equiv \beta_N \tag{2.37}$$

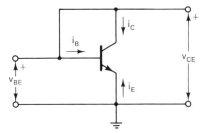

Figure 2.8 Common-emitter characteristic with base connected to collector.

Thus β_N is defined as the forward short-circuit current gain in the common-emitter connection for $v_{CB} = 0$ or $v_{CE} = v_{BE}$. Notice that this is identical with β_N arbitrarily defined in Eq. 2.18.

Equation 2.34 for zero-base current (open circuit) and v_{CE} large reduces to

$$i_C = \frac{I_{CO}}{1 - \alpha_N} \equiv I_{CEO} \tag{2.38}$$

Thus I_{CEO} is defined as the open-circuit reverse-bias collector saturation current in the common-emitter connection.

Regions of Operation

A plot of Eq. 2.34, which is $i_C = h_2(v_{CE}, i_B)$ for small positive values of v_{CE}, is made in Fig. 2.9(a) for a transistor for $\beta_N = 100$.

(a) The region to the right of the $v_{CE} = v_{BE}$ contour ($v_{CB} = 0$) represents the condition for reverse-bias collector–base junction and forward-bias base–emitter junction. This region and that extending to substantially larger values of v_{CE} is usually referred to as the *normal* or *forward active region*.

(b) The region to the left of $v_{CE} = v_{BE}$ has both junctions forward biased and is often referred to as the *forward saturation region*. It should be noted, however, that the form of the characteristic through most of this region is like that of the normal active

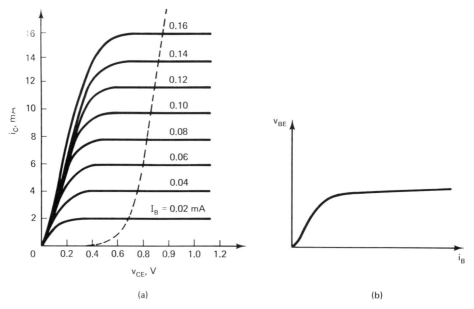

Figure 2.9 Common-emitter current–voltage characteristics. (a) Collector characteristic. (b) Base characteristic.

region, and what is usually thought of as saturation occurs only at very low voltages where all of the base current lines merge. That situation represents a more operational definition of saturation.

The base-emitter characteristic defined by Eq. 2.35 or 2.36 is a relatively insensitive function of v_{CE} or i_C and has the general form shown in Fig. 2.9 for positive values of i_B, given approximately from Eq. 2.35 for large values of v_{CE} by

$$e^{av_{BE}} = \frac{I_{CEO}}{I_{ES}} + (1 + \beta_N)\frac{i_B}{I_{ES}} \tag{2.39}$$

(c) The characteristics in the vicinity of the origin may be examined more closely as follows: If $i_C = 0$ in Eq. 2.34,

$$e^{av_{CE}} = \frac{I_{CEO} + \left(\dfrac{\beta_N}{1 + \beta_N} + \dfrac{I_{CEO}}{\beta_N I_{ES}}\right)\beta_N i_B}{I_{CEO} + \beta_N i_B} \tag{2.40}$$

For Eq. 2.40, for $I_{CEO} \ll \beta_N i_B$,

$$e^{av_{CE}} = \frac{\beta_N}{1 + \beta_N} + \frac{I_{CEO}}{\beta_N i_{ES}} \tag{2.41}$$

These results are plotted in Fig. 2.10(a), where Eq. 2.41 represents the flat portion of the curve.

Now, for $v_{CE} = 0$ in Eq. 2.34,

$$i_C = \frac{\beta_N/\{(1 + \beta_N)\}\{1 - [(1 + \beta_N)/\beta_N][I_{CEO}/I_{ES}]\}i_B}{1/(1 + \beta_N) + I_{CEO}/I_{ES}} \tag{2.42}$$

The i_C axis crossing is represented by Eq. 2.42 and the v_{CE} axis crossing by Eq. 2.40. These results are shown in Fig. 2.10(b). This region in the fourth quadrant identifies $v_{CE} > 0 < v_{BE}$ for $i_C < 0$. Both junctions are still forward biased but the collector current reverses direction. This region and the slight extension into the third quadrant is called the *inverse* or *reverse saturation region*.

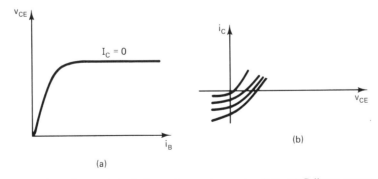

(a)

(b)

Figure 2.10 Transistor characteristics in low voltage and current regions. (a) Collector current = zero. (b) Low-voltage, low-current inverse saturation region.

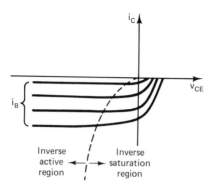

Figure 2.11 Transition from inverse saturation to inverse active region.

(d) As v_{CE} becomes negative in Eq. 2.34, operation is still in the third quadrant with both junctions still forward biased until a point is reached where $v_{BE} \to 0$ and the collector–base junction is simply that of a forward-biased diode. This is the limit of forward-bias conditions for both junctions and hence the limit of the inverse saturation region. The contour defining this limit is obtained by setting $v_{BE} = 0$ in Eq. 2.35 and using the result with Eq. 2.36 to obtain

$$i_C = -\left[\frac{I_{CEO}}{1 + \beta_N} + \left(\frac{\beta_N}{1 + \beta_N}\right)^2 I_{ES}\right][e^{-av_{CE}} - 1] \qquad (2.43)$$

This contour is the dashed line shown in Fig. 2.11. Then as v_{CE} becomes more negative (maintaining i_B positive), the base–emitter junction becomes reverse biased while the collector–base junction remains forward biased, and the region functions as an active region with roles of emitter and collector interchanged. This region to the left of the $v_{BE} = 0$ contour is referred to as the *inverse active region*.

In the inverse active region for $e^{-av_{CE}} \gg 1$, Eq. 2.34 reduces to

$$i_C = \frac{-I_{CEO} - \beta_N\{[\beta_N/(1 + \beta_N)] + (I_{CEO}/\beta_N I_{ES})\}i_B}{(I_{CEO}/I_{ES}) - \beta_N/(1 + \beta_N)} \qquad (2.44)$$

becoming independent of v_{CE} for even quite small values.

2.5 EXTRACTION OF BASIC PARAMETERS FROM TRANSISTOR MEASUREMENTS

In the forward and reverse (normal and inverse) saturation regions and in the relatively low voltage normal and inverse active regions, the Ebers–Moll equations and the model they represent, sometimes referred to as the diffusion model of the transistor, are reasonably valid and are extremely useful in modeling the switching characteristics of bipolar junction transistors.

In the active regions (primarily normal active) for higher voltages where transistors are usually used as small-signal amplifiers, most transistors depart somewhat from those defined by the simple model, and appropriate correction terms are required to accurately model the device, either by adding theoretical terms involving structural parameters to

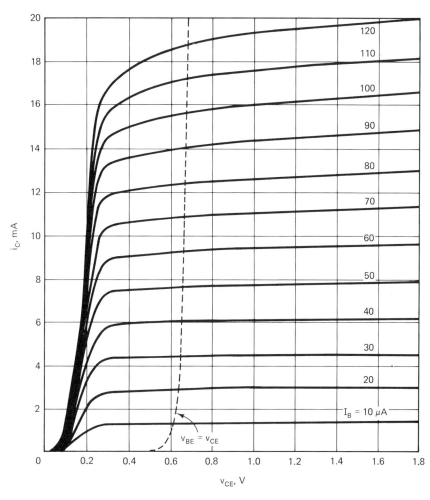

Figure 2.12 Common-emitter collector current–voltage characteristic of actual transistor at low voltages.

the basic model or by empirical modeling. In either case, it is possible by appropriate measurements to extract the basic parameters β_N, I_{CEO}, and I_{ES} from which the other Ebers–Moll parameters can be calculated and to which terms may be added to build a more accurate model. One manner in which this may be done is outlined as follows.

For the specific transistor having the collector characteristics at low voltages shown in Fig. 2.12, measurements were made along the $v_{BE} = v_{CE}$ (i.e., $v_{CB} = 0$) contour and plotted in semilog form as indicated in Fig. 2.13, where both collector current and base current are straight lines parallel to each other over several decades of current. The short-circuit current gain β_N is simply the ratio of i_C/i_B at some value of $v_{CE} = v_{BE}$ in the linear range. A value for $a = q/mkT$ can be obtained from the slope of the collector current curve as indicated and a value for I_{ES} obtained from the linear extrapolation of this curve to the i_C intersection for $v_{BE} = 0$. From a plot of $v_{CE} = f(i_B)$ for $i_C = 0$ (like Eq. 2.40), as shown in Fig. 2.14, I_{CEO} can be calculated from Eq. 2.41 as indicated.

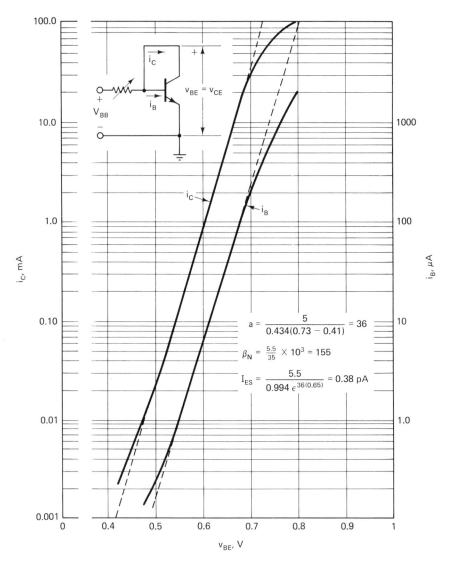

Figure 2.13 Collector and base currents for collector connected to base.

2.6 LINEAR AND NONLINEAR ANALYTICAL MODELS FOR TRANSISTORS IN THE NORMAL ACTIVE REGION

If Eq. 2.34 were valid, all the normal active region and a portion of the forward saturation region could be represented by

$$i_C = I_{CEO} + \beta_N i_B \simeq \beta_N i_B \qquad (2.45)$$

as shown in Fig. 2.15(a).

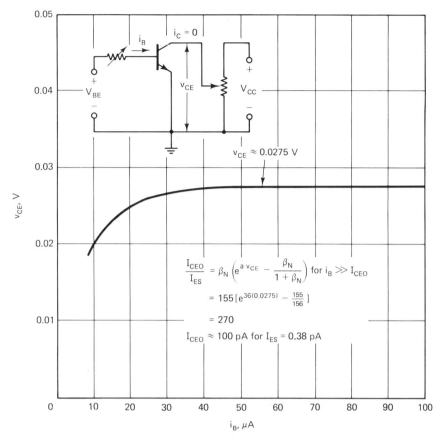

Figure 2.14 Collector–emitted voltage versus base current for zero collector current.

At the very least, but most simple approximation, an actual transistor departs from the diffusion model in a manner that can be approximated by the linear equation

$$i_C = I_{OS} + g_O v_{CE} + \beta_O i_B \qquad (2.46)$$

where, in terms of the diffusion model,

$$I_{OS} = I_{CEO} \pm \Delta I$$

$$\beta_O = \beta_N \pm \Delta\beta$$

$$g_O = \left. \frac{\Delta I}{\Delta V} \right]_{I_B = \text{constant}}$$

These characteristics are shown in Fig. 2.15(b).

Even when the departure from linearity is significant, as indicated in Fig. 2.15(c), Eq. 2.46 can still be used as a model in the neighborhood of a specified operating point;

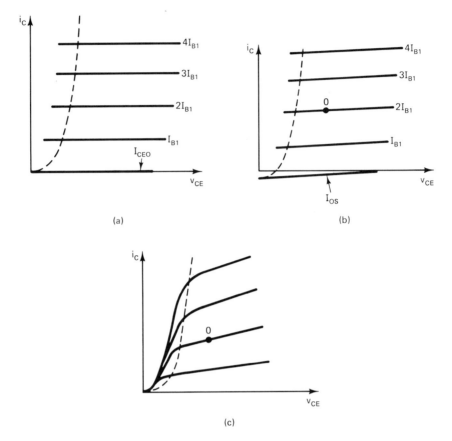

Figure 2.15 Approximate collector current–collector voltage characteristic for nonideal transistors.

however, over a wider range another term can be added to Eq. 2.46 to represent the characteristic more accurately as

$$i_C = I_{OS} + g_O v_{CE} + \beta_O i_B + f_1(i_B, v_{CE}) \tag{2.47}$$

where $f_1(i_B, v_{CE})$ is an empirical term. One possible and often useful form is

$$f_1(i_B, v_{CE}) = k_B i_B^x v_{CE}^y \tag{2.48}$$

where k_B, x, and y can be obtained from measured data.

For example, for the transistor shown in Fig. 2.16 (the same one previously referred to in Fig. 2.12), measurements at points 1, 2, 3, and 4 yield enough data to determine the constants. A variety of measurement sequences can be used. The base–emitter characteristic expressed as

$$v_{BE} = g(i_B, v_{CE})$$

(Eq. 2.35 for the diffusion model) is highly nonlinear but almost completely independent of v_{CE} in the normal active region. The actual transistor as shown in Fig. 2.17 is somewhat

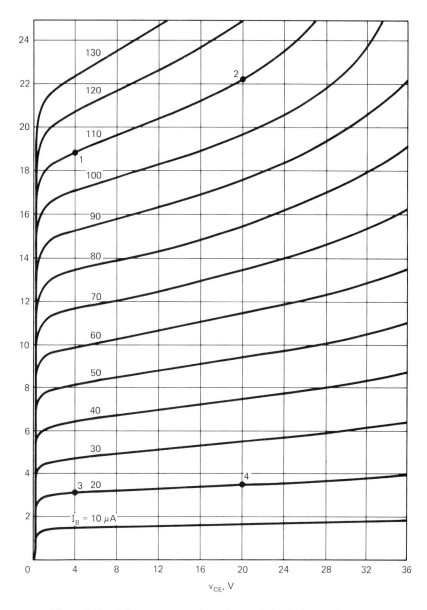

Figure 2.16 Collector current–voltage characteristics for large voltages.

less independent. This characteristic can be modeled somewhat more accurately as follows: Start with Eq. 2.36, which is of the form

$$e^{av_{BE}} = 1 + Bi_B + Ci_C \tag{2.49}$$

This equation is still reasonably accurate if the nonlinear model (Eq. 2.47) is used for i_C in Eq. 2.49 as

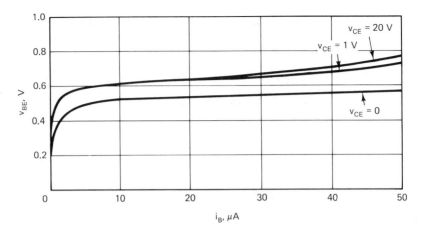

Figure 2.17 Base–emitter voltage–current characteristics.

$$e^{av_{BE}} \simeq Bi_B + C[I_{OS} + g_o v_{CE} + \beta_o i_B + f_1(i_B, v_{CE})] \qquad (2.50)$$

This allows us to write $v_{BE} = h_1(i_B, v_{CE})$ without the v_{BE} versus v_{CE} relationship being entirely lost.

2.7 INCREMENTALLY LINEAR (SMALL SIGNAL) CIRCUIT MODELS

Whether a transistor is described in its forward active region by an analytical model as discussed in the previous section or whether it is characterized only by a measured set of graphical characteristics, a linear model can be used to describe it in terms of deviation from a prescribed operating point, as indicated in Fig. 2.18. Any set of two-port network parameters theoretically may be used. However, those indicated by the functional relationships of Eqs. 2.31 and 2.32 more nearly fit the characteristics indicated.

Thus, starting with

$$i_C = h_2(v_{CE}, i_B) \qquad (2.31)$$

$$v_{BE} = h_1(v_{CE}, i_B) \qquad (2.32)$$

and noting an operating or quiescent point of $v_{CE} = V_{CE}$, $i_C = I_C$, $v_{BE} = V_{BE}$, and $i_B = I_B$, differential changes from this point may be written:

$$di_C = \frac{\partial i_C}{\partial i_B}\bigg]_{v_{CE} = V_{CE}} di_B + \frac{\partial i_C}{\partial v}\bigg]_{i_C = I_C} dv_{CE} \qquad (2.51)$$

$$dv_{BE} = \frac{\partial v_{BE}}{\partial i_B}\bigg]_{v_{CE} = V_{CE}} di_B + \frac{\partial v_{BE}}{\partial v_{CE}}\bigg]_{i_B = I_B} dv_{CE} \qquad (2.52)$$

Analytically, the partial derivatives can be evaluated from Eqs. 2.47 or 2.48.

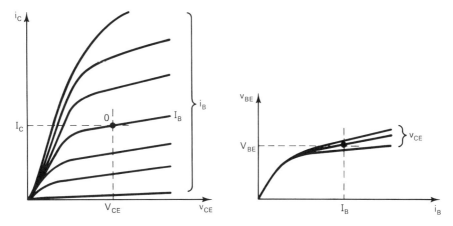

Figure 2.18 Transistor common-emitter characteristics for demonstration of development of incremental circuit models.

However, when (as is more often the case) the mathematical expressions for the nonlinearities have not been evaluated, linear approximations in the vicinity of the operating point as based on graphical or numerical data may be determined as follows: approximating Eqs. 2.51 and 2.52 by

$$\Delta i_C = \frac{\delta i_C}{\delta i_B}\bigg]_{v_{CE}=V_{CE}} \Delta i_B + \frac{\delta i_C}{\delta v_{CE}}\bigg]_{i_B=I_B} \Delta v_{CE} \tag{2.53}$$

$$\Delta v_{BE} = \frac{\delta v_{BE}}{\delta i_B}\bigg]_{v_{CE}=V_{CE}} \Delta i_B + \frac{\delta v}{\delta v}\bigg]_{i_B=I_B} \Delta v_{CE} \tag{2.54}$$

or in slightly simpler somewhat standardized notation

$$i_c = h_{fe}i_b + h_{oe}v_{ce} \tag{2.55}$$

$$v_{be} = h_{ie}i_b + h_{re}v_{ce} \tag{2.56}$$

where

$$i_c \equiv \Delta i_C, \qquad v_{ce} \equiv \Delta v_{CE}, \qquad v_{be} \equiv \Delta v_{BE}, \qquad i_b \equiv \Delta i_B$$

$$h_{fe} \equiv \frac{\partial i_C}{\partial i_B} \approx \frac{\delta i_C}{\delta i_B} \tag{2.57}$$

$$h_{oe} \equiv \frac{\partial i_C}{\partial v_{CE}} \approx \frac{\delta i_C}{\delta v_{CE}} \tag{2.58}$$

$$h_{ie} \equiv \frac{\partial v_{BE}}{\partial i_B} \approx \frac{\delta v_{BE}}{\delta i_B} \tag{2.59}$$

$$h_{re} \equiv \frac{\partial v_{BE}}{\partial v_{CE}} \approx \frac{\delta v_{BE}}{\delta v_{CE}} \tag{2.60}$$

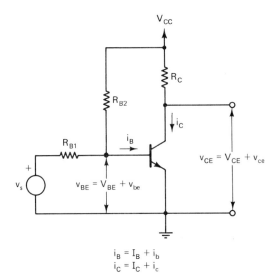

$$i_B = I_B + i_b$$
$$i_C = I_C + i_c$$

Figure 2.19 Transistor with simple bias network.

where the δ's may be obtained by measurements on the graphical characteristic or may be determined by small-signal ac measurements on actual devices.

The parameters h_{fe}, h_{oe}, h_{ie}, and h_{re} are defined as the incremental (small-signal) h parameters in the common-emitter connection. The first of the two subscripts have the normal identifications, f = forward, i = input, o = output, and r = reverse, while the second identifies the terminal that is common to both input and output.

The circuit of Fig. 2.19 illustrates a specific configuration that is consistent with the operating definitions that have been established and shows one method of establishing an operating point.

V_{CE}, I_C, V_{BE}, and I_B are established graphically for the signal input voltage $v_s = 0$. Then i_c, v_{ce}, i_b, and v_{be} are incremental changes as a result of application of v_s.

The device h parameters correspond to h-parameter designations of a two-port linear network (passive or active) in general, where as indicated in Fig. 2.20,

$$h_{21} = \frac{i_2}{i_1}\bigg]_{v_2=0} \qquad h_{22} = \frac{i_2}{v_2}\bigg]_{i_1=0}$$

$$h_{11} = \frac{v_1}{i_1}\bigg]_{v_2=0} \qquad h_{12} = \frac{v_1}{v_2}\bigg]_{i_1=0}$$

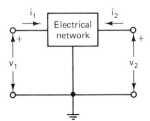

Figure 2.20 Sign conventions for h-parameter determination.

It is customary to use the device parameter notation when specifying characteristics of the device alone and the network parameter nomenclature when considering a circuit in which the device is only one component.

h-Parameter Circuit Model

The electrical circuit corresponding to Eqs. 2.55 and 2.56, known as the common-emitter h-parameter circuit model, is shown in Fig. 2.21(a). Inasmuch as the general h-parameter definitions may include frequency-dependent parameters, which will be discussed later, it is useful to devise a second set of symbols for the low-frequency or static small-signal h parameters as indicated in Fig. 2.21(b), where $\mu_{re} = h_{re}$ at low frequencies, $\beta = h_{fe}$ at low frequencies, $1/r_{ceo} = h_{oe}$ at low frequencies, and $r_{bes} = h_{ie}$ at low frequencies.

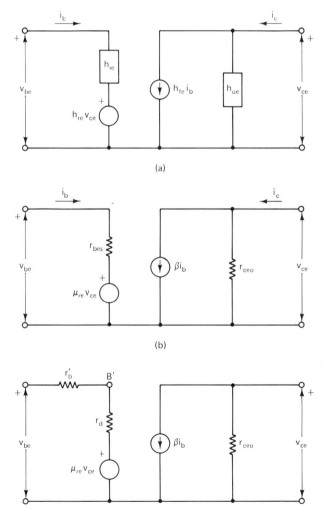

(a)

(b)

(c)

Figure 2.21 The h-parameter circuit model. (a) General model. (b) Low-frequency parameter model. (c) Modified low-frequency parameter model.

Furthermore, it is sometimes convenient to separate r_{bes} into two components, as indicated in Fig. 2.21(c), where $r_{bes} = r'_b + r_d$. Here r_d is defined as the diffusion component, which is principally the theoretical value of the input resistance as evolved from the Ebers–Moll relationships, and r'_b is the extrinsic or ohmic resistance external to the fundamental transistor processes, being largely a function of the resistivity of the base material itself.

Hybrid-π Model

It is sometimes useful to replace everything except r'_b by the π equivalent of the h-parameter model as indicated in Fig. 2.22. This is called the low-frequency hybrid-π circuit model.

The values of the hybrid-π parameters are obtained simply, *first* by solving for the h parameters of the designated portion of Fig. 2.22(b) according to their definitions. The results are

$$r_d = \frac{r'_{be} r'_{cb}}{r'_{be} + r'_{cb}} \tag{2.61}$$

$$\beta = r_{b'e} \left| \frac{g_m r_{cb'} - 1}{r_{b'e} + r_{cb'}} \right| \tag{2.62}$$

$$\frac{1}{r_{ceo}} = \frac{1 + g_m r_{b'e}}{r_{b'e} + r_{cb'}} + \frac{1}{r_{ce}} \tag{2.63}$$

$$\mu_{re} = \frac{r_{b'e}}{r_{b'e} + r_{cb'}} \tag{2.64}$$

This set of equations may then be solved simultaneously to obtain the hybrid-π parameters in terms of the h parameters. These results are

$$r_{b'e} = \frac{r_d}{1 - \mu_{re}} \simeq r_d \tag{2.65}$$

$$r_{cb'} = \frac{r_d}{\mu_{re}} \tag{2.66}$$

$$\frac{1}{r_{ce}} = \frac{1}{r_{ceo}} - \frac{\mu_{re}(1 + \beta)}{r_d} \tag{2.67}$$

$$g_m = \frac{\beta + \mu_{re}}{r_d} \simeq \frac{\beta}{r_d} \tag{2.68}$$

Although the hybrid-π circuit model is an equivalent circuit having only one dependent or controlled source rather than two as in the case of the h-parameter model, this source is a function of an internal voltage and, hence, in general does not result in

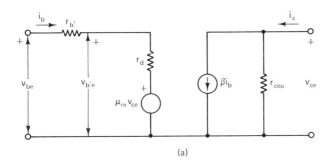

(a)

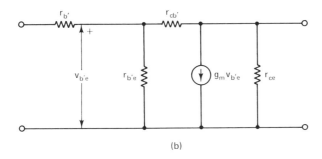

(b)

Figure 2.22 Hybrid-π circuit model from h-parameter model. (a) Modified h-parameter model. (b) Low-frequency hybrid-π model.

any significant simplification in the solution of actual circuit problems. Furthermore, the internal point is not accessible for parameter measurements.

It is also interesting to note that if we arbitrarily define

$$g'_m = \frac{\partial i_C}{\partial v'_{BE}} \bigg|_{v_{CE} = V_{CE}} \tag{2.69}$$

The result is

$$g'_m = \frac{\beta}{r_d} \tag{2.70}$$

which differs only minutely from Eq. 2.68. Hence it is convenient to think of Eq. 2.69 as the basic definition of g_m, which historically is referred to as the transconductance of a device; this is basically the rate of change of output current with respect to input voltage when the output voltage is maintained at a constant value. It is a most useful parameter when the device in question is voltage driven and exhibits a high input impedance relative to the source impedance.

T-Circuit Model

It is possible to replace the h-parameter model in Fig. 2.22(a) with a T equivalent, as illustrated in Fig. 2.23(a), excluding r'_b as was done for the hybrid-π model.

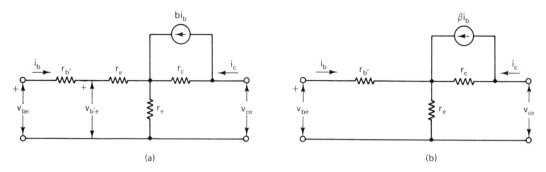

Figure 2.23 T-parameter circuit models. (a) General low-frequency T. (b) Approximate T-model.

The h parameters of the T model are shown from basic definitions to be

$$\beta = b\left(1 - \frac{r_e}{r_e + r_c}\right) - \frac{r_e}{r_e + r_c} \tag{2.71}$$

$$\mu_{re} = \frac{r_e}{r_e + r_c} \tag{2.72}$$

$$r_{ceo} = r_e + r_c \tag{2.73}$$

$$r_d = r_a + \frac{r_e r_c}{r_e + r_c}(1 + \beta) \tag{2.74}$$

These equations solved simultaneously yield the T parameters in terms of the h parameters, which are

$$b = \frac{\mu_{re} + \beta}{1 - \mu_{re}} \simeq \beta \tag{2.75}$$

$$r_e = \mu_{re} r_{ceo} \tag{2.76}$$

$$r_c = (1 - \mu_{re})r_{ceo} \simeq r_{ceo} \tag{2.77}$$

$$r_a = r_d - \mu_{re} r_{ceo}(1 + \beta) \tag{2.78}$$

Usually, $r_a \ll r_b'$ and the T model may be simplified to the form shown in Fig. 2.23(b). Also from Eqs. 2.78 and 2.76, $r_a = r_d - r_e(1 + \beta)$ and, for $r_a \to 0$,

$$r_e \simeq \frac{r_d}{1 + \beta} \simeq \mu_{re} r_{ceo} \simeq \frac{1}{g_m} \tag{2.79}$$

It will be useful to keep these approximate relationships arising from $r_a \to 0$ in mind, as well as their implication. For example, it is interesting to note that this approximation leads to a value of $r_{ce} = \infty$ in the hybrid-π model by substitution into Eq. 2.67.

2.8 HIGH-FREQUENCY CIRCUIT MODELS

Frequency-dependent circuit models for bipolar junction transistors, sufficiently accurate for many purposes, can be obtained by adding junction and diffusion capacitances to any of the incremental low-frequency models, as indicated in Fig. 2.24. The capacitance $C_{cb'}$ is primarily collector–base junction capacitance and has the voltage dependency discussed in Sec. 1.6 for the *pn* junction. The capacitance $C_{b'e}$ is composed of two components assumed to be acting in parallel:

$$C_{b'e} = C_{je} + C_b$$

where C_{je} is the junction capacitance of the forward-biased base–emitter junction, and C_b is the diffusion capacitance due to charge stored in the base as a result of minority carrier flow, also as discussed in Sec. 1.6 for the forward-biased *pn* junction.

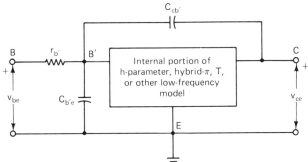

Figure 2.24 Capacitance additions to low-frequency circuit models.

The somewhat more complicated circuit model shown in Fig. 2.25, including the effect of extrinsic emitter resistance (small, but important in some circuit configurations) and additional peripheral capacitances, can be used for more accurate high-frequency solutions.

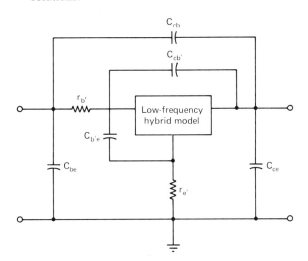

Figure 2.25 Additional capacitances for more accurate high-frequency representation.

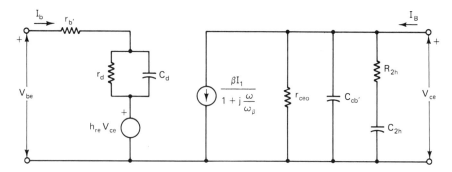

Figure 2.26 Complete frequency-dependent *h*-parameter circuit model.

Frequency-Dependent h-Parameter Model

The complete terminal-oriented *h*-parameter model may be obtained simply by solving for the parameters of one of the several circuits suggested in Fig. 2.24. The results in terms of the low-frequency *h* parameters are

$$h_{ie} = r'_b + \frac{r_d}{1 + j(\omega/\omega_\beta)} \tag{2.80}$$

where

$$\omega_\beta = \frac{1}{r_d C_d}, \quad \text{with } C_d = C_{cb'} + C_{b'e}$$

$$h_{fe} = \frac{\beta}{1 + j\omega/\omega_\beta} \tag{2.81}$$

$$h_{oe} = \frac{1}{r_{ceo}} + j\omega C_{cb'} + \frac{1}{R_{2h} + (1/j\omega C_{2h})} \tag{2.82}$$

where

$$R_{2h} = \frac{1}{\beta C_{cb} \, \omega_\beta} \quad \text{and} \quad C_{2h} = \beta C_{cb'}$$

$$h_{re} = \frac{\mu_{re} \{1 + j(\omega/\omega_\beta) \, C_{cb'}/[\mu_{re} \, (C_{cb'} + C_{b'e}]\}}{1 + j(\omega/\omega_\beta)} \tag{2.83}$$

The complete circuit model using these parameters is shown in Fig. 2.26.

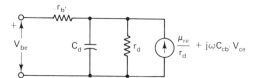

Figure 2.27 Modified input circuit for *h*-parameter model.

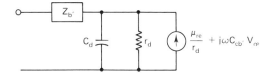

Figure 2.28 Alternate modified input circuit for h-parameter circuit model.

A minor simplifying modification is made of the input portion of the model by changing the series circuit between b' and e to the parallel equivalent shown in Fig. 2.27. It should be pointed out that for most transistors r_b' is not a pure resistance and varies with frequency at very high frequencies in a quite complicated manner. It can be replaced with an impedance $\mathbf{Z}_b'$, as shown in Fig. 2.28. Often an approximation using r_b' in parallel with C_b', where $C_b' = kC_b$ and $k < 1$, can be used as a more accurate representation.

The complete frequency-dependent h-parameter model is not often used in all its complexity because it is usually simpler to use one of the forms of Fig. 2.24. However, the h-parameter model illustrates well the high-frequency limitations of transistors. When connected to a source and load with very low values of load resistance, the output admittance and reverse voltage terms may be neglected, and the h-parameter model simplifies to that shown in Fig. 2.29. Then, under short-circuit conditions ($R_L \rightarrow 0$), ω_β is the angular frequency for which the current gain drops to 70.7% of its low-frequency value. Such a frequency in network usage is called the *cutoff frequency*. Hence ω_β or $f_\beta = \omega_\beta/2\pi$ is known as the β-cutoff frequency. At a frequency $\omega_T = \beta\omega_\beta$, known as the *transition frequency*, the short-circuit current gain drops to unity. This is also an important parameter because it relates to base transit time for minority carriers, as discussed in Sec. 1.6 for the diode.

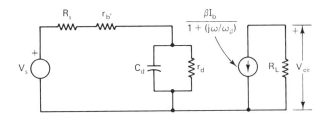

Figure 2.29 Simplified approximate high-frequency h-parameter circuit model.

2.9 VARIATIONAL NATURE OF INCREMENTAL PARAMETERS

It is useful first to examine the nature of bipolar transistor parameters assuming that the Ebers–Moll equations are operable and then to study the effects of additional nonlinearities as modifications of the basic characteristics.

In the normal active region, Eq. 2.34 reduces to Eq. 2.45, which is

$$i_C = I_{CEO} + \beta_N i_B \tag{2.45}$$

Rewriting Eq. 2.35 for $v_{B'E} = h_1(i_B, v_{CE})$ yields

$$
v_{B'E} = \frac{kT}{q} \left\{ \ln \left[I_{CEO} + \frac{I_{ES}}{1 + \beta_N} + (1 + \beta_N)i_B \right] \right.
$$

$$
\left. - \ln I_{ES} \left[1 + e^{-q(v_{CE}/kT)} \left(\frac{I_{CEO}}{I_{ES}} - \frac{\beta_N}{1 + \beta_N} \right) \right] \right\} \tag{2.84}
$$

Then the diffusion component of r_{bes} is

$$
r_d = \frac{\partial v_{B'E}}{\partial i_B} \bigg]_{i_B = I_B} = \frac{kT}{q} \frac{1 + \beta_N}{I_{CEO} + \dfrac{I_{ES}}{1 + \beta_N} + (1 + \beta_N)I_B} \tag{2.85}
$$

or in terms of the emitter current magnitude

$$
r_d = \frac{kT}{q} \frac{1 + \beta_N}{|I_E| + [I_{ES}/(1 + \beta_N)]} \tag{2.86}
$$

but since normally,

$$
|I_E| \gg \frac{I_{ES}}{1 + \beta_N}
$$

$$
r_d \simeq \frac{kT}{q} \frac{(1 + \beta_N)}{|I_E|} \tag{2.87}
$$

Then, for the hybrid-π model of Fig. 2.22, using Eq. 2.68,

$$
g_m = \frac{\beta_N}{1 + \beta_N} \frac{q}{kT} \left(|I_E| + \frac{I_{ES}}{1 + \beta_N} \right) \tag{2.88}
$$

or

$$
g_m \simeq \frac{\beta_N}{1 + \beta_N} \frac{q}{kT} |I_E| \tag{2.89}
$$

Also, using Eqs. 2.36 and 2.45 along with $i_B + i_C + i_E = 0$,

$$
e^{q(v_{B'E}/kT)} = \frac{1}{1 + \beta_N} - \frac{i_E}{I_{ES}} \tag{2.90}
$$

or

$$
v_{B'E} = \frac{kT}{q} \ln \left(\frac{1}{1 + \beta_N} - \frac{i_E}{I_{ES}} \right) \tag{2.91}
$$

Then the incremental emitter resistance in the T-equivalent circuit is

$$
r_e = \frac{\partial v_{B'E}}{\partial i_E} = \frac{kT}{q} \frac{1 + \beta_N}{(1 + \beta_N)|I_E| + I_{ES}} \tag{2.92}
$$

or
$$r_e \simeq \frac{kT}{q} \frac{1}{|I_E|} \tag{2.93}$$

for $(1 + \beta_N)|I_E| \gg I_{ES}$.

This result also could have been obtained theoretically from Eq. 2.76, $r_e = \mu_{re} r_{ceo}$, using

$$\frac{1}{r_{ceo}} = \frac{\partial i_C}{\partial v_{CE}} \bigg]_{i_B \,=\, I_B}$$

from Eq. 2.34, and

$$\mu_{re} = \frac{\partial v_{B'E}}{\partial v_{CE}} \bigg]_{i_B \,=\, I_B}$$

from Eq. 2.35.

Since $I_{ceo} \ll \beta_N i_b$, the quantity $h_{FE} = I_C/I_B$ can be used in the equations for r_e, r_d, and g_m. As an approximation, this substitution is often made for actual transistors, which may depart somewhat from the Ebers–Moll model.

However, in the linear region from the theoretical equation $\mu_{re} \to 0$, while simultaneously $r_{ceo} \to \infty$, and it is not convenient to use the preceding equations directly.

It is interesting to use Eqs. 2.86 and 2.92 to show that

$$r_a = r_d - r_e (1 + \beta) = 0$$

indicating the justification for using the circuit of Fig. 2.23(b).

The preceding relationships show that in the normal active region, if the Ebers-Moll equations are assumed to be valid, the current gain β_N is independent of operating point and temperature, whereas r_d in the h-parameter model is directly proportional to temperature and inversely proportional to operating emitter current, as is r_e in the T-equivalent circuit. The transconductance, g_m, is inversely proportional to temperature and directly proportional to operating current.

For the frequency-dependent circuit model of Fig. 2.24, the capacitances are voltage or current dependent like the capacitances of the junction diode discussed in Chapter 1. In the normal active region, the collector–base junction is reverse biased and $C_{cb'}$ is entirely junction capacitance, with a voltage dependence given in general by Eq. 1.81 with $v = -v_{cb'}$. The base–emitter capacitance $C_{b'e} = C_{je} + C_b$, where c_{je} is the base–emitter junction capacitance given by Eq. 1.81 where $v = v_{b'e}$, and C_b is the base diffusion capacitance in the manner discussed in Sec. 1.6(2). For an abrupt junction for an npn transistor, the result would be an equation like Eq. 1.108, but changed to

$$C_b = \frac{q}{kT} \frac{W_b'^2}{2D_{nB}} |i_E| \tag{2.94}$$

where W_B' is the effective width of the base, which for forward bias differs only little from the actual width at low collector voltage, and D_{nB} is the diffusion constant for electrons in the p-base region, and i_E is the electron current injected from the emitter into the base.

2.10 STRUCTURAL PARAMETERS AND NONLINEARITIES

The physical basis of the parameters of the Ebers–Moll equations developed in Sec. 2.2 and the incremental parameters discussed in Sec. 2.9, as well as extensions to include nonlinearities not included in the basic Ebers–Moll formulation, will be discussed in this section.

For convenience, Eqs. 2.20 through 2.22 are rewritten in common-emitter form, indicated in Fig. 2.30, separating out the base spreading resistance, r_b'. Added to these equations are additional components of current as a result of hole–electron recombinations occurring in the collector–base and base–emitter depletion regions. These modified equations are

$$i_C = I_s e^{av_{B'E}} (1 - e^{-av_{CE}}) + \frac{I_S}{\beta_I} \{1 - e^{-a(v_{CE} - v_{B'E})}\}$$

$$+ I_{CBR} \{1 - e^{-a(v_{CE} - v_{B'E})/n_c}\} \tag{2.95}$$

$$i_E = -I_s e^{av_{B'E}} (1 - e^{-av_{CE}}) - \frac{I_S}{\beta_N} (e^{av_{B'E}} - 1) - I_{EBF} [e^{(av_{B'E}/n_e)} - 1] \tag{2.96}$$

$$i_B = \frac{I_S}{\beta_N} (e^{av_{B'E}} - 1) - \frac{I_S}{\beta_I} \{1 - e^{-a(v_{CE} - v_{B'E})}\}$$

$$+ I_{EBF} (e^{av_{B'E}/n_e} - 1) - I_{CBR} \{1 - e^{-a(v_{CE} - v_{BE})/n_c}\} \tag{2.97}$$

In these above equations, n_e is called the emission coefficient for the emitter–base depletion region and n_c that for the collector–base region. Normally, $n_c \cong n_e \cong 2$, as suggested by Eq. 1.76. The various terms of reverse-bias current, I_S, I_{EBF}, I_{CBR}, β_N, and β_I, are made up of several components which will be discussed specifically for the *normal active region*.

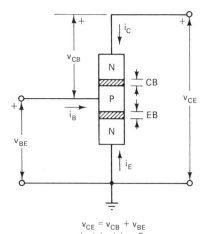

$$v_{CE} = v_{CB} + v_{BE}$$
$$i_B + i_C + i_E = 0$$

Figure 2.30 Transistor model for addition of depletion layer recombinations to Ebers–Moll model.

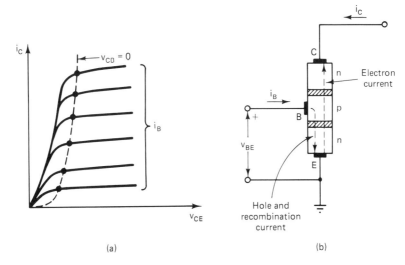

Figure 2.31 Nonlinear characteristic for normal active region of transistor. (a) Current voltage characteristics. (b) Current paths.

Normal Active Region

On the $v_{CE} = v_{BE}$ contour ($v_{CB} = 0$), as indicated in Fig. 2.31(a), the preceding equations reduce to

$$i_C = I_S e^{a v_{B'E}} \tag{2.98}$$

$$I_E \cong -I_S \left(1 + \frac{1}{\beta_N} \right) e^{a v_{B'E}} - I_{EBF} e^{a v_{B'E}/n_e} \tag{2.99}$$

$$i_B \cong \frac{I_S}{\beta_N} e^{a v_{B'E}} + I_{EBF} e^{a v_{B'F}/n_e} \tag{2.100}$$

However, a residual term

$$-\left(\frac{I_S}{\beta_N} + I_{EBF} \right)$$

in Eq. 2.99 and its complement in Eq. 2.100 are considered negligible. The equations are applicable well into the normal active region and somewhat into the forward saturation region.

As indicated in Fig. 2.31(b), the component reaching the collector is the electron current only, given by Eq. 2.98. Recombinations occurring in the base and emitter region and the depletion layer are replaced through the forward-biased base as indicated.

From Eq. 2.21 for $v_{CB} = 0$ and $e^{a v_{BE}} \to \infty$,

$$I_{ES} \cong I_S \left(1 + \frac{1}{\beta_N} \right) \tag{2.101}$$

where I_{ES} is the short-circuit reverse saturation current of the original Ebers–Moll model of Eqs. 2.13 and 2.14, which is composed of four components:

$$I_{ES} = I_{nF} + I_{pF} + I_{nBF} + I_{pEF} \qquad (2.102)$$

The terms I_{nF} and I_{pF} are the $e^{a v_{BE}}$ coefficients of electron and hole currents crossing the base–emitter junction analogous to those for any pn junction, while the terms I_{nBF} and I_{pEF} are the terms representing recombination of holes and electrons in the p-base and n-emitter regions, respectively, as discussed in Sec. 1.3. Furthermore, the term I_{NF} for the zero-biased collector junction is the sole component of I_S (i.e., $I_{NF} = I_S$).

For the normal transistor, the emitter is much more heavily doped than either the base or collector, so for the npn transistor $N_{DE} >> N_{AB}$, making $I_{nF} >> I_{pF}$. Also, the dimensions for a reasonably efficient transistor are such that base or emitter widths are small compared to the respective diffusion lengths, which makes the recombination currents small (e.g., $I_{NBF} << I_{NF}$) and because of relative dopings $I_{pEF} << I_{nBF}$ and can normally be neglected; hence Eq. 2.102 can normally be written as

$$I_{ES} \cong I_{nF} + I_{pF} + I_{nBF} \qquad (2.103)$$

where $I_{nF} = I_S$ is by far the largest term.

From Eqs. 2.101 and 2.103, the short-circuit common-emitter current gain in the Ebers–Moll model is

$$\beta_N = \frac{1}{(I_{pF}/I_{nF}) + (I_{nBF}/I_{nF})} \qquad (2.104)$$

This formulation does not include depletion layer recombinations which were later additions to the Ebers–Moll model.

If depletion layer recombinations are included, the complete short-circuit gain for $v_{CB} = 0$ can be obtained from the ratio of Eq. 2.98 to Eq. 2.100 as

$$\beta_N' = \left.\frac{I_C}{I_B}\right|_{v_{CB}=0} = \frac{1}{\dfrac{I_{pF}}{I_{nF}} + \dfrac{I_{nBF}}{I_{nF}} + \dfrac{I_{EBF}}{I_{nF}} \exp\left[-a\left(1 - \dfrac{1}{n_e}\right)v_{B'E}\right]} \qquad (2.105)$$

Thus β_N and β_N' can be related to structural parameters through the reverse bias components as follows:

For thin base and emitter regions such that $W_B' << L_n$ and $W_E' << L_p$, equations like Eqs. 1.68 and 1.67 may be written for the base–emitter region as

$$I_{nF} = \frac{qA_E D_{nB} n_i^2}{\displaystyle\int_0^{W_B'} N_{AB}(x_B')\, dx_B'} \qquad (2.106)$$

$$I_{pF} = \frac{qA_E D_{pE} n_i^2}{\displaystyle\int_0^{W_E'} N_{DE}(x_E')\, dx_E'} \qquad (2.107)$$

For convenience x' is defined as positive from the edge of each depletion layer boundary, as indicated in Fig. 2.32.

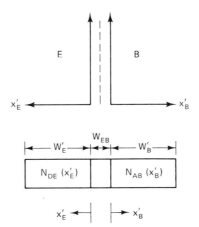

Figure 2.32 Base, emitter, and depletion regions for current component identification.

Although the equations on which Eqs. 2.106 and 2.107 are based were derived in Sec. 1.3 for the narrow p and n regions assuming no recombinations, they are *reasonably good approximations even allowing for a small component of recombination current,* which modifies the excess carrier distributions only slightly.

As a continuation of the charge–voltage–current relationships developed in Sec. 1.6, it was pointed out that minority carrier current was related to transit time. For electrons in the base

$$i_{nF} = \frac{\overline{q}_{nB}}{\tau_{FB}} \tag{2.108}$$

where $\overline{q}_{nB}$ is the total charge stored in the base as a result of the forward-biased junction and τ_{FB} is the forward transit time.

It can be argued analogously that, as an approximation, the charge lost by base recombinations that must be replaced by base current is given by

$$i_{nBF} = \frac{\overline{q}_{nB}}{\tau_n} \tag{2.109}$$

where τ_n is the electron lifetime in the base region.

Then, using the relationships of Secs. 1.3 and 1.6,

$$I_{nBF} \cong \frac{qAn_i^2}{\tau_n} \int_0^{W_B'} \frac{\int_x^{W_B'} N_{AB}(x')\,dx'}{N_{AB}(x')\int_0^{W_B'} N_{AB}(x')\,dx'} \tag{2.110}$$

$$= \frac{qA_E n_i^2}{\tau_n N_{AB}} \frac{W_B'}{2} \quad \text{for uniform impurity distributions.}$$

The validity of Eqs. 2.106, 2.107, and 2.110 depends on sufficiently small recombination current that total base charge determined under the assumption of no recombinations is a reasonably good approximation.

The depletion layer term can be approximated from Eq. 1.76 as

$$I_{EBF} = \frac{q n_i A_E W_{EB}}{2\tau_o} \tag{2.111}$$

where τ_o is the average lifetime in the depleted region.

For the special case of uniform impurity concentrations in the emitter and base regions, that is, $N_A(x_B') = N_A$ and $N_D(x_B') = N_D$, Eq. 2.105 becomes

$$\beta_N' = \frac{1}{\dfrac{N_{AB} W_B' D_{pE}}{N_{DE} W_E' D_{nB}} + \dfrac{W_B'^2}{2\tau_n D_{nB}} + \dfrac{1}{2\tau_o} \dfrac{W_{EB} W_B' N_{AB}}{D_{nB} n_i \exp[a(1 - 1/n_e) v_{B'E}]}} \tag{2.112}$$

The first group of terms in the denominator is the ratio of hole to electron current at the base–emitter junction and is sometimes called the *base injection factor*. Its inverse would be the value of β_N' in the absence of all recombinations. Thus the most important component in obtaining a large β_N' is the ratio of emitter-to-base doping. The second group accounts for base recombinations that reduce β_N' and is referred to as the *base recombination factor*. Considering these two terms only, $\beta_N' = \beta_N$ of the basic Ebers–Moll model. The third group accounts for depletion layer recombination, which was not included in the original Ebers–Moll model. It is seen to be voltage dependent and is important, if at all, only at very low forward voltages and currents where it acts to reduce β_N'. The terms are of course voltage dependent, but for all the other factors the voltage dependency is the same and cancels in the final assembly. The factor in the depletion layer term makes it different.

The preceding determination of current gain β_N' as related to structural parameters was based on equations assuming an effective base width W_B', using Eq. 2.98 through 2.100 for $v_{CB} = 0$, but it was inferred that the validity of these equations extended well into the normal active region. In reality, however, as v_{CE} is increased, the collector–base junction becomes more reverse biased, the depletion layer widens, and hence the effective base width W_B' narrows. The extent of base narrowing depends on the relative doping of the p-base and n-collector regions and on the grading of the junction.

As indicated in Sec. 1.2 for the general pn junction, the depletion width of the reverse-biased collector–base junction is given for the npn transistor by an equation of the form

$$d = K(\phi_B + v_{CB})^n \tag{2.113}$$

which may be expressed more explicitly as

$$d' \cong K'(v_{CE})^n \tag{2.114}$$

where d' is the portion extending into the base region, K' is the constant that relates to the portion extending into the base (i.e., Eq. 1.28 or 1.29 for the uniformly doped base) and V_{CE} is the collector–emitter voltage (i.e., $v_{CE} = v_{CB} + v_{BE}$ and v_{BE} approaches the value of a built-in voltage ϕ_B as an approximation).

Then the effective base width as a function of voltage can be written as

$$W_B' \cong W_B \left[1 - \frac{K'}{W_B} v_{CE}{}^n \right]$$ (2.115)

where W_B is the structural base width.

Furthermore, the widening of the depletion layer accounts for an excess of carriers generated in the depletion layer, resulting in an increase of collector current of the general form of Eq. 1.75 and given by

$$i_C' = \frac{q n_i A_{CB} W_{CB}'}{2 \tau_o}$$ (2.116)

where A_{CB} is the collector–base junction area and W_{CB}' is the total width of the collector–base depletion region.

Then from Eq. 2.113

$$i_C' = \frac{q n_i A_{CB}}{2 \tau_o} K v_{CE}{}^n$$ (2.117)

Therefore, if we were to use h_{FE} as defined by

$$h_{FE} = \frac{i_C}{i_B} \bigg|_{v_{CE} \, = \, v_{CE} \, > \, (v_{BE})}$$ (2.118)

in an equation like Eq. 2.105, the various terms in Eq. 2.105 would be modified with the appropriate narrowed W_B' used in Eqs. 2.106, 2.107, and 2.110. The excess carriers generated in the depletion region given by Eqs. 2.116 and 2.117 would not, in general, result in a further increase in h_{FE} because this component of current is balanced by hole current supplied through the base to preserve regional charge neutrality.

However, a further examination of h_{FE} using Eq. 2.112 modified to take base width variations into account would indicate that h_{FE} is not entirely independent of i_B, since W_B', W_E' and W_{EB} are all dependent upon forward bias base–emitter voltage which in turn is a function of base current. The net result of these effects is an increase in h_{FE} with both i_B and v_{CE} at normal operating levels.

Thus the general nonlinear equation

$$i_C = h_2 (v_{CE}, i_B)$$ (2.119)

can be written.

It is the overall complexity of this nonlinear relationship that led to the empirical modeling of the effect as suggested in Eqs. 2.47 and 2.48 to account for the various nonlinearities.

High Current and Voltage Effects

At high current levels, the models that have been proposed so far, structural or empirical, become less valid as current levels are increased, where a compression as indicated in Fig. 2.33 takes place. This suggests that low-level injection theory is no longer valid and

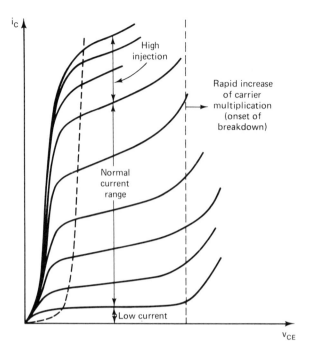

Figure 2.33 Collector characteristics at high currents and voltages.

that, for a given minority current density in the base region, the base current required to supply additional majority carriers must rise to preserve base charge neutrality. This results in a decreasing h_{FE} at higher collector currents. This effect is referred to as *conductivity* modulation. The onset of this effect is a function of transistor geometry. In a planer transistor, because of the distributed nature of the base resistance, r_b', the forward base–emitter bias is largest at the emitter edges, resulting in a nonuniform injection of carriers, which tend to concentrate at the emitter corners as indicated in Fig. 2.34. This effect, known as *emitter crowding,* causes high injection effects to become important at a lower collector current than might otherwise be predicted.

Also at high collector currents, particularly for relatively light collector doping in

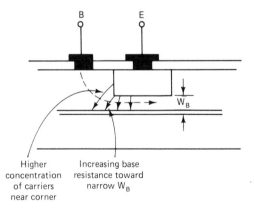

Figure 2.34 Emitter crowding phenomenon.

the vicinity of the collector–base junction, charge carriers flowing across the depletion region modify the electric field as previously established and cause the depletion layer to narrow as a function of injected current at a given collector–base voltage. The corresponding widening of the base region further decreases h_{FE} at high currents. This particular component of decrease is referred to as the *Kirk effect*.

As the collector voltage becomes higher (not necessarily at high currents), the collector current begins to increase more rapidly owing to the multiplication of charge carriers at the onset of reverse-bias collector–base junction breakdown, as discussed in Sec. 1.5 and modeled empirically by Eq. 1.78 for the general *pn* junction. This effect for the transistor can be modeled crudely by a multiplying factor of

$$\frac{1}{[1 - (v_{CE}/V_B]^n}$$

applied to Eq. 2.119.

For many purposes, the high-current and high-voltage effects can be modeled empirically by incorporating into the model suggested by Eq. 2.47 a term of the general form

$$- \beta_0 \frac{K_B i_B^q}{v_{CE}^m}$$

to account for various high-current effects as well as the term $k_B i_B{}^x v_{CE}{}^y$, as suggested in Eq. 2.48, which includes normal active-region nonlinearities even extending somewhat into the carrier multiplication region.

Then an overall empirical model derived from Eq. 2.47 can be written as

$$i_C = I_{OS} + g_O v_{CE} + \beta_O i_B + k_B i_B{}^x v_{CE}{}^y - \frac{\beta_O K_B i_B^q}{v_{CE}{}^m} \tag{2.120}$$

where all the constants may be determined by simultaneous solutions from measurements taken at points that define the various regions, as indicated in Fig. 2.35. This equation does not model accurately the very low current region where emitter–base depletion layer recombination might be significant, except that an offset of I_{OS} (positive or negative) will properly position the normal range curves on the characteristic, as indicated in Fig. 2.15.

2.11 EFFECTS OF TEMPERATURE VARIATIONS ON TRANSISTOR PARAMETERS

The most important effects of variation of junction temperature are the effects on the base–emitter voltage–current relationship and on the current gain, h_{FE}. The i_E, v_{BE} relationship will be analyzed first. Equation 2.99 using Eq. 2.103 can be written as

$$|i_E| = (I_{nF} + I_{pF} + I_{nBF}) e^{qv_B'E/kT} + I_{EBF} e^{qv_B'E/n_e kT} \tag{2.121}$$

The terms $I_{nF} + I_{pF}$ are electron and hole currents crossing the base–emitter junction, which have the temperature dependence discussed in Sec. 1.8 of the form

$$I_{nF} + I_{pF} = A_o' T^n e^{-(qV_{gO}/kT)} \tag{2.122}$$

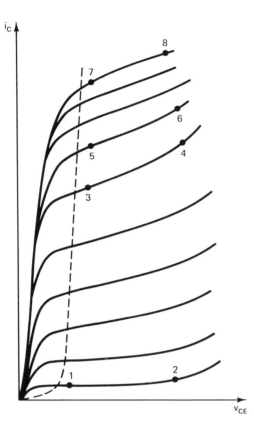

Figure 2.35 Collector current voltage characteristic at large currents and voltages, with suggested measurement points for parameter determination in nonlinear mathematical models.

where the exponent n is determined by the temperature dependence of n_i and the diffusion constants and v_{gO} is the extrapolated value of band gap voltage.

The recombination component I_{nBF} is of a slightly different form, being given by Eq. 2.110 for uniform dopings as

$$I_{nBF} = \frac{qAn_i^2}{\tau_n N_{AB}} \frac{W_B'}{2} \tag{2.123}$$

for the *npn* transistor. Since n_i^2 is proportional to T^3 and τ_n is proportional to $T^{-1/2}$, I_{nBF} can be expressed theoretically as

$$I_{nBF} = B_O' T^{3.5} e^{-(qv_{gO}/kT)} \tag{2.124}$$

The component I_{EBF} is still different, being given by Eq. 2.111, which can be written as

$$I_{EBF} = C_o' T^2 e^{-(qV_{gO}/2kT)} \tag{2.125}$$

since τ_o is proportional to $T^{-1/2}$.

In a normal transistor, the current injected from the emitter into the base is so completely dominant that the terms involving I_{nBF} and I_{EBF} are usually negligible, and the emitter current can be written as

$$|i_E| \cong A'_o T^n e^{-(qV_{gO}/kT)} e^{qv_{B'E}/kT} \qquad (2.126)$$

taken directly from Eq. 1.127 for the single *pn* junction. Any minor effect on I_{nBF} and I_{EBF} can be taken into account by slight modifications and A'_o and n.

Using this equation, relationships for base–emitter voltage $v_{B'E}$ may be expressed similarly to Eqs. 1.128 and 1.129 with $v_{B'E}$ replacing v_D and $|i_E|$ replacing i_D.

Alternatively, Eq. 2.126 may be used directly with

$$V_{B'E} = \frac{kT}{q} \left[\ln \frac{|i_E|}{A'_o} - n \ln T \right] + V_{gO}$$

with its temperature variation being

$$\frac{dv_{B'E}}{dT} \bigg]_{i_E = I_E} = -\frac{nk}{q} \left[1 + \ln T - \frac{1}{n} \ln \frac{|I_E|}{A'_o} \right] \qquad (2.127)$$

The temperature dependence of collector current as a function of base current can be established using the temperature-dependent h_{FE} derived from Eq. 2.112 modified to include base width variation with collector voltage. The result can be written for the *npn* transistor with uniform impurity concentrations as

$$h_{FE} = \cfrac{1}{\cfrac{N_{AE} W'_B D_{pE}}{N_{DE} W'_E D_{nB}} + \cfrac{W'^2_B}{2K_n K_{nB}} T^{\gamma + 1/2} + \cfrac{W'_{EB} W'_B N_{AB} T^{\gamma - 1}}{2K_o K_{nB} K_i \exp\left(\dfrac{-qV_{gO}}{2kT}\right) \exp\left[\dfrac{q}{kT}\left(1 - \dfrac{1}{n}\right)v_{B'E}\right]}}$$

where $\qquad (2.128)$

$$\tau_n = K_n T^{-1/2}$$

$$\tau_o = K_o T^{-1/2}$$

$$D_{nB} = K_{nB} T^{-\gamma}$$

with γ being the magnitude of the temperature exponent of the diffusion constant ($\frac{1}{2}$ for lattice scattering and 3/2 for the more commonly determined experimental value for doped silicon).

As a result of the base recombination factor alone, h_{FE} would be a decreasing function of temperature, and although the equation does not show it directly, the base injection factor is also slightly temperature dependent particularly at low collector voltages. Noting that W'_B narrows with increasing reverse-bias base–collector voltage, that the voltage function involved is $\phi_o + |v_{CB}|$, and that $\phi_o =$ constant (kt/q), therefore W'_B is a slightly decreasing function of temperature resulting in a slightly increasing component of h_{FE} with temperature at low collector voltages. This base narrowing also minimizes the decrease in h_{FE} due to the temperature-dependent base recombination factor. The overall combination results in an h_{FE} that is only slightly temperature dependent.

The depletion layer recombination term is a much more complex function of temperature involving $T^{\gamma - 3/2}$, V_g, and v_{BE}, with the net result being a component of h_{FE}, which is an increasing function of temperature of low values of current. This effect decreases as larger collector currents resulting from larger values of v_{BE} are approached. This effect, however, is negligible except at very low values of collector current.

The collector current is also dependent on reverse-bias collector–base voltage according to Eq. 2.116, where the added component can be written as

$$i_C' = \frac{qA_{CB}KK_iT^2e^{-qv_g/kT}v_{CE}^n}{2K_o} \tag{2.129}$$

where $v_{CE} >> v_{CB}$. The voltage-dependent term is negligible at low values of collector voltage but increases rapidly with temperature at higher values.

This equation pinpoints a particular problem: As v_{CE} is increased, the transistor power dissipation, which is proportional to the $(V_{CE})(i_C)$ product is increased, which in turn raises the junction temperature and results in a further increase in collector current. The compounding of this effect may result in destruction of the transistor. This effect is referred to as *thermal runaway*.

2.12 NOISE SOURCES AND MODELS

There are both thermal- and current-related noise sources associated with bipolar transistors, which are mostly similar to those discussed in Sec. 1.9 for the junction diode. Noise sources associated with the various noise-producing mechanisms in bipolar transistors may be added to any of the various small-signal circuit models that have been discussed. In general, these sources can be represented as indicated in Fig. 2.36.

The equivalent noise voltage

$$\overline{v_{rb'}^2} = 4kTr_B\,\Delta f \tag{2.130}$$

is thermal noise related to the ohmic base resistance using the Ohm's law computation for dc base resistance r_B'.

The equivalent noise current

$$\overline{i_{b'e}^2} = 2qI_B\,\Delta f + \text{excess noise} \tag{2.131}$$

is composed of at least two components: The first component, proportional to the dc base current, is pure shot noise. The second component, excess noise, may contain flicker noise and burst noise as represented by terms like Eq. 1.138 and has been found to be primarily related to base current. Hence such components are reasonably modeled as additions to the pure shot noise component. Flicker noise and burst noise can be minimized by careful processing methods.

The shot noise current

$$\overline{i_c^2} = 2qI_C'\,\Delta f \tag{2.132}$$

represents random arrived times of carriers produced in the emitter and reaching the collector terminal through the successive processes of diffusion and drift. The dc component I_C' in this equation is the collector current that would exist were there no excess carrier components generated in the base–collector depletion region.

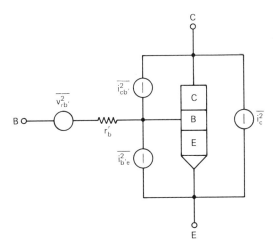

Figure 2.36   Noise sources in a transistor.

The collector–base shot noise component

$$\overline{i_{cb'}^2} = 2qM(\Delta I_c)\ \Delta f \tag{2.133}$$

is a representation of random variations due to the increased current ΔI_c as a result of depletion layer widening with reverse-bias voltage, as indicated by Eq. 2.117. The factor M is a carrier multiplication factor that is voltage dependent and becomes most significant as the base–collector breakdown voltage is approached. This results in the spike or avalanche noise discussed in Sec. 1.9. For low-voltage transistor operation, the term $\overline{i_{cb'}^2}$ is usually neglected and the total I_C rather than I_C' is used in Eq. 2.132.

PROBLEMS

2.1 Assume that the Ebers–Moll equations for an *npn* transistor are valid and that $\alpha_N I_{ES} = \alpha_I I_{CS}$.
  (a) Write as many equations as you can involving relationships among the parameters I_S, I_{ES}, I_{CS}, I_{EO}, I_{CO}, I_{CEO}, α_N, α_I, β_N, and β_I.
  (b) Determine the approximations for which

$$\beta_I \cong \frac{I_{ES}}{I_{CEO}}\ \beta_N$$

  Keep your results for future reference.

2.2 There are a number of ways in which a transistor may be used as a diode, one of which is shown here where the diode characteristic is the v_{BE} versus i_1 relationship.
  (a) Write the input current–voltage relationship as $i_1 = f(v_{BE}, I_{ES})$.
  (b) If $I_1 = 1.5$ mA when $V_{BE} = 0.72$ V, determine I_{ES}.
  (c) Under the same conditions, the base current is $I_B = 1\ 5\ \mu$A; determine β_N and α_N.

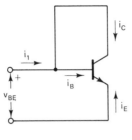

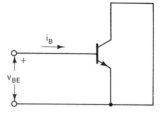

Prob. 2.2   Prob. 2.3

2.3 A transistor is used as a diode using the i_B, v_{BE} relationship for the connection shown.
   (a) Write an equation for $i_B = f(v_{BE})$ involving I_{CEO}, I_{ES}, and β_N. (Use Eq. 2.35 for $v_{CE} = 0$.) Simplify your result for

$$I_{CEO} \gg \frac{I_{ES}}{1 + \beta_N}$$

2.4 Consider the diode-connected transistor shown using the i_C, v_{CE} relationship.
   (a) Write an equation for $v_{CE} = f(i_C)$ in terms of Eq. 2.16 with appropriate conditions.
   (b) Expand the result of part (a) involving the parameters I_{CEO}, β_N, and I_{ES}.

2.5 Using results obtained from Probs. 2.2 through 2.4, determine which of the three diodes yields the largest terminal current for the same applied voltage. To illustrate your answer, use a transistor with parameters of $\beta_N = 100$, $I_{ES} = 0.5$ pA, and $I_{CEO} = 100$ pA.

2.6 Assume that the basic Ebers–Moll equations are valid and that measurements are made at 300 K on a transistor connected as shown.
   (a) With $V = +5$ V, a current of 5 pA is found to flow; then V is reversed and adjusted to make a current of 5 mA flow in the opposite direction. What would you expect the voltage at the base–emitter terminals to be?
   (b) Independent of part (a), the actual value of v_{BE} is found to be 0.65 V at a current $I = -5$ mA. What would you expect I_{ES} to be?
   (c) Discuss whatever discrepancies you find between the results of parts (a) and (b).

2.7 With the *npn* transistor connected as shown, the following three sets of measurements are made.

   1. For $I_C = 0.1$ mA, $I_B \cong 1.25$ µA, $V_{BE} \cong 0.6$ 0V.
   2. For $I_C = 1.0$ mA, $I_B \cong 12.5$ µA, $V_{BE} \cong 0.66$ V.
   3. For $I_C = 10.0$ mA, $I_B \cong 125$ µA, $V_{BE} \cong 0.72$ V.

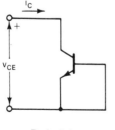

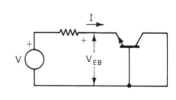

Prob. 2.4   Prob. 2.6

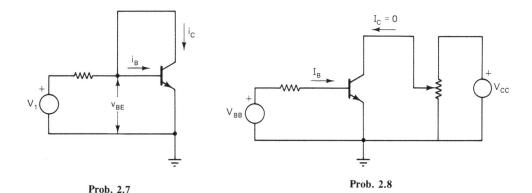

Prob. 2.7

Prob. 2.8

(a) Determine a value for $V_T \cong 1/a$ starting with the ratio of two collector currents (i.e., I_{C2}/I_{C1} or I_{C3}/I_{C2}), noting that they are the same.

(b) Find a value for β_N.

(c) Find I_S and I_{ES} in the Ebers–Moll model.

2.8 Using the transistor of Prob. 2.7, the circuit shown is adjusted to maintain $i_C = 0$ as i_B is varied (see Fig. 2.14). It is found that the resultant collector–emitter voltage has an almost constant value of $V_{CE} \cong 0.032$ V for values of $I_B > 50$ μA.

(a) Determine I_{CEO}.

(b) From part (a), and the results of Prob. 2.7, find the values of I_{CS} and β_I.

2.9 Refer to the transistor with the characteristics shown in Figs. 2.16 and 2.17.

(a) Determine approximately the small-signal h parameters, β, r_{ceo}, r_{bes}, from the graphical characteristics at an operating point of $V_{CE} = 10$ V and $I_C = 5$ mA.

(b) Calculate a value for r_e from theoretical considerations at a junction temperature of 300 K and find values for r_d and μ_{re}.

(c) What is your estimate for r_b'?

(d) Determine g_m, $r_{b'e}$, and $r_{cb'}$ for the transistor at the prescribed operating point.

2.10 It is suggested that the most significant parameters of the high-frequency h-parameter circuit model can be obtained by a few direct measurements. For example, consider the simplified model of Fig. 2.29, where R_L is either zero or sufficiently small that the shunt admittances of Fig. 2.26 can be neglected. The following measurements are made at a dc collector current of 2.0 mA; $\beta = 100$, $f_\beta = 10$ MHz, $r_{bes} = 1.8$ kΩ. Then holding I_1 constant in the input circuit shown by varying the amplitude of V_S as its frequency is varied, the value of the magnitude of V_1 is found to be one-half of its very low-frequency value at a frequency of 18 MHz.

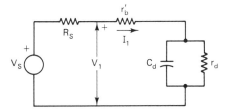

Prob. 2.10

(a) Determine r_b' and r_d.

(b) Determine the capacitance, C_d.

(c) Compare part (a) with values you would obtain using a value of r_d calculated at the dc collector current of 2.0 mA.

(d) What is g_m using both experimentally determined and calculated values for r_d?

2.11 An *npn* transistor has uniform doping concentrations in all three regions of values

$$N_{DE} = 10^{18} \text{ cm}^{-3}, \qquad N_{AB} = 10^{16} \text{ cm}^{-3}, \qquad N_{DC} = 10^{15} \text{ cm}^{-3}$$

Also,

$$W_B = 1.2 \times 10^{-4} \text{ cm}, \qquad A_E = 1.0 \times 10^{-5} \text{ cm}^2$$

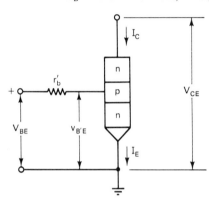

Prob. 2.11

(a) At $T = 300$ K and an emitter current $I_E = 2$ mA, determine approximately the diffusion capacitance, C_b. Repeat for $I_E = 1$ mA and $I_E = 0.1$ mA.

(b) For the conditions of part (a), determine approximately the internal base–emitter junction voltage $v_{B'E}$, the built-in voltage, and the emitter–base junction capacitance, C_{je}.

(c) If the collector–base junction is reverse biased at $V_{CB} = 5$ V, determine the collector–base junction capacitance, C_{jc}. Repeat for $V_{CB} = 10$ V.

2.12 A *pnp* transistor has uniform impurity concentrations in the base, emitter, and collector regions as follows: $N_{AE} = 4.5 \times 10^{17}$, $N_{DB} = 1.2 \times 10^{16}$, and $N_{AC} = 1 \times 10^{15}$. Also, $W_E = 2.0 \times 10^{-4}$, $W_B = 1.0 \times 10^{-4}$, and $W_C = 4.0 \times 10^{-4}$.

(a) Write an approximate equation for h_{FE} at very low collector voltage including base recombinations current but neglecting depletion layer recombinations.

(b) With $h_{FE} = 50$ (measured), determine approximately the lifetime of holes injected into the base.

(c) Determine the base transit time for injected holes.

2.13 A particular transistor with all three regions uniformly doped has the following structural and process parameters:

$$N_{DE} = 1.5 \times 10^{17} \text{ cm}^{-3}, \qquad N_{AB} = 1.0 \times 10^{16} \text{ cm}^{-3}, \qquad N_{DC} = 1.5 \times 10^{15} \text{ cm}^{-3}$$

$$W_B = 1.5 \times 10^{-4} \text{ cm}, \qquad W_E = 2.0 \times 10^{-4} \text{ cm}, \qquad W_C = 2.5 \times 10^{-4} \text{ cm}$$

$$A_C = A_E = 1.5 \times 10^{-5} \text{ cm}^2$$

The lifetime of minority carriers (electrons) in the base is estimated to be $\tau_n = 5 \times 10^{-8}$ s.

(a) Determine the dc common-emitter current gain h_{FE}, assuming negligible depletion layer recombinations.

(b) Calculate the base transit time for electrons.

(c) Using your answer from part (a) as an approximation for the incremental current gain, $h_{FE} \cong h_{fe} = \beta$, and that the base transit time is given approximately by $\tau_{FB} \cong 1/\omega_T$, determine the frequency f_β.

2.14 An integrated-circuit *npn* transistor with uniformly doped regions has the parameters as shown. Assume that carrier flow, as shown, is directly controlled by the emitter–base junction area.

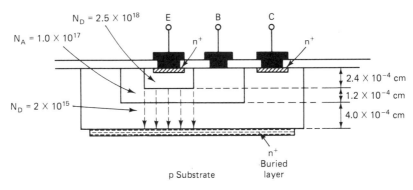

Prob. 2.14

(a) Calculate the dc common-emitter current gain neglecting all recombination currents and base-width versus voltage variations.

(b) In the transistor shown, the actual value of current gain was measured and found to be $h_{FE} = 100$. Neglecting depletion layer recombinations, determine the minority carrier lifetime in the base.

2.15 Consider a *pnp* transistor whose base has an exponential impurity profile given by

$$N_D(x) = N_D(0)e^{-(x/W_B)}$$

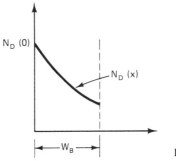

Prob. 2.15

(a) Derive an equation for the excess minority carrier distribution and plot it on the accompanying in figure.

(b) In the equation

$$i = I_{PF}(e^{qV_{EB}/kT} - 1)$$

determine I_{PF} for the particular doping profile given.

(c) Derive an expression for base transit time involving only D_{pn}, $N_D(x)$, and W_B.

2.16 Consider the circuit shown and the validity of the Ebers–Moll model in the normal active region.

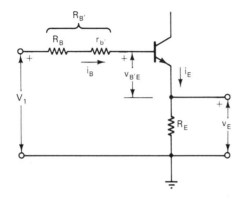

Prob. 2.16

(a) Show that from Eq. 2.126 and

$$v_{B'E} = V_1 - i_B R_B' - i_E R_E$$

and

$$i_B = \frac{i_E}{1 + h_{FE}}$$

the following equation can be written.

$$i_E \exp\left[\frac{q}{kT}\left(1 + \frac{R_B'}{R_E}\frac{1}{1 + h_{FE}}\right)i_E R_E\right] = A_o' T^n e^{-(qV_{gO}/kT)} e^{(q/kT)V_1}$$

(b) From the preceding relationship obtain $di_E/dT = f(dV_1/dT)$ independent of A_o'.
(c) Then obtain an expression for dV_1/dT to make $di_E/dT = 0$.
(d) For

$$\frac{R_B'}{R_E}\frac{1}{1 + h_{FE}} << 1$$

and for $T = 300$ K, $V_{gO} = 1.21$ V, $V_1 = 2.8$ V, and $n = 1.5$, $R_E = 2$ kΩ and $I_E = 1$ mA, determine the value of dV_1/dT. Repeat for $R_B' = 10$ kΩ.

REFERENCES

EBERS, J. J. AND J. L. MOLL. "Large Signal Behavior of Junction Transistors," *Proc. IRE,* vol. 42, no. 12, Dec. 1943, pp. 1761–1772.

GLASER A. B. AND G. E. SUBAK-SHARPE. *Integrated Circuit Engineering,* Addison-Wesley, Reading, Mass., 1979.

GLASFORD, G. M. *Linear Analysis of Electronic Circuits,* Addison-Wesley, Reading, Mass., 1965.

———. "Nonlinear Distortion Models for Bipolar Junction Transistors," *Conf. Record, Seventh Asilomar Conference on Circuits, Systems, and Computers,* Pacific Grove, Calif., Nov. 1973, pp. 548–552; Western Periodicals, North Hollywood, Calif.

————. "Comparisons of Nonlinear Distortion Models for Bipolar Transistors," *Conf. Record, Eighth Asilomar Conference on Circuits, Systems, and Computers,* Dec. 1974, pp. 351–355; Western Periodicials, North Hollywood, Calif.

————. "Transistor Model Parameters as Related to Structural Parameters for Geometrical Dissymmetry and Nonuniform Impurity Concentrations," *Conf. Record, Ninth Asilomar Conference on Circuits, Systems, and Computers,* Nov. 1975, pp. 359–364; Western Periodicials, North Hollywood, Calif.

GROVE, A. S. *Physics and Technology of Semiconductor Devices,* Wiley, New York, 1967.

GUMMEL, H. K., AND H. C. POON. "An Integral Charge Control Model of Bipolar Transistors," *Bell Syst. Tech. J.* vol. 49, May 1970, pp. 827–852.

LINDMAYER J., AND C. Y. WRIGLEY. *Fundamentals of Semiconductor Devices,* Van Nostrand Reinhold, New York, 1964.

MULLER R. S. AND T. I. KAMINS. *Device Electronics for Integrated Circuits,* Wiley, New York, 1977.

NANAVATI, R. P. *Semiconductor Devices,* Intext Educational Publisher, New York, 1975.

SANSEN, W. M. C., AND R. G. MEYER. "Characterization and Measurement of the Base and Emitter Resistances of Bipolar Transistors," *IEEE J. Solid State Circuits,* vol. SC-7, no. 6, pp. 492–498.

SZE, S. M. *Physics of Semiconductor Devices,* Wiley, New York, 1981.

WHITTIER, R. J., AND D. A. TREMERE. "Current Gain and Cutoff Frequency Falloff at High Currents," *IEEE Trans. Electron. Devices,* vol. Ed-16, no. 1, Jan. 1969, pp. 39–57.

Chapter 3

Junction
and Insulated Gate
Field-Effect Transistors

INTRODUCTION

This chapter discusses in some detail the basic characteristics of field-effect transistors using both *pn* junction and insulating film technologies. The treatment roughly parallels that accorded the bipolar junction transistor discussed in Chapter 2.

The principles involved are basically quite simple. In the case of the junction field-effect transistor (JFET), a single *pn* junction is used to establish a two-port characteristic by using the variation in depletion layer width, as discussed in Chapter 1 for the diode, to modulate the width of a semiconductor "channel."

In the case of the insulated gate type (IGFET), very similar characteristics are obtained by using the voltage applied between a conduction layer (gate) and the channel, which is insulated from it, to modify the charge distribution within the channel, and hence its conductivity. The insulating layer is usually an oxide and the gate a metal; hence the IGFET is alternatively referred to as a metal oxide semiconductor field-effect transistor, or MOSFET, although the gate may often consist of a polysilicon crystal film that has the essential characteristics of the metal gate.

3.1 BASIC n-CHANNEL JUNCTION GATE FET STRUCTURE

The junction FET (JFET) is essentially a single junction device, usually realized as some variation of the diffused planar form, as indicated in Fig. 3.1(a) for an *n*-channel FET. The two highly doped *n* diffusions labeled source and drain only ensure good ohmic

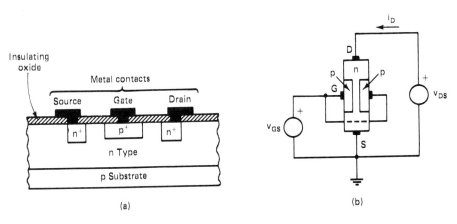

Figure 3.1 Junction FET geometries. (a) Planar structure. (b) Symmetrical structure.

contact at the terminals, as discussed in Sec. 1.4, and are not germane to the basic processes.

For conceptual purposes and for the development of the basic equations, the symmetrical structure shown in Fig. 3.1(b) is usually analyzed. The two separate gate diffusions are indicated for symmetry only, and the structure is considered as having a single gate junction. When the gate is reverse biased (negative with respect to the source), the effective width of the channel under the gate becomes a function of the gate–source voltage and to a lesser extent the positive drain–source voltage. A set of characteristics like those shown in Fig. 3.2 might be anticipated.

3.2 BASIC EQUATIONS FOR SYMMETRICAL STRUCTURE

In the symmetrical structure shown in Fig. 3.3, the *p* region is assumed to be highly doped relative to the *n* region, which causes the depletion layer to extend mostly into

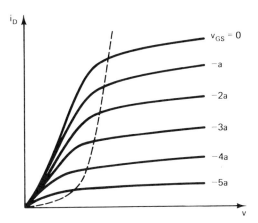

Figure 3.2 Drain current–drain voltage characteristic of FET.

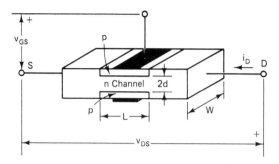

Figure 3.3 FET geometry for derivation of basic equations.

the n region as the gate becomes more negative. The equations that may be derived for the

$$i_D = f(v_{DS}, v_{GS}) \tag{3.1}$$

relationship based on variation of depletion layer width depends on the impurity distribution in the channel.

Uniformly Doped Channel

The equations for the "built-in" voltage of a pn junction and the depletion layer width as a function of applied voltage were given in Sec. 1.2. Relating these equations to the structure shown in Fig. 3.3, the conductivity of the channel depends on the effective channel width as constricted by the widening of the depletion layer into the relatively lightly doped n channel. This, in turn, is a function of distance along the channel as a function of the combined effect of gate and drain voltages.

Based on the preceding considerations, the following equation relating the components can be obtained:

$$i_D = 3I_{DSS} \left[\frac{V_{DS}}{V_P} - \frac{2}{3} \left(\frac{v_{DS} + \phi_0 - v_{GS}}{V_P} \right)^{3/2} + \frac{2}{3} \left(\frac{\phi_0 - v_{GS}}{V_P} \right)^{3/2} \right] \tag{3.2}$$

where

$$I_{DSS} = \frac{d^3 W q^2 \mu_n N_D^2}{3 L \varepsilon_o \varepsilon_r} \tag{3.3}$$

$$V_P = \frac{q N_D d^2}{2 \varepsilon \varepsilon_o}, \qquad \text{for } N_A \gg N_D \tag{3.4}$$

$$\phi_0 = \frac{kT}{q} \left(\ln \frac{N_D}{n_i} + \ln \frac{N_A}{n_i} \right) \tag{3.5}$$

In these equations it is assumed that the impurity concentration, N_D, in the channel is small compared to that, N_A, in the gate region. The electron mobility in the channel is μ_n. It is further assumed that $L \gg d$, which permits "end effects" to be neglected.

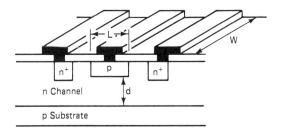

n Channel

p Substrate

Figure 3.4 Planar JFET.

If we define $v'_{GS} = v_{GS} - \phi_0$, the current equation may be written as

$$i_D = 3I_{DSS}\left[\frac{v_{DS}}{V_P} - \frac{2}{3}\left(\frac{v_{DS} - v'_{GS}}{V_P}\right)^{3/2} + \frac{2}{3}\left(\frac{-v'_{GS}}{V_P}\right)^{3/2}\right] \qquad (3.6)$$

For the planar structure shown in Fig. 3.4, Eqs. 3.4 and 3.5 are still valid; but

$$I_{DSS} = \frac{d^3 W q^2 \mu_n N_D^2}{6L\varepsilon_o\varepsilon_r} \qquad (3.7)$$

which is half the value for the symmetrical structure.

Equations 3.2 and 3.6 are based on the variation of depletion layer width as a function of v_{DS} and v_{GS}; hence there should be a contour on the i_D, v_{DS} characteristic for which

$$\frac{\partial i_D}{\partial v_{DS}} = 0$$

corresponding to the conditions for which the depletion layer width extends across the entire channel, thus reducing the effective channel width to zero. Using Eq. 3.6,

$$\frac{\partial i_D}{\partial v_{DS}} = \frac{3I_{DSS}}{V_P}\left[1 - \left(\frac{v_{DS} - v'_{GS}}{V_P}\right)^{1/2}\right] \qquad (3.8)$$

and $\qquad \dfrac{\partial i_D}{\partial v_{DS}} = 0 \qquad$ when $v_{DS} - v'_{GS} = V_P \qquad (3.9)$

The drain current corresponding to the preceding condition, called $i_{D\text{sat}}$, is obtained by using Eq. 3.9 in Eq. 3.6 in terms of v_{DS}:

$$i_{D\text{sat}} = I_{DSS}\left[(-2) + 3\frac{v_{DS}}{V_P} + 2\left(1 - \frac{v_{DS}}{V_P}\right)^{3/2}\right] \qquad (3.10)$$

Or, in terms of V'_{GS},

$$i_{D\text{sat}} = I_{DSS}\left[1 + 3\frac{v'_{GS}}{V_P} + 2\left(\frac{-v'_{GS}}{V_P}\right)^{3/2}\right] \qquad (3.11)$$

Equation 3.10 plots as the contour in Fig. 3.5, and Eq. 3.11 identifies specific values of v'_{GS}. The contour is called the *pinchoff line* because it represents the combination of v_{DS} and v'_{GS}, which reduces the channel width to zero. Equation 3.6 is valid everywhere to the left of the pinchoff line.

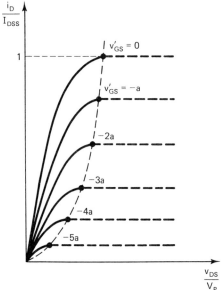

Figure 3.5 Current–voltage characteristic of FET from basic theoretical model.

This region to the left of the pinchoff contour is known as the *triode region,* whereas the region beyond pinchoff is referred to as the *saturation region.* The derivation of Eq. 3.6 does not predict any further change of i_D with v_{DS} in the saturation region; hence the lines are assumed to have zero slope pending further analysis.

In the triode region the small-signal conductance of the channel is defined by

$$g_C = \frac{\partial i_D}{\partial v_{DS}}\bigg]_{v_{GS} = \text{const.}} = \frac{3I_{DSS}}{V_P}\left[1 - \left(\frac{v_{DS} - v'_{GS}}{V_P}\right)^{1/2}\right] \qquad (3.12)$$

The value of g_c becomes zero when $v_{DS} - v'_{GS} = V_P$.

In the triode region for $v'_{GS} = 0$, we define a particular value of g_c as

$$g_o = \frac{\partial i_D}{\partial v_{DS}}\bigg]_{\substack{v_{GS} = 0 \\ v_{DS} < V_D}} = \frac{3I_{DSS}}{V_P}\left[1 - \left(\frac{v_{DS}}{V_P}\right)^{1/2}\right] \qquad (3.13)$$

As v_{DS} decreases from V_P, g_o increases and approaches a limiting value called the open-channel conductance:

$$G_O = \frac{\partial i_D}{\partial v_{DS}}\bigg]_{\substack{v'_{GS} = 0 \\ V_{DS} \to 0}} = \frac{3I_{DSS}}{V_P} \qquad (3.14)$$

Thus, in the triode region, the JFET can be thought of as the channel being a variable resistance whose value can be controlled by the gate voltage.

The most significant two-port parameter of the FET may be identified as its transconductance, g_m, defined as

$$g_m = \frac{\partial i_D}{\partial v_{GS}}\bigg]_{v_{DS}} = \text{const.}$$

In the triode region, from Eq. 3.6

$$g_m = \frac{3I_{DSS}}{V_P}\left[\left(\frac{v_{DS} - v'_{GS}}{V_P}\right)^{1/2} - \left(\frac{V'_{GS}}{V_P}\right)^{1/2}\right] \tag{3.15}$$

On the pinchoff contour

$$g_{mp} = \frac{3I_{DSS}}{V_P}\left[1 - \left(\frac{-v'_{GS}}{V_P}\right)^{1/2}\right] \tag{3.16}$$

Its maximum value occurs when $v'_{GS} = 0$, given by

$$g_{mpM} = \frac{3I_{DSS}}{V_P} \tag{3.17}$$

which may be observed to be identical to the open-channel conductance G_O defined in Eq. 3.14.

In terms of the maximum value along the pinchoff contour

$$g_{mp} = G_O\left[1 - \left(\frac{-v'_{GS}}{V_P}\right)^{1/2}\right] \tag{3.18}$$

Theoretically, the g_m in the saturation region is the same as that along the pinchoff contour, and the equations hold everywhere in saturation to a reasonable approximation, subject to modifications that will be discussed later.

Nonuniformly Doped Channel

Equations similar to Eq. 3.2 and those following may be derived for channel impurity profiles other than uniform across the channel, although all such general solutions are much more complicated, and I_{DSS}, V_P, and the exponents in the equations will differ. For example, for the limit of an exponentially rising doping density across the channel approaching a "spike" of doping at the channel edge, it can be shown that for the planar structure

$$I_{DSS} = \frac{d^3 W q^2 \mu_n N_D{}^2}{2L\varepsilon_o\varepsilon_r} \tag{3.19}$$

and

$$V_P = \frac{qN_D d^2}{\varepsilon_o\varepsilon_r} \tag{3.20}$$

Furthermore, for the "spike" profile it can be shown that in the triode region the drain current can be written in the form

$$i_D = K \left[(v'_{GS} + V_P)v_{DS} - \frac{v^2_{DS}}{2} \right] \qquad (3.21)$$

At pinchoff, using Eq. 3.9, $v_{DS} = v'_{GS} + V_P$ and

$$i_{Dsat} = \frac{K}{2} (V'_{GS} + V_P)^2$$

$$= \frac{K}{2} V^2_P \left(1 + \frac{v'_{GS}}{V_P} \right)^2 \qquad (3.22)$$

or

$$i_{Dsat} = I_{DSS} \left(1 + \frac{v'_{GS}}{V_P} \right)^2 \qquad (3.23)$$

where

$$K = \frac{2I_{DSS}}{V^2_P} \qquad (3.24)$$

or

$$K = \frac{\mu_n W \varepsilon_o \varepsilon_r}{dL} \qquad (3.25)$$

For such a "spike" distribution, it is logical to identify the channel capacitance as

$$C_o = \frac{\varepsilon_o \varepsilon_r WL}{d} \qquad (3.26)$$

which is equivalent to the capacitance of a fully depleted region at a *pn* junction as given by Eq. 1.86. Then K may be expressed as

$$K = \frac{\mu_n C_O}{L^2} \qquad (3.27)$$

Equation 3.21 may be written in terms of channel capacitance as

$$i_D = \frac{\mu_n C_o}{L^2} \left[(v'_{GS} + V_P)v_{DS} - \frac{v^2_{DS}}{2} \right] \qquad (3.28)$$

and

$$i_{Dsat} = \frac{\mu_n C_O}{2L^2} (v'_{GS} + V_P)^2 \qquad (3.29)$$

It is sometimes useful to model i_{Dsat} for all JFETs regardless of channel impurity distribution by an equation of the form

$$i_{Dsat} = I_{DSS} \left(1 + \frac{v'_{GS}}{V_P} \right)^n \qquad (3.30)$$

For the spike distribution, $n = 2$ (Eq. 3.23), and for the uniform distribution of Eq. 3.22, n might be chosen to make Eqs. 3.30 and 3.11 match at some specific point.

Example

The transconductance at pinchoff from Eq. 3.11 is

$$g_{mp} = \frac{3I_{DSS}}{V_P} \left[1 - \left(\frac{-v'_{GS}}{V_P} \right)^{1/2} \right] \tag{3.16}$$

while, using Eq. 3.30,

$$g_m = \frac{nI_{DSS}}{V_P} \left(1 + \frac{v'_{GS}}{V_P} \right)^{n-1} \tag{3.31}$$

Thus, if it is desired to make the maximum value of g_{mM} agree for the two cases, $n = 3$ and

$$i_{D\text{sat}} = I_{DSS} \left(1 + \frac{v'_{GS}}{V_P} \right)^3 \tag{3.32}$$

and

$$g_m = \frac{3I_{DSS}}{V_P} \left(1 + \frac{v'_{GS}}{V_P} \right)^2 \tag{3.33}$$

Then the plot shown in Fig. 3.6 shows how the g_m for the two models compares over the entire range as a plot of g_{mp}/g_{mM}.

For the "spike" distribution ($n = 2$), Eq. 3.30 is

$$g_{mp} = \frac{2I_{DSS}}{V_P} \left(1 + \frac{v'_{GS}}{V_P} \right) \tag{3.34}$$

which represents a linear variation with gate voltage.

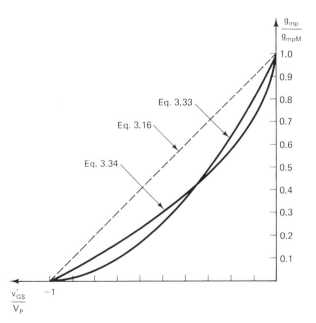

Figure 3.6 Nonlinear FET transductance from comparative models.

3.3 SEMIEMPIRICAL MODELS FOR THE SATURATION REGION

The equations developed so far for the *n*-channel JFET apply only in the triode region and imply that there are no further changes in i_{DS} as v_{DS} is increased (beyond the value for $\partial i_{DS}/\partial v_{DS} = 0$), and hence that g_m remains constant at its value $g_m = g_{mp}$ in the saturation region. This is not exactly true and the equations predicting the characteristics precisely are quite complex, but the properties upon which they are based are quite simple.

In the structure shown in Fig. 3.7(a), the pinchoff value of g_m is predicted when the effective channel width is reduced to zero at the drain end.

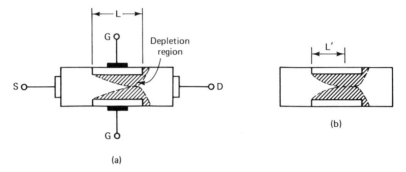

(a)

(b)

Figure 3.7 Symmetrical structure showing depletion layer progression at increased drain voltage.

As the drain voltage is further increased beyond the saturation value, the effect is to extend the depletion region back toward the source, as shown in Fig. 3.7(b), which in turn shortens the effective channel length, thereby increasing the conductivity of the channel beyond that predicted at pinchoff, as indicated in Fig. 3.8.

The effects of channel shortening in the saturation region can be modeled in a semiempirical fashion as follows: Starting with the approximate equation (3.30)

$$i_{D\text{sat}} = I_{DSS}\left(1 + \frac{v'_{GS}}{V_P}\right)^n \tag{3.35}$$

and assuming that v_{DS} continues to influence i_D beyond pinchoff, but by a factor $1/\mu_A$ less than that of v_{GS}, Eq. 3.35 is modified as follows:

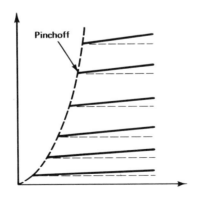

Figure 3.8 Simple nonlinear model for JFET in saturation region.

$$i_D = I_{DSS} \left[1 + \frac{1}{V_P} \left(v'_{GS} + \frac{v_{DS}}{\mu_A} \right) \right]^n \tag{3.36}$$

where μ_A is a factor that indicates the relative effectiveness of drain and gate voltages in controlling the increase of drain current in the saturation region. Then in the saturation region

$$g_m = \frac{\partial i_D}{\partial v_{GS}} \bigg]_{v_{DS} = V_{DS}} = \frac{n I_{DSS}}{V_P} \left[1 + \frac{1}{V_P} \left(v'_{GS} + \frac{v_{DS}}{\mu_A} \right) \right]^{n-1} \tag{3.37}$$

Furthermore, the incremental drain resistance is defined by

$$\frac{1}{r_d} = \frac{\partial i_D}{\partial v_{DS}} \bigg]_{v_{GS} = V_{GS}} = \frac{n I_{DSS}}{\mu_A V_P} \left[1 + \frac{1}{V_P} \left(v'_{GS} + \frac{v_{DS}}{\mu_A} \right) \right]^{n-1} \tag{3.38}$$

From Eqs. 3.37 and 3.38,

$$\mu_A = g_m r_d \tag{3.39}$$

or for a basic definition of μ, Eq. 3.35 may be written as

$$v_{DS} = \mu_A V_P \left[\left(\frac{i_D}{I_{DSS}} \right)^{1/n} - 1 - \frac{v_{GS}}{V_P} \right] \tag{3.40}$$

Then

$$\frac{\partial v_{DS}}{\partial v_{GS}} \bigg]_{i_D = I_D} = -\mu_A \tag{3.41}$$

The factor μ_A is termed the *voltage amplification factor*, not to be confused with the reverse voltage μ_r defined for the bipolar junction transistor discussed in Chapter 2, or with hole and electron mobilities, which also use μ_n or μ_p for their symbols.

A slightly different model for the saturation region, which relates somewhat more closely to the structural parameters of short-channel JFETs, is given by

$$I_{DSS} = I_{DSS} \left(1 + \frac{V'_{GS}}{V_P} \right)^n (1 + \lambda v_{DS}) \tag{3.42}$$

where λ is a structurally related parameter. For this model, g_m and μ_A are both functions of both v_{DS} and v_{GS}, and n may be a number different from 2, but the relationship $\mu = g_m r_d$ still holds.

3.4 INCREMENTAL (SMALL SIGNAL) CIRCUIT MODELS

An n-channel JFET is shown schematically in Fig. 3.9 as a two-port circuit element with source and load connected. The characteristics of a specific discrete JFET are shown in Fig. 3.10. The following equations may be written for the circuit shown.

$$v_{GS} = V_{GG} + v_{gs}$$

$$v_{DS} = V_{DD} - i_D R_L$$

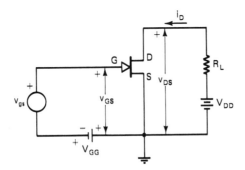

Figure 3.9 Symbols and terminology for JFET incremental models.

A specific operating point may be established by plotting the preceding equation on the characteristic of a typical JFET shown in Fig. 3.10. Such a plot is known as the *load line*. For this example, $V_{DD} = 24$ V, I_D for $V_{DD} = 0$ is 4.25, $V_{GG} = -1$, and $R_L = 5.65$ kΩ. The selected operating point is $V_{DS} = 10$ V, $V_{GS} = -1.0$ V, and $I_D = 2.5$ mA.

Deviations from the selected operating point are v_{ds}, v_{gs}, and i_d, as indicated. All deviations actually take place along the path defined as the load line.

Whether the drain characteristics are described analytically as discussed in Sec. 3.3 or graphically as shown in Fig. 3.10, using the functional relationship

$$i_D = f(v_{GS}, v_{DS}) \tag{3.43}$$

differential variations from an established operating point may be written as

$$di_D = \frac{\partial i_D}{\partial v_{GS}}\bigg]_{v_{ds} = V_{DS}} dv_{GS} + \frac{\partial i_D}{\partial v_{DS}}\bigg]_{v_{GS} = V_{GS}} dv_{DS} \tag{3.44}$$

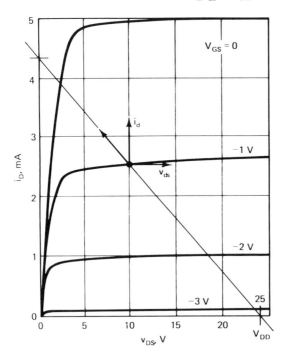

Figure 3.10 Drain current–voltage characteristic of typical JFET.

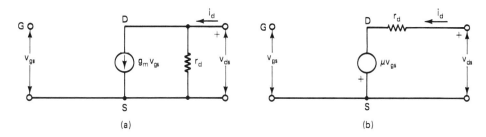

Figure 3.11 Incremental JFET low-frequency circuit models. (a) Current source model. (b) Voltage source model.

Noting from Sec. 3.3 that

$$\frac{\partial i_D}{\partial v_{GS}} \equiv g_m \qquad (3.45)$$

and

$$\frac{\partial i_D}{\partial v_{DS}} \equiv \frac{1}{r_d} \qquad (3.46)$$

and using small finite values as approximations in the vicinity of the established operating point, Eq. 3.44 may be approximated by

$$i_d = g_m v_{gs} + \frac{1}{r_d} v_{ds} \qquad (3.47)$$

which is represented by the equivalent circuit of Fig. 3.11(a).

Alternatively, the circuit of Fig. 3.11(b) is equivalent for the equation

$$v_{DS} = i_d r_d - \mu V_{gs} \qquad (3.48)$$

where $\mu = g_m r_d$.

3.5 p-CHANNEL JUNCTION FIELD-EFFECT TRANSISTOR

The JFET can be fabricated as a *p*-channel device in all the structure variations indicated in Sec. 3.2, with all *n* and *p* regions interchanged. The *p*-channel JFET shown schematically in Fig. 3.12 in an actual circuit has characteristics similar to those of the *n*-channel type except for voltage polarities and current directions.

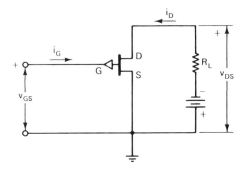

Figure 3.12 *p*-channel JFET symbols and external connections.

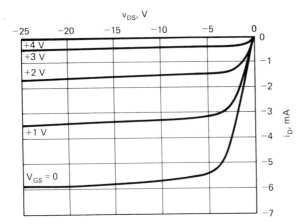

Figure 3.13 Typical *p*-channel JFET drain current–voltage characteristic.

If the sign conventions used for the *n*-channel structure (in accordance with standard two-port network terminology) are as indicated in Fig. 3.12, the actual directions and polarities for normal FET operation are v_{GS} positive, v_{DS} negative, and i_D negative.

Then, for the same equations to hold, the signs of these quantities should be changed accordingly. This causes the resultant characteristics comparable to those of the *n*-channel device to be plotted as shown in Fig. 3.13 for a typical device. Once the operating point is established, the incremental circuit models are identical to those for the *n*-channel FET.

3.6 HIGH-FREQUENCY INCREMENTAL MODELS

The major frequency dependency of the JFET is due to the effective capacitance of the depletion layer under the gate. In the triode region, this capacitance is voltage dependent in the manner of the reverse-biased junction. At pinchoff for the uniform channel impurity distribution, where the depletion region extends across the channel at the extreme end, the capacitance of the resultant "wedge-shaped" region is roughly twice that of the fully depleted channel. As the drain voltage increases beyond its saturated value, the total channel capacitance decreases as the portion of the channel length that is fully depleted increases, with the capacitance ultimately reaching that of the fully depleted channel. The relative portion of the total channel capacitance allocated to that between gate and source and that between gate and drain is then also a function of increased drain voltage beyond saturation.

For the "spike" impurity distribution for the planar structure, the channel capacitance given by Eq. 3.26 at pinchoff is appropriate, and changes more slowly with increased drain voltage because there is no marked change in effective channel width.

With the total channel capacitance apportioned between gate and drain and between gate and source, a simple approximate high-frequency incremental circuit model may be obtained by adding the appropriate capacitances to the incremental model of Fig. 3.11, as indicated in Fig. 3.14(a). For completeness, an incremental resistance has been added to the input circuit which was not included in the model of Fig. 3.11 because, being that

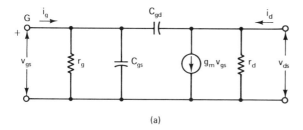

(a)

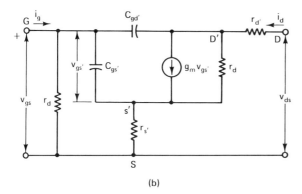

(b)

Figure 3.14 High-frequency circuit models for JFETs.

of a reverse-biased diode, it is extremely high; but in some cases it may be important when frequency dependencies are considered.

This model is sufficiently accurate for most purposes; however, a more accurate model can be obtained if small series resistances of the source and drain ends of the channel are included, as shown in Fig. 3.14(b). These are bulk resistances extrinsic to the processes defining FET operation and are like the base-spreading resistance discussed for the bipolar junction transistor.

The resistance, r_s', in the common source lead is the most important of the two extrinsic resistances because it causes the effective g_m (as obtained from external measurements) to be less than that (which we will now call g_m') as calculated using previous equations. The external g_m in terms of g_m' can be shown to be

$$g_m = \frac{g_m'}{1 + g_m' r_s'} \tag{3.49}$$

The frequency characteristics of specific amplifier configurations will be discussed in subsequent chapters. However, in a general sort of way, the voltage gain at low frequencies will be proportional to g_m, whereby at high frequencies it will be gradually reduced by shunt capacitance related to the channel capacitance. Therefore, it is convenient to define a figure of merit for an FET as

$$\text{FM} = \frac{g_m}{C_o} \tag{3.50}$$

Sec. 3.6   High-Frequency Incremental Models   **111**

In terms of structural parameters this can be determined as follows for the "spike" distribution. From Eqs. 3.19, 3.20, and 3.34, the maximum value of g_m at pinchoff is

$$g_{mpM} = \frac{qN_D W d \mu_n}{L} \tag{3.51}$$

and from Eq. 3.26

$$C_o = \frac{\varepsilon_o \varepsilon_r W L}{d} \tag{3.26}$$

$$\text{FM} = \frac{qN_D \mu_n}{\varepsilon_o \varepsilon_r} \frac{d^2}{L^2} \tag{3.52}$$

These equations can be used as a guide in optimizing the structure of the FET. In general, it would be desirable to keep V_P low to minimize the voltage supply required. From Eq. 3.20 this would indicate keeping d, the channel thickness, as small as practical. At the same time, it is generally desirable to maintain g_m high, which can be done by increasing the width W and minimizing the channel length L. Increasing the width also increases C_o, which is undesirable, although it does not affect the figure of merit; but decreasing L also decreases C_o, which *is* desirable. Thus the one single parameter change that increases g_m and decreases C_o simultaneously without affecting V_p is the decrease in channel length L; hence most high-frequency design is based on making L as small as possible, while maximizing the W/L ratio, at the same time maintaining V_P low by making d small.

3.7 INSULATED GATE FIELD-EFFECT TRANSISTOR

The insulated gate field-effect transistor (IGFET), also referred to as the metal oxide field-effect transistor (MOSFET), has characteristics similar in form to those for the JFET. The structure and general principles of operation of one particular form of IGFET may be described as follows: Two highly doped n^+ diffusions provide ohmic contact between the source and drain terminals to a relatively lightly doped p substrate, as indicated in Fig. 3.15(a). These contacts are made through the insulating oxide film of depth d. A metal surface of length L bonded to this insulating layer is the gate.

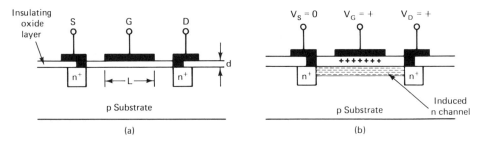

Figure 3.15 Cross section of insulated-gate FET. (a) Structure. (b) Induced channel for positive gate voltages.

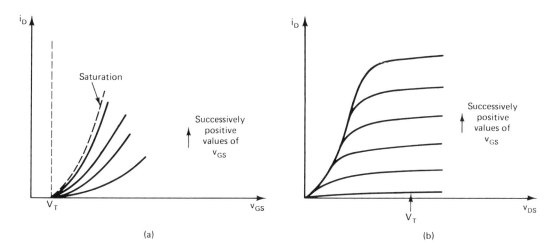

Figure 3.16 Suggested drain current–voltage characteristics for induced channel IGFET.

If the drain is made positive relative to the source, electron current flows from source to drain; however, if the substrate is very lightly doped, and with no voltage applied to the gate, the current would be small. (As a matter of fact, there would be a reverse-biased *pn* junction formed at the drain–substrate diffusion.)

Now if a positive voltage is applied to the gate, the gate–insulator–substrate region acts as a capacitor with + charges appearing at the gate–insulator surface and − charges in the substrate adjacent to the insulator, as indicated in Fig. 3.15(b). These excess electrons form an "induced" *n* channel in the region between the source and drain, as indicated. Such a layer is referred to as an *inversion layer*. The more positive the gate becomes, the wider the effective channel becomes. If the drain is made increasingly positive, more current will flow through the channel for a particular value of drain voltage, as indicated in Fig. 3.16(a).

Regardless of the drain voltage, it would be expected that very little current would flow until some threshold voltage, V_T, is reached, sufficient to induce a significant *n* channel. Also, it would be expected that some positive value of v_{GS} would finally be reached (saturation) when i_D would no longer be a function of v_{DS}.

The effects indicated in Fig. 3.16(a) translate into another set of characteristics indicated in Fig. 3.16(b) of $i_D = f(v_{DS})$ for increasing values of v_{GS}. These properties may be recognized as similar to those of the *n* channel JFET, except that the gate is forward rather than reverse biased with respect to the source.

In the modified structure in Fig. 3.17, there is an *n* channel diffused into the substrate at the same time the drain and source contact diffusions are formed. If now the gate is reverse biased relative to the source, an increasingly negative gate voltage induces positive charges within the channel, which partly neutralizes the negative charges due to impurities, thus tending to deplete the region of free carriers. The resultant characteristic, shown in Fig. 3.17(b), is similar to that of the JFET with "spike" impurity distribution and is referred to as a *depletion-mode* IGFET, whereas the induced channel structure of Fig. 3.15 with characteristics shown in Fig. 3.16 is referred to as an *enhancement mode* IGFET.

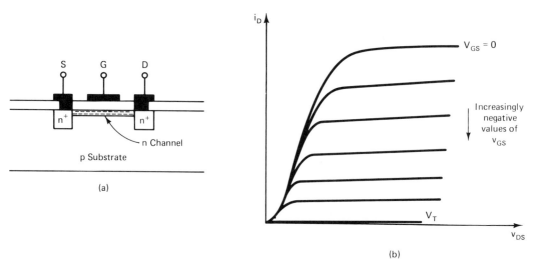

Figure 3.17   Insulated-gate FET with diffused channel. (a) Cross section. (b) Drain current–voltage characteristic.

In the case of the depletion-mode structure, V_T is the negative value of v_{GS}, which effectively depletes the region of carriers and reduces the conductivity of the channel to zero. It is possible by use of a very lightly doped channel to achieve both enhancement- and depletion-mode operation in the same device.

3.8 BASIC EQUATIONS FOR IGFET OPERATION IN THE LOW-VOLTAGE (TRIODE) REGION

The drain current–drain voltage relationship for the n-channel IGFET in the triode region can be modeled by the approximate relationship

$$i_D = K\left[(v_{GS} - V_T)v_{DS} - \frac{v_{DS}^2}{2}\right] \tag{3.53}$$

where

$$K = \frac{\mu_n C_{ox}}{L^2}$$

with C_{ox} being the capacitance of the insulating oxide given by

$$C_{ox} = \frac{\varepsilon_o \varepsilon_i WL}{d_i} \tag{3.54}$$

or $C_{ox} = WLC_i$ with C_i being the capacitance per unit area. The thickness of the oxide is d_i and ε_i is its dielectric constant (approximately 3.9 for SiO_2).

The mobility μ_n is derived on the basis of electrons in the lightly doped substrate, but it is a surface mobility that is approximately half the value obtained in the bulk.

Equation 3.53 models, although somewhat crudely, both enhancement- and depletion-mode operation, where V_T is positive for enhancement mode and highly negative for depletion mode. This is the most critical parameter in defining IGFET operation and will be considered in some detail as it relates to basic structural and process parameters. A rough approximation will first be made for depletion-mode V_T, followed by a more detailed expression for enhancement-mode devices and a return to a more accurate representation for the depletion-mode device.

Buried Channel Depletion-Mode IGFET

Shown in Fig. 3.18 is an IGFET with an n channel implanted at the edge of the p substrate. The thickness of the insulating oxide is d_i and that of the n channel is d_n. Assuming that $n \cong N_D$ is distributed uniformly in the channel, the total charge per unit area in the region under the gate is $Q_n = -qN_Dd_n$.

The potential across the channel when the channel is completely depleted of mobile carriers can be expressed approximately as

$$\phi_n = -\frac{qN_Dd_n}{2C_n}$$

where

$$C_n = \frac{\varepsilon_o\varepsilon_r}{d_n}$$

The factor of 2 in the denominator is based on the approximation that on the average the work done in transferring charges out of the layer (voltage) is roughly as though the charges were all at the center of the layer rather than being uniformly distributed. The applied negative voltage between gate and substrate must include Q_n plus the negative charge at the gate surface required to equal the ion charge left in the depletion layer. Thus the threshold voltage is very approximately

$$V_T = -\frac{qN_Dd_n}{2C_n} - \frac{qN_Dd_i}{C_i}$$

or

$$V_T = -\frac{qN_Dd_i}{C_i}\left(1 + \frac{C_i}{2C_n}\right) \tag{3.55}$$

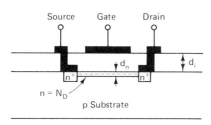

Figure 3.18 Buried channel depletion-mode IGFET.

In this equation, if $C_i << C_n$, which is the usual case since $\varepsilon_i << \varepsilon_r$, and if $d_n << d_i$, it may be written as

$$V_T \cong -\frac{q\,N_D d_i^2}{\varepsilon_o \varepsilon_i} \tag{3.56}$$

This value compares with $-V_p$ given by Eq. 3.20 for the spike channel JFET, except for the difference in dielectric constant of the insulator and the semiconductor. This result should not be entirely unexpected and could have been carried out in a similar manner for the JFET.

However, this value for V_T neglects the built-in potential of the pn junction at the channel–substrate interface, the effect of fixed charges in and at the oxide surface as a result of processing imperfections, and an offset voltage due to differences in work functions of the metal and semiconductor. These factors all combine to make V_T somewhat more negative than the value given by Eq. 3.55 or 3.56. These will be reconsidered after the enhancement-mode IGFET is discussed, where they become the more important factors.

Also, V_T was derived as being independent of distance along the channel, which would not be true for other than very small values of v_{DS}.

Enhancement-Mode IGFET with Channel Inversion

For the enhancement-mode IGFET, the threshold voltage focuses on conditions involving creation of the inverted channel. The components comprising V_T may be determined approximately with reference to Fig. 3.19, which shows the distances under the gate in expanded form.

In the p region, free electrons are available in the concentration of

$$n \cong \frac{n_i^2}{N_A}$$

A potential at the surface of approximately

$$\phi_s = 2\phi_F = 2\,\frac{kT}{q}\ln\frac{N_A}{n_i} \tag{3.57}$$

is required to create a strong inversion, that is, to bring a large number of free electrons to the vicinity of the surface.

The potential ϕ_F in this case is approximately the potential corresponding to the Fermi level, which is an energy level approximately midway between the upper edge of

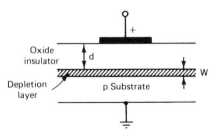

Figure 3.19 Enhancement-mode IGFET structure.

the valence band and the lower edge of the conduction band. Thus roughly twice this value is required to create substantial numbers of conduction electrons. Such a potential creates a depletion layer of width W, which by using the depletion approximation discussed in Chapter 1 is approximately

$$W = \left(\frac{2\varepsilon_o\varepsilon_r\phi_s}{qN_A} \right)^{1/2} \tag{3.58}$$

The depletion layer charge (which resides near the surface) is given by $Q_d = -qN_AW$, or using Eqs. 3.57 and 3.58,

$$Q_d = -(2\varepsilon_o\varepsilon_r qN_A^2 2\phi_F)^{1/2} \tag{3.59}$$

The additional potential across the insulating oxide, assuming that ϕ_d is at or almost at the surface of the substrate, is

$$V_i = \frac{|Q_d|d}{\varepsilon_o\varepsilon_i} = \frac{Q_d}{C_i} \tag{3.60}$$

where C_i is defined as the capacitance per unit area of the oxide with ε_i its dielectric constant. Then from the preceding relationships

$$V_i = \frac{\sqrt{2\varepsilon_o\varepsilon_r qN_A(2\phi_F)}}{C_i} \tag{3.61}$$

From the preceding discussion, V_T would be the sum of Eqs. 3.57 and 3.61, which would be the minimum applied voltage needed to produce the required inverted channel at the $p\text{-}i$ surface. However, there are other necessary modifications.

Inasmuch as at one side of the insulating oxide is a metal and at the other is the surface of the substrate, the threshold voltage determination must take into account the difference in work functions of the metal and the semiconductor. This difference in work function is designated as ϕ_{ms} and is also approximately the potential required to equalize the Fermi levels of the metal and the semiconductor. The value of ϕ_{ms} is normally negative, inasmuch as the work function of the metal is less than that of the semiconductor. In addition, there are surface charges and charges due to trapped oxide ions not previously accounted for. The sum of the two factors is called the *flat-band* voltage given by

$$V_{FB} = \phi_{ms} - \frac{Q_i}{C_i} \tag{3.62}$$

where Q_i is a net positive charge referred to as the *interface charge*. The flat-band voltage is negative because ϕ_{ms} is also negative.

The threshold voltage combining all the preceding factors is given by

$$V_T = V_{FB} + 2\phi_F + \frac{\sqrt{2\varepsilon_o\varepsilon_r qN_A(2\phi_F - V_{su})}}{C_i} \tag{3.63}$$

The additional term V_{su} in the V_i component accounts for any bias voltage applied to the substrate relative to the source. If V_{su} is negative, the V_i component and hence the threshold voltage is increased.

In a normal enhancement-mode n-channel FET, it is intended that V_T be positive; hence the impurity concentration in the p substrate would be made sufficiently high to achieve the desired value. Other adjustments to V_T can be made by controlling $C_i = \varepsilon_o \varepsilon_i / d_i$ through control of the oxide and its relative dielectric constant by using compounds of silicon other than S_iO_2. Also, substrate bias may be added where circuit configurations permit. In addition, the negative V_{FB} can be substantially reduced by using a very highly doped polycrystalline silicon gate, rather than a metal, for a better match of gate and substrate work functions. On the other hand, if desired, V_T can actually be made zero or even negative. Normally, ϕ_{ms} for the metal–insulator–semiconductor (MIS) system is on the order of -0.9 V. Q_i/C_i is normally small (perhaps 0.1 v), but can be made larger through control of fixed surface and oxide charges. Normally, the flat-band voltage lies between -1 and -1.5 V for a metal gate and much less for a polysilicon gate.

More Detailed Representation for the Buried Layer IGFET

The model discussed for the buried layer IGFET with Eqs. 3.55 and 3.56 as first-order approximations for the threshold voltage did not include some important correction terms. The flat-band voltage, V_{FB}, is also present and is of the same polarity as it is for the enchancement-mode structure, and hence makes V_T more negative than that given by Eq. 3.55. Also, it did not take into account the built-in voltage and depletion layer widening into the p region, as indicated in Fig. 3.20.

Since the built-in voltage $\phi_B = \phi_p - \phi_n$ is positive, it reduces the negative threshold voltage by a term that can be shown to be approximately

$$\frac{\sqrt{2\varepsilon_o \varepsilon_r\, q\, [N_A N_D/(N_A + N_D)]}}{C_i} \left(1 + \frac{C_i}{C_n}\right) \sqrt{(\phi_B - V_{su})}$$

Then the total threshold voltage is given by

$$V_T = V_{FB} - \frac{qN_D d_i}{C_i}\left(1 + \frac{C_i}{2C_n}\right)$$
$$+ \frac{\sqrt{2\varepsilon_o \varepsilon_r q[N_A N_D/(N_A + N_D)]}}{C_i}\left(1 + \frac{C_i}{C_n}\right)\sqrt{\phi_B - V_{su}} \quad (3.64)$$

In this equation, the flat-band voltage, V_{FB}, can be made to have a reasonably wide range (largest for a metal such as aluminum or very heavily doped polysilicon, to almost zero for more lightly doped polysilicon). This value can be adjusted to balance the term involving ϕ_B, which would leave Eq. 3.55 as a reasonably good approximation for V_T.

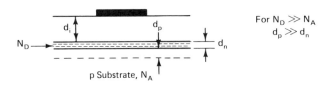

Figure 3.20   Details of buried layer IGFET structure.

Enhancement–Depletion Mode IGFETS

IGFETS can be fabricated to operate as combined enhancement–depletion mode devices by employing a very lightly doped buried layer n channel. This maintains V_T at a negative value required to deplete the channel of carriers as in the depletion-mode structure, and then, as the gate voltage becomes positive, an inversion layer is formed as in the case of the enhancement-mode structure. However, the model equations are not as simple as they are for either structure alone, and the threshold voltage is highly dependent on the relative dopings of the buried channel and the substrate. Furthermore, the electron mobility for the depletion mode region is the bulk mobility in the channel, whereas, as enhancement-mode operation is reached, it becomes the surface mobility. These factors render Eq. 3.53 a less accurate representation for the combined E–D operation. However, it can still be used as a guide to construct approximate characteristic curves that can later be empirically modified to take various factors into account.

An actual E–D IGFET is illustrated in Fig. 3.21, which reveals the enhancement–depletion dissymmetry due to the factors that have been discussed.

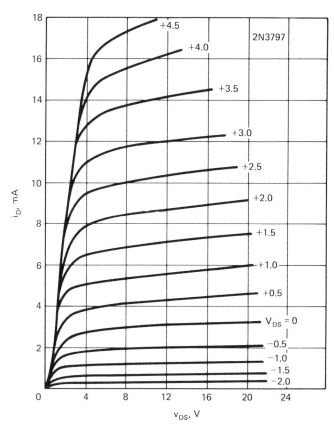

Figure 3.21 Characteristics of typical depletion–enhancement mode IFGET.

3.9 APPROXIMATE IGFET MODELS FOR THE SATURATION REGION

If the starting point in the construction of reasonably accurate saturation region models is to assume the validity of Eq. 3.53 at the edge of saturation (i.e., for $\partial i_D/\partial v_{DS} = 0$), the two separate equations for $i_D = f_1(v_{DS})$ alone and $i_D = f_2(v_{GS})$ alone may be written as

$$i_{D\text{sat}} = \frac{K}{2} v_{DS}^2 \tag{3.65}$$

and

$$i_{D\text{sat}} = \frac{K}{2}(v_{GS} - V_T)^2 \tag{3.66}$$

The dashed line in Fig. 3.22 is a plot of Eq. 3.65, while points on the plot are for specific values of equal increments of v_{GS}.

The transconductance at the edge of saturation defined by $g_m = \partial i_D/\partial v_{GS}$ obtained from Eq. 3.66 is

$$g_{ms} = K(v_{GS} - V_T)$$

where the appropriate K factor as related to mobility and V_T is determined for the particular transistor (i.e., whether enhancement or depletion mode). In addition to reasons previously cited, Eq. 3.53, which leads to Eqs. 3.65 and 3.66, is only approximate at the edge of saturation and does not model the transistor fully in the saturation region.

For one thing, the threshold voltage V_T was determined on the basis of the assumption that the channel charge from which it was determined is uniformly distributed along the channel, which would be true only for very small values of v_{ds}. Also, for FETs of small dimensions (for enhancement-mode devices), the surface mobility is dependent on v_{GS} because the transverse electric field $\mathscr{E}_t = v_{GS}/d$ is larger for a thin-channel device, and it can be shown that mobility is a decreasing function at such transverse electric fields.

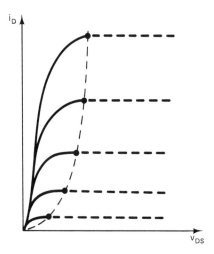

Figure 3.22 Extension of i_D–v_{DS} characteristic into the saturation region.

The axial field $\mathscr{E}_a = v_{DS}/L$ is greater for shorter-channel devices with the following effect: The electron drift velocity is normally given by $\eta = \mu_n \varepsilon$. However, at high electric fields the drift velocity tends to saturate at a constant value for very high electric fields. This makes the effective mobility a decreasing function of the longitudinal field as well as the transverse field. The net result is a shift of the i_D versus v_{DS} characteristic away from the square-law characteristic expressed by Eq. 3.66 toward a more linear i_D versus v_{GS} relationship. In addition to this compression at saturation, there is a continuing positive slope for each v_{GS} contour as a result of effective channel shortening due to the extension of the depletion layer back into the channel as v_{DS} is increased.

These effects can be modeled at the edge of and well into the saturation region by modifying Eq. 3.66 to the form

$$i_D = \frac{\mu_n C_o}{2L^2 a}(v_{GS} - V_T)^n (1 + \lambda v_{DS}^m) \tag{3.67}$$

where a is a thin-channel field-related conductance degradation constant with dimensions $v^{n/2}$ and n for $n < 2$ accounts for short-channel velocity saturation effects. The constant λ with dimension at v^{-m} accounts for channel length modulation at higher values of v_{ds}. At the same time, $m > 1$ models excess carrier generation at higher voltages.

From Eq. 3.67, the small-signal parameters may be written as

$$g_m = \frac{nK}{2a}(v_{GS} - V_T)^{n-1}(1 + \lambda v_{DS}^m) \tag{3.68}$$

$$\frac{1}{r_d} = \frac{K\lambda m}{2a} v_{DS}^{m-1} (v_{GS} - V_T)^n \tag{3.69}$$

$$\mu_A = \frac{n}{\lambda} \frac{1 + \lambda v_{DS}^m}{m v_{DS}^{m-1}} \frac{1}{v_{GS} - V_T} \tag{3.70}$$

$$= g_m r_d \tag{3.71}$$

In the simplest case, using what is referred to as the first-order model, $m = 1$, $a = 1$, and $n = 2$, making

$$g_m = K(v_{GS} - V_T)(1 + \lambda v_{DS}) \tag{3.72}$$

$$r_d = \frac{2}{\lambda K} \frac{1}{(v_{GS} - V_T)^2} \tag{3.73}$$

$$\mu_A = \frac{2}{\lambda} \frac{(1 + \lambda v_{DS})}{v_{GS} - V_T} \tag{3.74}$$

A simple alternative first-order model based on an assumed constant amplification factor uses $v_{GS} + V_{DS}/\mu_A$ in Eq. 3.66 in the same manner as was suggested by Eq. 3.36 for the JFET. Then

$$i_D = \frac{K}{2}\left(v_{GS} - V_T + \frac{V_{DS}}{\mu}\right)^2 \tag{3.75}$$

A still different empirical equation that accounts for higher degrees of nonlinearity may be written as

$$i_D = \frac{K}{2}(v_{GS} - V_T) + (G_o^{1/n} + kv_{GS})v_{DS}^n \qquad (3.76)$$

The constants n and k can be chosen to model varying degrees of nonlinearity.

p-Channel IGFET

Both enhancement- and depletion-mode IGFET can be fabricated using an n substrate with p^+ source and drain contact diffusions and with a p-channel buried layer for the depletion-mode device. Equations similar to those that have been developed can be applied to the p-channel IGFET with appropriate polarity changes. A typical discrete p-channel enhancement-mode MOSFET is shown in Fig. 3.23. This particular transistor is designed for a relatively large value of V_T. For many applications, threshold voltages of V_T in the neighborhood of 1 or 2 volts are more typical.

3.10 IGFET (MOSFET) SYMBOLS

Schematically, several alternative symbols are used to identify MOSFET in electronic circuit schematic diagrams. Some of these are shown for n-channel devices in Fig. 3.24.

In Fig. 3.24(a), the direction of the arrow indicates the direction of current flow and its position the source terminal. The labels G, S, and D are redundant. The symbol of Fig. 3.24(b) differs only in that the lead from the gate terminal is positioned to be closest to the source terminal. For this symbol, if it is already understood that n-channel devices are being described, the arrow on the source lead is not needed. The solid fill-

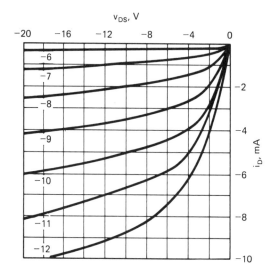

Figure 3.23 Characteristic of typical p-channel enhancement-mode IGFET.

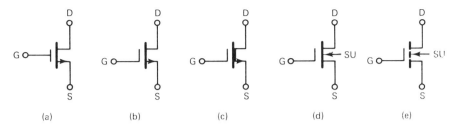

Figure 3.24 IGFET (MOSFET) symbols.

in shown in Fig., 3.24(c) indicates specifically a depletion-mode device. If both symbols as in parts (b) and (c) are used on the same circuit schematic, the ones without the fill-in would be assumed to be enhancement-mode devices. The slightly more elaborate symbol shown in Fig. 3.24(d) separately identifies the substrate conection and indicates a depletion-mode device.The split line in the symbol of Fig. 3.24(e) indicates an enhancement-mode n-channel device. The symbols in part (d) or (e) are most often used when it is necessary to indicate substrate connections.

A similar set of symbols with all arrow directions changed identifies p-channel devices in the same manner.

3.11 TEMPERATURE DEPENDENCE OF FIELD-EFFECT TRANSISTORS

There are temperature factors for both JFETs and MOSFETs that are common to both structures and some that differ. The K factor is temperature dependent through temperature-dependent mobility, as discussed in Sec. 1.8. For the JFET, the built-in voltage ϕ_0 of the pn junction is temperature dependent through the temperature-dependent intrinsic carrier concentration as suggested by Eq. 1.124. For the JFET, the pinchoff voltage V_p is not temperature dependent, but for the MOSFET, the corresponding threshold voltage V_T is temperature dependent through ϕ_F, as indicated in Eq. 3.57, and through ϕ_{FB}, which is slightly temperature dependent through the work-function differential, as indicated by Eq. 3.62.

The threshold voltage for the enhancement-mode MOSFET as it is presented in Eq. 3.63 contains the temperature-dependent V_{FB} and ϕ_F. For the buried layer depletion-mode MOSFET, the built-in voltage of the channel-substrate junction as indicated by Eq. 3.64 is also involved.

3.12 NOISE SOURCES IN FIELD-EFFECT TRANSISTORS

It would be expected that the dominant source of noise in the FET would be thermal noise as related to the resistance of the channel. As the saturation range is approached from the triode region, this resistance tends toward an almost constant value at the boundary of and into the saturation region.

The thermal noise voltage of such a channel is given by $\overline{v_R^2} = 4kTR\Delta f$, which can be expressed as an equivalent thermal noise drain current given by

$$\overline{i_{dt}^2} = 4KG_c\,\Delta f \tag{3.77}$$

where G_c is the channel conductance $G_c = 1/R_c$ as determined on the saturation contour given by

$$G_c = \frac{i_{D\text{sat}}}{v_{DS}} \tag{3.78}$$

where for the simple square-law FET

$$i_{D\text{sat}} = \frac{K}{2}v_{DS}^2$$

Hence for this particular device model

$$G_c = \frac{K}{2}\,v_{DS}$$

This value can be conveniently expressed in terms of transconductance given by $g_m = KV_{DS}$, making $G_c = g_m/2$.

Therefore, Eq. 3.77 may be written as

$$\overline{i_{dt}^2} = 4k\left(\frac{g_m}{2}\right)\Delta f \tag{3.79}$$

Sometimes the value $\frac{2}{3}g_m$ rather than $\left(\frac{g_m}{2}\right)$ is suggested. This would be more accurate for the uniformly doped channel JFET.

There may also be present excess low-frequency noise, such as flicker and burst noise in the channel. Equation 3.79 can be modified to include these effects, allowing a total mean-square noise current to be written as

$$\overline{i_d^2} = 4kT\left(\frac{g_m}{2}\right)\Delta f + \text{excess noise}$$

where excess noise contains terms of the general nature of Eq. 1.138.

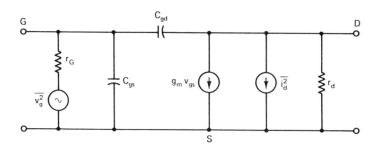

Figure 3.25 Noise model for FETs.

There is another source of thermal noise arising from the gate leakage resistance given by

$$\overline{v_R^2} = 4kTr_G\,\Delta f \qquad (3.80)$$

A complete circuit model including both channel and gate leakage effects is shown in Fig. 3.25. The noise voltage $\overline{v_g^2}$ would normally be negligible unless the transistor is driven from an extremely high impedance source.

PROBLEMS

3.1 A symmetrical n-channel JFET with uniformly doped channel has $N_A = 10^{18}$, $N_D = 5 \times 10^{15}$, $d = 0.4\ \mu\mathrm{m}$, $L = 8\ \mu\mathrm{m}$, and $w = 500\ \mu\mathrm{m}$.
  (a) Determine ϕ_o, V_p, and I_{DSS} at 300 K.
  (b) Calculate the maximum value of g_m.
  (c) Write an equation for I_{DSS}/V_P and note which parameters yield the highest I_{DSS} while minimizing the pinchoff voltage.

3.2 A planar-type JFET has the same impurity concentrations and dimensions as in Prob. 3.1 (note that d is the distance from the gate to the p-channel substrate).
  (a) What are the values for V_P and I_{DSS}?
  (b) Repeat part (a) for $d = 0.8\ \mu\mathrm{m}$.

3.3 A planar p-channel JFET with uniformly doped p and n regions has $N_D = 10^{18}$, $N_A = 5 \times 10^{15}$, $d = 0.4\ \mu\mathrm{m}$, $L = 8\ \mu\mathrm{m}$, and $w = 500\ \mu\mathrm{m}$.
  (a) Determine ϕ_o, V_p, and I_{DSS}.
  (b) Calculate g_m at pinchoff for the value of I_{DSS} determined in part (a).
  (c) What is the ratio of I_{DSS} to that of a comparable n-channel JFET with all structural and process parameters the same?

3.4 For the circuit shown, using a JFET known to have approximately a "spike" channel impurity distribution the gate voltage, v_{GS} is derived from a supply voltage, v_{GG}, in series with a large resistance, R_G. V_{GG} is gradually increased from negative values through zero and up to a point for which $v_{GS} \to 0.8$ V and does not increase further with increasing v_{GG}. Simultaneously, R_D is decreased from a very large value to the point for which $i_D \to 10$ mA for $v_{DS} = 4.0$ V. As R_D is decreased further, no further change takes place in i_D.

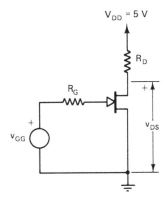

Prob. 3.4

(a) Determine K and V_P in the defining equation for the JFET.

(b) What is the maximum value of transconductance at pinchoff?

3.5 A planar JFET, as shown, has impurities concentrated in a narrow region at the edge of the p substrate.

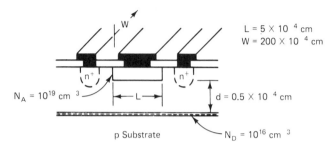

$$L = 5 \times 10^{-4} \text{ cm}$$
$$W = 200 \times 10^{-4} \text{ cm}$$

$N_A = 10^{19} \text{ cm}^{-3}$

$d = 0.5 \times 10^{-4} \text{ cm}$

p Substrate

$N_D = 10^{16} \text{ cm}^{-3}$

Prob. 3.5

(a) Determine V_P, I_{DSS}, and ϕ_0.

(b) Determine the maximum value of g_m at pinchoff.

(c) Calculate the total channel capacitance.

3.6 Assume that a particular n-channel JFET can be modeled in the saturation region by the equation

$$i_D = I_{DSS} \left(1 + \frac{v'_{GS}}{V_P} \right)^2 (1 + \lambda v_{DS})$$

(a) Write equations for g_m, μ_A, and r_d according to their basic definitions at an operating point V_{GS}, V_{DS}, and I_D.

(b) Determine the value of $\mu_A = g_m r_d$ and compare it with the value determined directly in part (a).

3.7 Refer to the n-channel JFET whose characteristics are shown in Fig. 3.10.

(a) From the graphical characteristics, make estimates of g_m, r_d, and μ_A at the operating point $v_{DS} = 10$ V, $V_{GS} = -1.0$ V, and $I_D = 2.5$ mA.

(b) From the measurements of part (a), find a value for λ as defined in Prob. 3.6.

3.8 Determine the maximum value of transconductance at pinchoff, the total channel capacitance, and the figure of merit for an n-channel JFET (spike channel) having the following parameters: $W = 200$ μm, $d = 1.0$ μm, $L = 8$ μm, and $N_D = 1 \times 10^{16}$. If d and L are both reduced by a factor of 2, what is the effect on g_m? on C_o? on FM and on V_P and I_{DSS}?

3.9 The scale for the determination of the parameters in the basic equation for an MOS transistor is shown in the sketch.

(a) Determine V_T and write the equation for i_{Dsat} (with numbers); plot the curve for $i_{Dsat} = f(v_{GS}, v_{DS})$ on the sketch.

(b) Write the complete equation for $i_D = f(v_{GS}, v_{DS})$ in the triode region and plot a family of characteristics.

3.10 It is desired that two MOSFETs, one an enhancement-mode type with its gate and drain connected and the other a depletion-mode type with its gate connected to its source, have the same drain current I_{DS} at $V_{DS} = 5$ V. Channel depth, d, is the same for both devices, but other factors are such that $V_T = -2.5$ V for the depletion-mode structure and $V_T = +1$ V for the enhancement-mode structure.

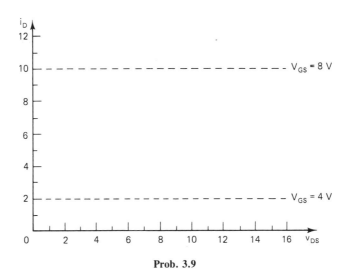

Prob. 3.9

(a) Determine L_E/L_D for the described condition, where L_E is the gate length for the enhancement-mode device and L_D that for the depletion-mode device, assuming that the channel widths are the same.

(b) Assuming equal channel lengths, what should be the ratio of channel widths?

3.11 A buried n-layer MOSFET with metal gate has the process and structural parameters indicated.

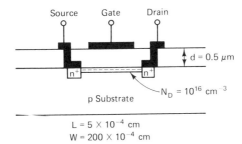

L = 5 × 10⁻⁴ cm
W = 200 × 10⁻⁴ cm   Prob. 3.11

(a) Assuming that the thickness of the buried channel is much less than d, determine an approximate value for V_T (assume $\varepsilon_i = 3.9$ for the insulating oxide).

(b) What is the total channel capacitance?

(c) Compare your results with those of Prob. 3.5, indicating major reasons for any differences.

3.12 An enhancement-mode silicon MOSFET has an S_iO_2 insulating film of thickness $d = 0.2$ μm, and the impurity concentration in the p substrate is $N_A = 1.45 \times 10^{16}$ cm⁻³. The gate dimensions are $L = 5$ μm and $W = 200$ μm.

(a) Assuming that the substrate is grounded (0 V) and that the flat-band voltage is approximately -1.0 V, determine the threshold voltage V_T.

(b) What is the value of the K coefficient and the channel capacitance?

(c) What is the transconductance at a saturation drain current of $I_D = 10$ mA?

3.13 The MOSFET having the characteristics shown in Fig. 3.21 is connected to external sources as shown.

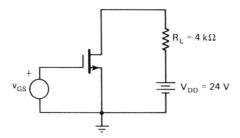

Prob. 3.13

(a) Determine the operating point, V_{GS}, V_{DS}, I_D.

(b) At the operating point, make an estimate of g_m, r_d, and μ_A from the graphical data.

(c) Determine a new value for g_m for $V_{DS} = 12$ V and $V_{GS} = +1.5$ V. Repeat with V_{DS} unchanged but with V_{GS} changed to -1.5 V.

(d) From the values of g_m found, would you or would you not conclude that g_m is approximately proportional to gate voltage, for this particular transistor?

3.14 Consider the *p*-channel MOSFET shown in Fig. 3.23, connected as shown, and try to model it by the equation

$$i_D = -\frac{K}{2}(v_{GS} - V_T)^2(1 - \lambda v_{DS})$$

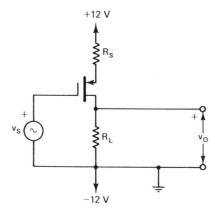

Prob. 3.14

(a) For a specific operating point, $I_D = -5$ mA at $V_{DS} = -10$ V and $V_{GS} = -10$ V, find the required values of R_S and R_L and make an estimate of g_m, r_d, and μ_A at the operating point.

(b) From the result of part (a), find a value for λ to be used in the preceding equation.

(c) Determine K from values of parameters at the operating point.

(d) Make a plot of i_D, v_{DS} curves over a wide range in the neighborhood of the operating point and observe how closely they model the characteristics of Fig. 3.23.

(e) At the operating point, determine the small-signal voltage gain, v_o/v_s.

3.15 For the built-in junction voltage of a JFET, show that its variation with temperature can be expressed as

$$\frac{d\phi_O}{dT} = \frac{k}{q}(\ln N_A N_D - 2\ln K_i) - \frac{3k}{q}(1 + \ln T)$$

Then for an n-channel JFET with $N_A = 5 \times 10^{18}$ and $N_D = 5 \times 10^{15}$, determine a value for $d\phi_O/dT$ at 300 K after first determining K_i at 300 K.

3.16 Show that the variation in transconductance with temperature of a JFET at a prescribed operating point can be written as

$$\frac{dg_m}{dT} = \frac{C_O}{L^2} \left[(V_{GS} - \phi_O + V_P) \frac{d\mu_n}{dT} - \mu_n \frac{d\phi_O}{dT} \right]$$

Then, using $\mu_n = K_\mu T^{-5/2}$, determine dg_m/dT for the FET of Prob. 3.15 at an operating temperature of 300 K with the added information that $W = 500$ μm, $L = 5$ μm, and $d = 0.2$ μm.

3.17 Under what conditions can the threshold voltage of an n-channel enhancement-mode MOSFET be expressed as

$$V_T \cong \frac{[\varepsilon_o \varepsilon_r q N_A (k/q)]^{1/2}}{C_i} T^{1/2} \left(\ln N_A - \ln K_i - \frac{3}{2} \ln T + \frac{qV_g}{2kT} \right)$$

Then derive an expression for dV_T/dT. For an n-channel MOSFET with $N_A = 2 \times 10^{16}$cm^{-3}, $L = 5$ μm, $w = 200$ μm, and $d = 0.15$ μm, determine V_T at 300 K and dV_T/dT in that temperature range.

REFERENCES

FUKUMA, M., AND Y. OKUTO. "Analysis of Short Channel MOSFETS with Field Dependent Carrier-Drift Mobility," *IEEE Trans. Electron. Devices*, vol. ED-27, no. 11, Nov. 1980, pp. 2109–2114.

GROVE, A. S. *Physics and Technology of Semiconductor Devices*, Wiley, New York, 1967.

LIU, S., AND L. W. NAGEL. "Small-Signal MOSFET Models for Analog Circuit Design," *IEEE J. Solid State Circuits*, vol. SC-17, no. 6, Dec. 1982, pp. 983–998.

SZE, S. M. *Physics of Semiconductor Devices*, 2nd ed., Wiley, New York, 1982.

VAN DER ZIEL, A. *Solid State Physical Electronics*, 3rd ed., Prentice-Hall, Englewood Cliffs, N.J., 1976.

WORDEMAN, M. R., AND R. H. DENNARD. "Threshold Voltage Characteristics of Depletion Mode MOSFETS," *IEEE Trans. Electron. Devices*, vol. ED-28, no. 9, Sept. 1981, pp. 1025–1030.

YAMAQUACHI, T., AND S. MORIMOTO. "Analytical Model and Characterization of Small Geometry MOSFETS," *IEEE Trans. Electron. Devices*, vol. ED-30, no. 6, June 1983, pp. 559–566.

Chapter 4

Basic Properties
of Single-Device
Amplifier Structures

INTRODUCTION

The incremental (small-signal) circuit models developed in Chapter 2 for the bipolar junction transistor and in Chapter 3 for the field-effect transistor suggest their application to small-signal linear amplifiers where a voltage or current applied at the input terminals results in a magnified voltage or current at the output terminals.

The most descriptive incremental parameter of the BJT was found to be its current gain, β or h_{fe}, and used specifically in this mode it functions as a current amplifier, whereas the most significant parameter of the FET was found to be its transconductance, g_m. When the input–output voltage relationship is specified, the amplifier is referred to as a *voltage amplifier*, and when the voltage–current relationship is emphasized, it is referred to as a *transconductance amplifier*.

Both devices, when used with appropriate external circuits, function as voltage, current, or transconductance amplifiers; but when used in other than their natural modes, their input and output impedance characteristics become important, because the overall transfer function will depend on how the input of the device loads the source to which it is connected, and also the output impedance it presents to a load to which the output is connected.

The BJT was characterized in detail in Chapter 2 primarily by its h parameters in common-emitter form, and the FET in Chapter 3 by its g_m and r_d which are the low-frequency output components of the y parameter system in common-source form. Both devices function as amplifiers in other configurations, the BJT in the common-collector or common-base mode, and the FET in the common-gate or common-drain mode.

In this chapter it is intended to derive expressions which best describe the fundamental properties of voltage or current gain and input and output impedances for both BJTs and FETs to be used as amplifiers in their various circuit configurations, and in terms of the basic parameters that are most often used to describe the devices, as well as in terms of other related parameters.

Although amplifiers are rarely composed of a single transistor, usually containing two or more in a composite or compound configuration, such configurations can best be analyzed when the properties of the single-device structures comprising them are thoroughly understood.

4.1 AMPLIFIER CHARACTERIZATIONS AND DEFINITIONS

A voltage amplifier containing one or more active devices with one or two inputs, the output being inverting with respect to one input and noninverting with respect to the other, but with one input common to input and output, is shown in Fig. 4.1(a). Such an amplifier is generally characterized by its input impedance with load connected as

$$Z_i \equiv \frac{V_1}{I_1}\bigg]_{Z_L} \tag{4.1}$$

Its output impedance with its source impedance connected with Z_L replaced by a voltage as shown in Fig. 4.1(b), is given by

$$Z_o = \frac{V_2}{I_2}\bigg]_{Z_s} \tag{4.2}$$

And its voltage gain is given by

$$A = \frac{V_2}{V_1} \tag{4.3}$$

with load impedance in place.

The overall circuit model is shown in Fig. 4.1(c), where A_o is the open-circuit voltage gain given by

$$A_o = \frac{V_2}{V_1}\bigg]_{Z_L \to \infty} \tag{4.4}$$

The overall gain with source and load connected is

$$\frac{V_2}{V_s} = \frac{V_1}{V_s}\frac{V_2}{V_1} \tag{4.5}$$

where

$$\frac{V_1}{V_s} = \frac{Z_i}{Z_s + Z_i}$$

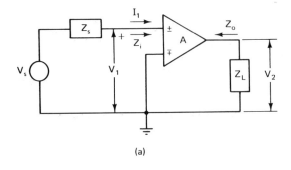

(a)

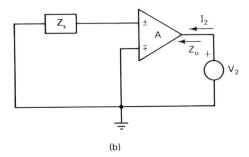

(b)

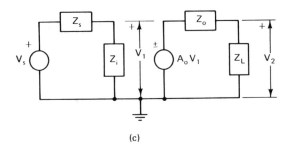

(c)

Figure 4.1 Voltage amplifier terminology.
(a) Input impedance. (b) Output impedance.
(c) Circuit model.

Voltage gain is often expressed in logarithmic units derived from the concept of power gain of an amplifier,

$$A_P = \frac{P_2}{P_1}$$

where P_2 is the power delivered from the output of an amplifier to its load impedance Z_L, and P_1 is the power delivered from the signal source to the input impedance of the amplifier. For $Z_s = R_s$, $Z_i = R_i$, $Z_o = R_o$, and $Z_L = R_L$ in the circuit model of Fig. 4.1(c),

$$A_P = \left(\frac{V_2}{V_1}\right)^2 \frac{R_i}{R_L}$$

The power gain in decibels (dB) expressed as a logarithmic ratio is defined as

$$A_P \text{ (dB)} = 10 \log_{10} \frac{P_2}{P_1}$$

$$= 20 \log_{10} \frac{V_2}{V_1} + 10 \log_{10} \frac{R_i}{R_L}$$

The first term of this expression is a logarithmic voltage ratio, and voltage gain in decibels is commonly defined by it as

$$A_v \text{ (dB)} = 20 \log_{10} \frac{V_2}{V_1} \tag{4.6}$$

irrespective of the relative values of R_i and R_L.

The overall voltage gain from source to load given by Eq. 4.5 but expressed in decibels is

$$\frac{V_2}{V_s} \text{ (dB)} = 20 \log \frac{V_1}{V_s} \frac{V_2}{V_1}$$

$$= 20 \log \frac{V_1}{V_s} + 20 \log_{10} \frac{V_2}{V_1} \tag{4.7}$$

This illustrates the important concept that a gain function that is the product of separate cascaded components is expressed in decibels as the sum rather than the product of individual components.

If the gain of an amplifier is a function of frequency, its magnitude $|A(j\omega)|$ relative to its magnitude in a nonfrequency-dependent range (usually zero or very low frequencies) $|A_0|$ is expressed in decibels as

$$A_{\text{REL}} \text{ (dB)} = 20 \log_{10} \frac{|A(j\omega)|}{|A_0|}$$

$$= 20 \log |A(j\omega)| - 20 \log|A_0| \tag{4.8}$$

Single-Device Amplifiers

Single devices such as bipolar or field-effect transistors may be utilized as voltage amplifiers in various circuit configurations with various combinations of input and output terminals. For example, for the bipolar transistor, the input may be applied to the base, with the emitter common to the input and output circuits (common-emitter connection as analyzed in Chapter 2), to the base with the collector common, or to the emitter with the base common. Very often an external impedance Z_T is included in the common lead, as indicated in Fig. 4.2. This affects the input impedance and gain in various ways depending on the current flowing in Z_T. Because of the importance of this common impedance, it is included in the circuits to be analyzed in this chapter.

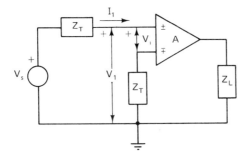

Figure 4.2 Two-input amplifier with common input impedance.

4.2 BIPOLAR TRANSISTORS AS VOLTAGE AMPLIFIERS

The bipolar junction transistor is commonly utilized as a voltage amplifier with the device itself being characterized by the various circuit models described in Secs. 2.7 and 2.8. However, the device internally functions as an almost linear current amplifier described by its most important parameter, h_{fe} of β, which is relatively independent of bias current levels and temperature variations.

The generalized h-parameters represent a unified approach to the characterization of bipolar transistors in various amplifier configurations by direct substitution into a universal set of h parameter equations.

The circuit of Fig. 4.3 illustrates a device model described by h parameters used as an amplifier with an impedance in the common lead, as described in the previous section. The solution of this circuit yields the following equations for input impedance, output impedance, and overall voltage gain.

$$Z_i = h_i - \frac{h_r Z_L[(h_f/h_o) - Z_T]}{Z_L + (1/h_o) + Z_T} + Z_T(1 - h_r)\frac{Z_L + [(1 + h_f)/h_o]}{Z_L + (1/h_o) + Z_T} \quad (4.9)$$

$$Z_o = \frac{(1/h_o)\{1 + [Z_T/(Z_s + h_i)]\,[(1 - h_r)(1 + h_f) + h_o(Z_s + h_i)]\}}{1 - (h_r h_f/h_o)[1/(Z_s + h_i)] + [Z_T/(Z_s + h_i)]} \quad (4.10)$$

$$A_v =$$

$$\frac{-(h_f - Z_T h_o)Z_L}{(Z_s + h_i)(1 + Z_L h_o)\left\{1 - \dfrac{h_f h_r Z_L}{(Z_s + h_i)(1 + h_o Z_L)}\right\} + Z_T h_o\left[Z_L + Z_s + h_i + \dfrac{(1 - h_r)(1 + h_f)}{h_o}\right]}$$

$$(4.11)$$

The components of these equations are segregated to show readily the effects of the common impedance Z_T and the conditions under which the internal feedback parameter h_r involved in various terms can be neglected. Also, these general equations can be used to describe the voltage gain properties of the transistor in various configurations simply by appropriate subscript additions (e.g., h_{fe} for the common-emitter configuration, h_{fb} for the common-base configuration, and h_{fc} for the common-collector configuration).

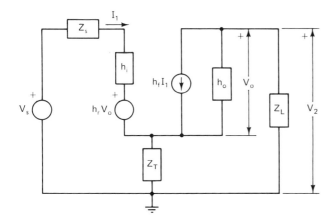

Figure 4.3 Transistor circuit using *h*-parameters with impedance in common lead.

The most important transistor amplifier parameters for various configurations will be evaluated within the framework of the preceding equations.

4.3 COMMON-EMITTER AMPLIFIER

The common-emitter amplifier including the presence of an impedance in the common lead is shown in Fig. 4.4. This fits the equivalent small-signal circuit of Fig. 4.3 with

$$h_i = h_{ie} \qquad h_o = h_{oe} \qquad h_f = h_{fe} \qquad h_r = h_{re}$$

substituted directly into Eqs. 4.9, 4.10, and 4.11.

The most significant properties of the common-emitter amplifier may be examined by using approximations that are almost always valid in practical circuits. These are $h_{re} \ll 1, Z_T \ll 1/h_{oe}, Z_L \ll 1/h_{oe}$, and $Z_s \ll (1 + h_{fe})/h_{oe}$. Under these conditions, Eqs. 4.9, 4.10, and 4.11 reduce to

$$Z_i \cong h_{ie} - h_{re}h_{fe}Z_L + (1 + h_{fe})Z_T \tag{4.12}$$

$$Z_o \cong \frac{(1/h_{oe})\{1 + [Z_T(1 + h_{fe})/(Z_s + h_{ie})] + h_{oe}(Z_s + h_{ie})\}}{1 - [(h_{fe}h_{re}/h_{oe})/(Z_s + h_{ie})] + [Z_T/(Z_s + h_{ie})]} \tag{4.13}$$

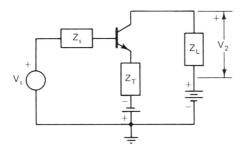

Figure 4-4 Common-emitter transistor amplifier with impedance in emitter lead.

$$\frac{V_2}{V_s} \cong \frac{-h_{fe}Z_L}{(Z_s + h_{ie})[1 - h_{re}h_{fe}Z_L/(Z_s + h_{ie})] + Z_T(1 + h_{fe})} \tag{4.14}$$

The terms involving h_{re} illustrate a general property of internal feedback inherent in the common-emitter amplifier, whereas the terms containing Z_T represent negative feedback derived in series with the load and source, that is, reduction in gain and increase of input and output impedances, assuming that the freqency of operation is sufficiently low that neither Z_T nor any of the device parameters cause excessive phase shift. These properties can best be illustrated using resistances only, $Z_s = R_B$, $Z_L = R_L$, and $Z_T = R_E$, and the low-frequency device parameters β, μ_{re}, r_{bes} and r_{ceo}. Using this terminology, Eqs. 4.12, 4.13, and 4.14 may be written as

$$R_i \cong r_{bes} - \mu_{re}\beta R_L + (1 + \beta)R_E \tag{4.15}$$

$$R_o \cong \frac{r_{ceo}\{1 + [R_E(1 + \beta)/(R_B + r_{bes})] + [(R_B + r_{bes})/r_{ceo}]\}}{1 - [\mu_{re}\beta r_{ceo}/(R_B + r_{bes})] + [R_E/(R_B + r_{bes})]} \tag{4.16}$$

$$\frac{v_2}{v_s} = \frac{-\beta R_L}{(R_B + r_{bes})\{1 - \mu_{re}\beta R_L/(R_B + r_{bes})\} + R_E(1 + \beta)} \tag{4.17}$$

Inasmuch as the presence of the common impedance R_E reduces the gain or "degenerates" the output, its use is commonly referred to as *emitter degeneration*. It is interesting to note that the effect of internal positive feedback involving μ_{re} can be neutralized at a particular operating point with respect to input resistance and gain by making

$$R_E = \frac{\mu_{re}\beta R_L}{1 + \beta}$$

Since a value for μ_{re} often does not appear in published specifications, since it is difficult to measure and is highly dependent on operating point and temperature, it may be difficult to assess the magnitude of the effect of terms involving it and to determine the conditions under which it can be neglected. However, this can be done indirectly by using the relationship between the h-parameter model and the T-model discussed in connection with Fig. 2.23, using $r_e = \mu_{re}r_{ceo}$. Then

$$\frac{\mu_{re}\beta R_L}{R_B + r_{bes}} = \frac{r_e}{r_{ceo}} \frac{\beta R_L}{R_B + r_b' + r_d} \tag{4.18}$$

in Eq. 4.17.

Then using the approximations developed in Sec. 2.9 of

$$r_e \cong \frac{kT}{q} \frac{1}{|I_E|} \tag{2.93}$$

$$r_d \cong \frac{kT}{q} \frac{(1 + \beta)}{|I_E|} \tag{2.87}$$

making an estimate of $r_b' \ll r_d$, and measuring or otherwise determining r_{ceo}, the magnitude of Eq. 4.18 can be approximated.

Similarly, for the input and output impedance relationships of Eqs. 4.15 and 4.16,

$$\frac{\mu_{re}\beta r_{ceo}}{R_B + r_{bes}} = r_e \frac{\beta}{R_B + r_{bes}} \qquad (4.19)$$

Where it can be determined that the internal feedback terms can be neglected, Eqs. 4.15, 4.16, and 4.17 reduce to

$$R_i \cong r_{bes} + (1 + \beta)R_E \qquad (4.20)$$

$$R_o \cong \frac{r_{ceo}\{1 + [R_E(1 + \beta)/(R_B + r_{bes})] + [(R_B + r_{bes})/r_{ceo}]\}}{1 + [R_E/(R_B + r_{bes})]} \qquad (4.21)$$

$$\frac{v_2}{v_s} \cong \frac{-\beta R_L}{R_B + r_{bes} + (1 + \beta)R_E} \qquad (4.22)$$

These equations involve neglecting the internal feedback term involving μ_{re} which is proportional to R_L. This leads to a word of caution about the approximation $R_L \ll r_{ceo}$ used in all the preceding equations. There may be unusual circumstances under which this and the approximation of $R_E \ll r_{ceo}$ may not be valid. Then the more exact equations are available.

4.4 COMMON-COLLECTOR AMPLIFIER: EMITTER FOLLOWER

The common-collector amplifier as shown in Fig. 4.5 appears to be identical with that of Fig. 4.4 except for an interchange of Z_L and Z_T. The general equations, 4.9, 4.10, and 4.11, can be used for the common-collector amplifier simply by substituting $h_i = h_{ic}$, $h_o = h_{oc}$, $h_f = h_{fc}$, and $h_r = h_{rc}$.

However, inasmuch as the common-emitter parameters are more significant and more generally available, it is preferable to make the further substitution of common-collector parameters in terms of common-emitter parameters as

$$h_{ic} - h_{ie}, \qquad h_{fc} = -(1 + h_{fe}), \qquad h_{oc} = h_{oe}, \qquad h_{rc} = 1 - h_{re}$$

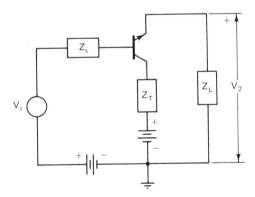

Figure 4.5 Common-collector amplifier with impedance in emitter lead.

In these terms, Eqs. 4.9, 4.10 and 4.11 may be written as

$$Z_i = h_{ie} + (1 - h_{re})Z_L \frac{[(1 + h_{fe})/h_{oe}] + Z_T}{Z_L + (1/h_{oe}) + Z_T} + h_{re}Z_T \frac{Z_L - (h_{fe}/h_{oe})}{Z_L + (1/h_{oe}) + Z_T} \quad (4.23)$$

$$Z_o = \frac{(1/h_{oe})\{1 + [Z_T/(Z_s + h_{ie})][(Z_s + h_{ie})h_{oe} - h_{re}h_{fe}]\}}{1 + [(1 + h_{fe})(1 - h_{re})/h_{oe}][1/(Z_s + h_{ie})] + [Z_T/(Z_s + h_{ie})]} \quad (4.24)$$

$$\frac{V_2}{V_s} =$$

$$\frac{[(1 + h_{fe}) + Z_T h_{oe}]Z_L}{[(Z_s + h_{ie})(1 + Z_L h_{oe})]\left[1 + \dfrac{(1 - h_{re})(1 + h_{fe})Z_L}{(Z_s + h_{ie})(1 + Z_L h_{oe})}\right] + Z_T h_{oe}\left[Z_L + Z_s + h_{ie} - \dfrac{h_{re}h_{fe}}{h_{oe}}\right]}$$

$$(4.25)$$

Examination of Eq. 4.25 shows that at relatively low frequencies the internal feedback term involving $h_{rc} = 1 - h_{re}$ is large compared with a similar term in the common-emitter amplifier, and thus represents a reduction in the input voltage effective at the base–emitter terminals, which reduces the gain and hence is described as *negative feedback*. The overall gain is also very insensitive to the common term Z_T. However, since the effect of the Z_T term in the denominator dominates its effect in the numerator, to the extent that it has any importance, it also causes the gain to be reduced and is also negative feedback. For the same approximations that were considered for the common-emitter amplifier, the governing equations reduce to

$$Z_i = h_{ie} + Z_L(1 + h_{fe}) - h_{re}h_{fe}Z_T \quad (4.26)$$

$$Y_o = \frac{1}{Z_o} = \frac{1 + h_{fe}}{Z_s + h_{ie}} + h_{oe}\left(1 + \frac{Z_T}{Z_s + h_{ie}}\right) \quad (4.27)$$

$$\frac{V_2}{V_s} = \frac{(1 + h_{fe})Z_L}{(Z_s + h_{ie})\{1 + [(1 + h_{fe})Z_L/(Z_s + h_{ie})]\}} \quad (4.28)$$

A rearrangement of Eq. 4.28 in the form

$$\frac{V_2}{V_s} \simeq \frac{1}{1 + [(Z_s + h_{ie})/(1 + h_{fe})Z_L]} \quad (4.29)$$

illustrates more clearly the most significant property of the common-collector amplifier: the voltage gain is less than unity but approaches unity for reasonably large values of Z_L. For this reason it is called an *emitter follower*, which is a specific example of a class of circuits with similar properties referred to as voltage followers.

Equation 4.26 for input impedance is identical in form to Eq. 4.12 for the common-emitter amplifier except that Z_T and Z_L are interchanged. However, in the case of the emitter follower the load impedance itself is in the feedback loop and increases the effective input impedance. This property is consistent with the general properties of voltage feedback as derived in shunt with the load voltage but applied in series with the input circuit.

The nature of the Z_T term in the collector in reducing the effective input impedance is consistent with the concept of positive feedback applied in series with the source internally through the $h_{re}v_{ce}$ feedback voltage.

For convenience, the output impedance is expressed in admittance form in Eq. 4.27. This equation defines two parallel branches of admittance looking back from the load into the device, one through the base–emitter resistance into the source impedance and the other through the common-collector circuit and the common impedance, Z_T. The second of these two terms is usually relatively small, illustrating again the relative independence of circuit performance upon the common impedance. Under such conditions

$$Z_o \cong \frac{Z_s + h_{ie}}{1 + h_{fe}}$$

For convenience, approximate equations in terms of the low-frequency device parameters and resistive circuit elements $Z_L = R_L$, $Z_s = R_B$, and $Z_T = R_C$ may be written as

$$R_i \cong r_{bes} + (1 + \beta)R_L - \mu_{re}\beta R_C \tag{4.30}$$

$$G_o \cong \frac{1}{R_o} \cong \frac{(1 + \beta)}{R_B + r_{bes}} + \frac{1}{r_{ceo}}\left(1 + \frac{R_C}{R_B + r_{bes}}\right) \tag{4.31}$$

$$\frac{v_2}{v_s} \cong$$

$$\frac{1}{1 + [(R_B + r_{bes})/(1 + \beta)R_L] + (R_C/r_{ceo})[(R_L + R_B + r_{bes} - \mu_{re}\beta r_{ceo})]/(1 + \beta)R_L]} \tag{4.32}$$

Under most conditions, the internal feedback term can be neglected, and these equations further reduce to

$$R_i \cong r_{bes} + (1 + \beta)R_L \tag{4.33}$$

$$R_o \cong \frac{R_B + r_{bes}}{1 + \beta} \tag{4.34}$$

$$\frac{v_2}{v_s} \cong \frac{1}{1 + [(R_B + r_{bes})/(1 + \beta)R_L]} \cong \frac{(1 + \beta)R_L}{(R_B + r_{bes})\{1 + [(1 + \beta)R_L/(R_B + r_{bes})]\}} \tag{4.35}$$

These equations demonstrate that a very useful property of the emitter follower is that of an impedance-matching device involving large R_s/R_L ratios. Although it has less than unity voltage gain, it has a higher input impedance than the common-emitter amplifier, and hence loads the source less, and a lower output impedance, which provides a better driving source to a low value of R_L. It therefore does provide appreciable *insertion* gain when used with a large source resistance and a small load resistance. It thus functions as a step-down impedance-matching transformer without the attendant voltage reduction of the impedance-matching transformer.

4.5 COMMON-BASE AMPLIFIER

The controlling h-parameter equations for the common-base amplifier shown in Fig. 4.6 may be written by simply using

$$h_i = h_{ib} = \frac{h_{ie}}{1 + h_{ie}}$$

$$h_o = h_{ob} = \frac{h_{oe}}{1 + h_{fe}}$$

$$h_f = h_{fb} = -\frac{h_{fe}}{1 + h_{fe}}$$

$$h_r = h_{rb} \cong \frac{h_{ie}h_{oe}}{1 + h_{fe}} - h_{re}$$

in Eqs. 4.9, 4.10, and 4.11.

Using the approximations $Z_L \ll 1/h_{ob}$, $Z_T \ll 1/h_{ob}$, and $h_{rb} \ll 1$, the controlling equations in terms of common-emitter parameters (with the exception of h_{rb}) may be expressed as

$$Z_i \simeq \frac{h_{ie}}{1 + h_{fe}} + \frac{h_{rb}h_{fe}Z_L}{1 + h_{fe}} + \frac{Z_T}{1 + h_{fe}} \tag{4.36}$$

$$Z_o \simeq \frac{1}{h_{oe}} \frac{1 + h_{fe}}{1 + \{[(h_{rb}h_{fe}/h_{oe}) + Z_T]/Z_s + [h_{ie}/(1 + h_{fe})]\}} \tag{4.37}$$

$$\frac{V_2}{V_s} \simeq \frac{[h_{fe}/(1 + h_{fe})]Z_L}{\left(Z_s + \dfrac{h_{ie}}{1 + h_{fe}}\right)\left\{1 + \dfrac{[h_{rb}\, h_{fe}/(1 + h_{fe})]Z_L}{Z_s + [h_{ie}/(1 + h_{fe})]}\right\} + \dfrac{Z_T}{1 + h_{fe}}} \tag{4.38}$$

Assuming sufficiently low frequencies and that there is very little phase shift, these equations identifty a noninverting amplifier with less than unity current gain. The internal feedback involving h_{rb} is negative voltage feedback applied in series with the input signal, which has the effect of decreasing the output impedance and voltage gain while increasing

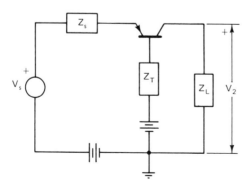

Figure 4.6 Common-base amplifier with impedance in base lead.

the input impedance. External feedback involving the common impedance Z_T is in series with both the output load impedance and the input signal. It is identifiable as positive current feedback, which by itself should increase the voltage gain, which, however, is overridden by the direct effect of Z_T.

In terms of resistive circuit elements $Z_s = R_E$, $Z_L = R_L$, and $Z_T = R_B$ and low-frequency parameters, Eqs. 4.36, 4.37, and 4.38 are

$$R_i \cong \frac{r_{bes}}{1 + \beta} + \frac{R_B}{1 + \beta} + \frac{\mu_{rb}\beta R_L}{1 + \beta} \tag{4.39}$$

$$R_o \cong \frac{r_{ceo}(1 + \beta)\{1 + [R_B]/[r_{bes} + R_E(1 + \beta)]\}}{1 + [(1 + \beta)(\mu_{rb}\beta r_{ceo} + R_B)]/[(r_{bes} + R_E(1 + \beta)]} \tag{4.40}$$

$$\frac{v_2}{v_s} \cong \frac{[\beta/(1 + \beta)]R_L}{\left(R_E + \dfrac{r_{bes}}{1 + \beta}\right)\left[1 + \dfrac{\mu_{rb}[\beta/(1 + \beta)]R_L}{(R_E + r_{bes})/(1 + \beta)}\right] + \dfrac{R_B}{1 + \beta}} \tag{4.41}$$

In these equations, low-frequency common-emitter parameters are used with the exception of μ_{rb}, which is retained because of the complexity of the $\mu_{rb} - \mu_{re}$ relationship, which will now be examined in more detail.

Starting with

$$\mu_{rb} \cong \frac{r_{bes}/r_{ceo}}{1 + \beta} - \mu_{re}$$

and using the relationships $r_e = \mu_{re}r_{ceo}$ and $r_d = r_a + r_e(1 + \beta)$, from Eqs. 2.76 and 2.78, along with $r_{bes} = r_b' + r_d$, μ_{rb} can be written approximately as

$$\mu_{rb} \cong \frac{r_b' + r_a}{r_{ceo}(1 + \beta)} \tag{4.42}$$

with r_a from the T-circuit model being very small and usually assumed to be negligible.

This relationship is useful when μ_{rb} needs to be calculated from parameter values that have been determined in common-emitter form. The result also shows that μ_{rb} is a very small number, in most cases being $\mu_{rb} \ll \mu_{re}$.

The low-frequency equations, 4.39, 4.40, and 4.41, further identify an amplifier with less than unity current gain, a very low input resistance, and a very high output resistance compared to the common-emitter circuit (by the factor $1 + \beta$ in each case). The resistance R_B in the common lead is important as it affects both the input resistance and the gain. A value of voltage gain of $V_2/V_s > 1$ can be achieved only by resistance ratios $R_L > R_S$.

As an amplifier, the common-base stage will be found most useful in cascade with amplifier stages of other configurations, where its particular input–output impedance relationships can be utilized to advantage.

A further simplification, neglecting the internal feedback terms, results in the approximate equations

$$R_i \cong \frac{r_{bes} + R_B}{1 + \beta} \tag{4.43}$$

$$R_o \cong \frac{r_{ceo}(1 + \beta)}{1 + [R_B(1 + \beta)]/[r_{bes} + R_E(1 + \beta)]} \tag{4.44}$$

$$\frac{v_2}{v_s} \cong \frac{\beta R_L}{(r_{bes} + R_B) + R_E(1 + \beta)} \cong \frac{\beta R_L}{(r_{bes} + R_B)\{1 + [R_E(1 + \beta)/(r_{bes} + R_B)]\}} \tag{4.45}$$

Except for being noninverting, this gain is identical to that for the common-emitter amplifier given by Eq. 4.22 for the same values of R_B and R_E, except that it is noninverting. This would not be true if R_E and R_B were chosen to optimize the gain for each configuration.

4.6 TWO-INPUT TRANSISTOR AMPLIFIER

In the circuit shown in Fig. 4.7, signals are applied to both the emitter and base inputs, so it is simultaneously a common-emitter and common-base amplifier. The approximate gain with respect to v_s is given by Eq. 4.22 and with respect to v_{s2} by Eq. 4.45. Therefore, by linear superposition,

$$v_2 \cong \frac{\beta R_L(v_{s2} - v_{s1})}{(r_{bes} + R_B) + R_E(1 + \beta)} \tag{4.46}$$

Within the limits of the approximations used in determining the gains, the two-input bipolar transistor amplifier is a true difference amplifier but with appreciable gain only for relatively large R_L/R_E ratios.

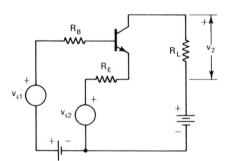

Figure 4.7 Two-input transistor amplifier.

4.7 FIELD-EFFECT TRANSISTOR AMPLIFIERS

Because of the natural function of the common-source FET as a voltage amplifier (very high input impedance) and the relatively simple frequency-dependent parameters (capacitive in nature), it is simpler to proceed directly to consideration of the low-frequency characteristics of FET amplifiers and factor in, as additions, the capacitances involved, rather than to begin with the generalized equations using a complete set of parameters and simplify them to the low-frequency model, as was done for the bipolar transistor.

However, were we to proceed on a general basis it would be most appropriate to work from the frequency-dependent y parameters discussed in Chapter 3, rather than the

h parameters for the bipolar transistor. In particular, a current gain parameter, β, is useless, because at low frequencies the input current is insignificant, whereas the low-frequency transconductance g_m, is the most significant natural parameter, with the incremental drain resistance r_d being of less but of some importance, because it is usually large compared with the load resistance, although it sometimes needs to be considered.

4.8 COMMON-SOURCE AND COMMON-DRAIN AMPLIFIER

A simple gate-driven JFET amplifier is shown in Fig. 4.8(a). It may be looked on as a common-source amplifier with source degeneration where, as a byproduct, R_s may be chosen to bias the gate negative with respect to the source (known as self-bias), or it may be considered as a common-drain amplifier or source follower with common resistance in the drain circuit.

The two circuits are analogous to the common-emitter and emitter-follower circuits.

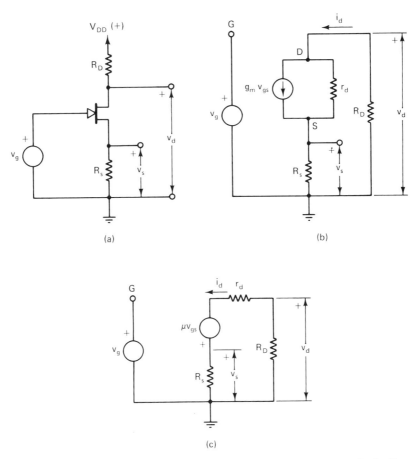

Figure 4.8 Common-source and common-drain FET amplifiers. (a) Actual circuit. (b) Current source circuit model. (c) Voltage source circuit model.

However, an important difference lies primarily in the input characteristic, with the input resistance being so high that its loading effect on the signal source can usually be ignored. For this reason, the simple circuit models shown in Fig. 4.8(b) and (c) are sufficient to characterize the low-frequency performance as an amplifier under most source and load conditions.

Using Fig. 4.8(c), the controlling equations are

$$\mu v_{gs} = i_d(r_d + R_s + R_D) \tag{4.47}$$

$$v_{gs} = v_g - i_d R_s \tag{4.48}$$

The solution for drain current is

$$i_d = \frac{\mu v_g}{r_d + R_D + (\mu + 1)R_s} = \frac{\mu v_g}{(r_d + R_D)\{1 + (\mu + 1)R_s/[r_d + R_D]\}} \tag{4.49}$$

or in terms of $g_m = \mu/r_d$

$$i_d = \frac{g_m v_g}{1 + [(R_D + R_s)/r_d] + g_m R_s} \tag{4.50}$$

Then for $r_d \gg (R_D + R_s)$

$$i_d \cong \frac{g_m v_g}{1 + g_m R_s} \tag{4.51}$$

This current is common to both drain and source circuits since any current to or from the gate terminal is negligible. Hence this drain current applies both to the common-source amplifier and the source follower.

Common-Source Amplifier Gain Equations

The small-signal output voltage at the drain is given by

$$v_d = -i_d R_D \tag{4.52}$$

Hence, from the alternative equations for i_d,

$$\frac{v_d}{v_s} = \frac{-\mu R_D}{r_d + R_D + (\mu + 1)R_s} \tag{4.53}$$

$$\frac{v_d}{v_s} = \frac{-g_m R_D}{1 + [(R_D + R_s)/r_d] + g_m R_s} \tag{4.54}$$

$$\frac{v_d}{v_s} \cong \frac{-g_m R_D}{1 + g_m R_s} \tag{4.55}$$

The output resistance, R_o, for the common source amplifier can be determined from the circuit of Fig. 4.9 using $v_{gs} = -i_d R_s$ for $v_g = 0$. The result is

$$R_o = \frac{v_d}{i_d} = r_d + R_s(\mu + 1) \tag{4.56}$$

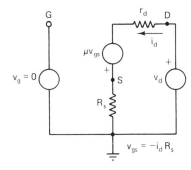

Figure 4.9 Circuit for determination of output resistance of common-source amplifier.

or for

$$\mu = g_m r_d \gg 1$$

$$R_o \cong r_d(1 + g_m R_s) \tag{4.57}$$

From examination of the gain and output resistance equations, source degeneration may be interpreted as negative current feedback, whereby R_s decreases the gain and increases the output impedance.

Common-Drain Amplifier: Source Follower

With the output voltage in the circuit of Fig. 4.8 taken from the source terminal using $v_s = i_d R_s$, gain equations similar to those of Eqs. 4.53, 4.54, and 4.55 may be written.

$$\frac{v_s}{v_g} = \frac{+\mu R_s}{r_d + R_D + (\mu + 1)R_s} \tag{4.58}$$

$$\frac{v_s}{v_g} = \frac{g_m R_s}{1 + [(R_D + R_s)/r_d] + g_m R_s} \tag{4.59}$$

$$\frac{v_s}{v_g} \cong \frac{g_m R_s}{1 + g_m R_s}, \qquad \text{for } R_d \gg (R_D + R_s) \tag{4.60}$$

These equations show that the voltage gain is less than unity, as it was for the emitter follower, and approaches unity for $g_m R_s \gg 1$ with $r_d \gg (R_D + R_s)$.

The output resistance, R_o, may be determined from the circuit of Fig. 4.10 and may be written as

$$R_o = \frac{v_s}{i_s} = \frac{r_d + R_D}{\mu + 1} \tag{4.61}$$

or

$$R_o \cong \frac{1}{g_m}, \qquad \text{for } r_d \gg R_D \tag{4.62}$$

Similar to the emitter follower, the output resistance is decreased. This, together with the decrease in gain, exhibits the properties of negative voltage feedback. However,

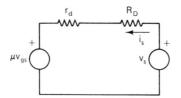

Figure 4.10 Circuit for determination of output resistance of common-drain amplifier.

the primary current path is through the drain–source path because of the extremely high resistance of the gate–source path, whereas in the case of the emitter follower it was through the base–emitter circuit and the signal source resistance.

4.9 COMMON-GATE AMPLIFIER

The source-driven common-gate FET has a relatively low input resistance, being that of the output resistance of the source follower, given by Eqs. 4.61 and 4.62; it must be taken into account when driven from a signal voltage with a finite but even low value of series resistance, as indicated in the circuit of Fig. 4.11(a). The input terminal voltage of this circuit is given by

$$\frac{v_i}{v_s} = \frac{(r_d + R_D)/(\mu + 1)}{R_s + [(r_d + R_D/(\mu + 1)]} \tag{4.63}$$

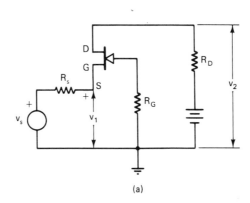

(a)

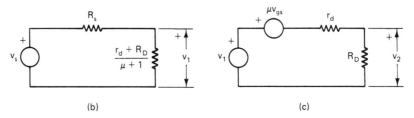

(b)   (c)

Figure 4.11 The common-gate FET amplifier. (a) Actual circuit. (b) Input impedance characteristic. (c) Circuit model for gain determination.

or for $r_d \gg R_D$

$$\frac{v_i}{v_s} \simeq \frac{1}{1 + g_m R_s} \tag{4.64}$$

The gain, v_2/v_1, of the amplifier itself may be determined from Fig. 4.11, with $v_{gs} = -v_1$ as

$$\frac{v_2}{v_1} = \frac{(\mu + 1)R_D}{r_d + R_D} \tag{4.65}$$

or

$$\frac{v_2}{v_1} \simeq g_m R_D, \qquad \text{for } r_d \gg R_D \tag{4.66}$$

The overall gain

$$\frac{v_2}{v_s} = \frac{v_1}{v_s} \frac{v_2}{v_1} \tag{4.67}$$

obtained by combining Eqs. 4.63 and 4.65 is

$$\frac{v_2}{v_s} = \frac{(\mu + 1)R_D}{r_d + R_D + R_s(\mu + 1)} \tag{4.68}$$

This is identical to Eq. 4.53 for the common-source amplifier except for the factor $(\mu + 1)$ rather than μ, and for the fact that it is noninverting.

Alternatively, for $\mu \gg 1$,

$$\frac{v_2}{v_s} \cong \frac{g_m R_D}{1 + (r_d + R_D)/r_d + g_m R_s} \tag{4.69}$$

which except for being noninverting is identical to Eq. 4.54 for the common-source amplifier.

Then for $r \gg R_D$

$$\frac{v_2}{v_s} \simeq \frac{g_m R_D}{1 + g_m R_s} \tag{4.70}$$

analogous to Eq. 4.55 for the common-source amplifier except for the non-polarity inversion.

Another important difference is that of the loading of the signal source by the input resistance of the transistor, which was shown to be

$$R_i \simeq \frac{r_d + R_D}{\mu + 1} \quad \text{or} \quad R_i \cong \frac{1}{g_m}, \qquad \text{for } r_d \gg R_D$$

Also, the output resistance obtained by replacing R_D with a voltage source and setting $v_s = 0$ is found to be

$$R_o = r_d + R_s(\mu + 1)$$

4.10 DUAL-INPUT FET AMPLIFIER

The gate and source of the FET may be driven simultaneously with separate signal sources as indicated in Fig. 4.12. Then v_2/v_{s1} for $v_{s2} = 0$ is given by Eq. 4.53, Eq. 4.54, or approximately by Eq. 4.55.

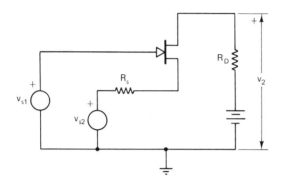

Figure 4.12 Two-input FET amplifier.

Also v_2/v_{s2} for $v_{s1} = 0$ is given by Eq. 4.68 or approximately by Eq. 4.69 or 4.70.

For sufficiently small input signals such that device nonlinearities need not be considered, the output by linear superposition may be written by combining Eqs. 4.53 and 4.68 as

$$v_2 = \frac{(\mu + 1)R_D v_{s2} - \mu R_D v_{s1}}{r_d + R_D + (\mu + 1)R_s} \tag{4.71}$$

or, by combining Eqs. 4.54 and 4.69, as

$$v_2 \cong \frac{g_m R_D (v_{s2} - v_{s1})}{1 + [(R_D + R_s)/r_d] + g_m R_s} \tag{4.72}$$

or, for $r_d \gg R_D + R_s$, combining Eqs. 4.55 and 4.70:

$$v_2 \simeq \frac{g_m(v_{s2} - v_{s1})}{1 + g_m R_s} \tag{4.73}$$

These equations demonstrate that, with source and gate both driven, the FET functions as an amplifier for both inputs, the only differences with respect to the two inputs being the signal source loading by the input resistance of the source-driven input and the polarity inversion with respect to the gate signal source.

4.11 INTRODUCTION TO FREQUENCY- AND TIME-DOMAIN RESPONSES

The time-varying response of complete amplifiers may be analyzed by incorporating the frequency-dependent models of single devices (see Sec. 2.8 for the BJT and Sec. 3.6 for the FET) into two-port networks including whatever RC components might be associated with sources and loads. However, prior to consideration of specific amplifiers, it is useful to review the characteristics of simple RC circuits.

Single-Pole High-Pass Network

Sometimes a source $v_s(t)$ with its output resistance R_s is coupled to a load resistance R_L through a series capacitance C_c as indicated in Fig. 4.13(a) or its current source equivalent, Fig. 4.13(b). For example, the source might represent the output of a simple amplifier with $v_s(t) = A_o v_i(t)$, where $v_i(t)$ is the input to the amplifier, A_o is its open-circuit voltage gain, and R_s is its output resistance. The capacitance C_c might be a dc "blocking" capacitance to decouple the dc offset voltage arising from amplifier biasing levels.

The equation for overall steady-state frequency response can be expressed as

$$\frac{V_2}{V_s} = \frac{R_L}{R_L + R_s} \frac{1}{1 - j(\omega_1/\omega)} \tag{4.74}$$

where

$$\omega_1 = \frac{1}{(R_L + R_s)C_c} \tag{4.75}$$

The response, alternatively, can be written in terms of magnitude and phase as

$$\left|\frac{V_2}{V_s}\right| = \frac{R_L}{R_L + R_s} \frac{1}{[1 + (\omega_1/\omega)^2]^{1/2}} \tag{4.76}$$

and

$$\theta = \tan^{-1}\frac{\omega_1/\omega}{1} \tag{4.77}$$

These responses are plotted in normalized form in Fig. 4.14(a) and (b).

Alternatively, Eq. 4.74 may be written as

$$\frac{V_2}{V_s} = \frac{R_L}{R_L + R_s} \frac{s}{s + \omega_1} \tag{4.78}$$

where $s = j\omega$.

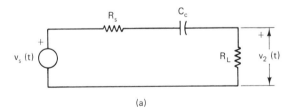

(a)

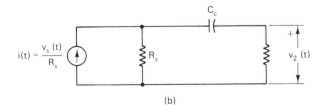

(b)

Figure 4.13 Single-pole high-pass *RC* networks.

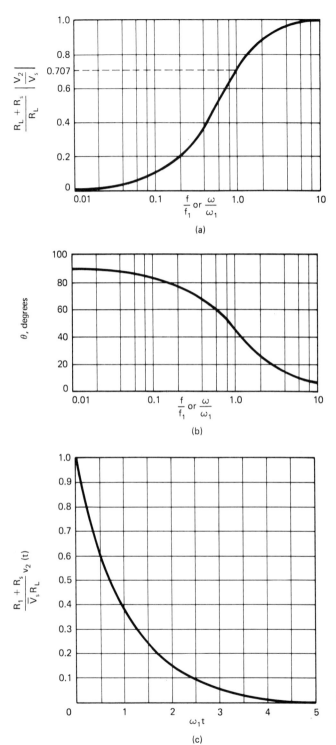

Figure 4.14 Frequency characteristics of single-pole high-pass network. (a) Amplitude response. (b) Phase response. (c) Transient (step function) response.

If a step function described by $v_s(t) = \overline{V}_s U(t)$; defined as zero for $t < 0$ and V_s for $t > 0$, whose Laplace transform is $V_s(s) = \overline{V}_s/s$ is applied, the response transforms to

$$\frac{V_2}{V_s} = \frac{R_L}{R_L + R_s} \frac{s}{s(s + \omega_1)} \tag{4.79}$$

The resultant time-domain response from the inverse Laplace transform is

$$v_2(t) = \overline{V}_s \frac{R_L}{R_L + R_s} e^{-\omega_1 t} \tag{4.80}$$

which is plotted in Fig. 4.14(c).

After the initial transient, the remaining dc level at the input is not sustained at the output.

Single-Pole Low-Pass Network

A voltage source coupled directly to a load that consists of R_L shunted by a capacitance C_2 is shown in Fig. 4.15(a), and its equivalent network forms (from the standpoint of the load terminals) in Fig. 4.15(b) and (c).

The steady-state frequency response, V_2/V_s, is

$$\frac{V_2}{V_s} = \frac{R_L}{R_L + R_s} \frac{1}{1 + j(\omega/\omega_2)} \tag{4.81}$$

where

$$\omega_2 = \frac{1}{R_2 C_2} \quad \text{with} \quad R_2 = \frac{R_s R_L}{R_s + R_L} \tag{4.82}$$

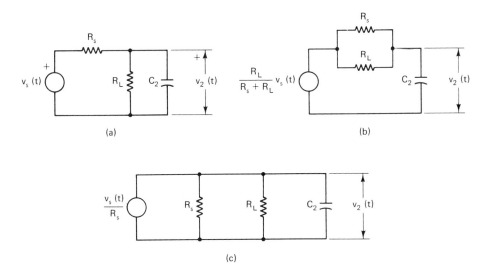

(a)

(b)

(c)

Figure 4.15 Single-pole low-pass RC networks.

In terms of magnitude and phase,

$$\frac{V_2}{V_s} = \frac{R_L}{R_L + R_s} \frac{1}{[1 + (\omega/\omega_2)^2]^{1/2}} \tag{4.83}$$

$$\theta = -\tan^{-1} \frac{\omega/\omega_2}{1} \tag{4.84}$$

These quantities are plotted in Fig. 4.16 normalized to the reference frequence ω_2. The time constant R_2C_2 limits the high-frequency response of a circuit, which is finally limited by C_2, the presence of which can never be entirely eliminated in an actual circuit.

The limitation on the steady-state high-frequency response can also be related to the finite rise time in response to a step function. The overall network response of Eq. 4.81 can be written as

$$\frac{V_2}{V_s} = \frac{R_L}{R_L + R_s} \frac{\omega_2}{s + \omega_2} \tag{4.85}$$

and, in response to the step function whose transform is $\overline{V}_s/s$,

$$\frac{V_2}{V_s} = \frac{R_L}{R_L + R_s} \frac{\omega_2}{s(s + \omega_2)} \tag{4.86}$$

The time-domain response is

$$v_2(t) = \frac{\overline{V}_s R_L}{R_L + R_s} (1 - e^{-\omega_2 t})\overline{V}_s \tag{4.87}$$

which is plotted in Fig. 4.16(c).

It is customary to define, somewhat arbitrarily, the rise time T_r as the time required for the response to progress from 0.1 to 0.9 of full amplitude. From Eq. 4.87,

$$T_r = \frac{2.19}{2\pi f_2} = \frac{0.35}{f_2}$$

whereas the bandwidth, arbitrarily defined as the 0.707 response point from Fig. 4.16, is f_2. Hence the product at rise time and bandwidth, known as the time–bandwidth product, is

$$T_r B = 0.35$$

It can be shown that more complex low-pass networks, which may include inductances, have time–bandwidth products that tend toward a constant value, but exceeding somewhat that of the simple RC circuit. As a rule of thumb, it can be generally stated that $0.35 < T_r B < 0.45$.

Combined Low-Pass, High-Pass Network

A two-port network that contains limiting shunt capacitances at both source and load together with dc decoupling capacitance C_c is shown in Fig. 4.17. The combined steady-state response is

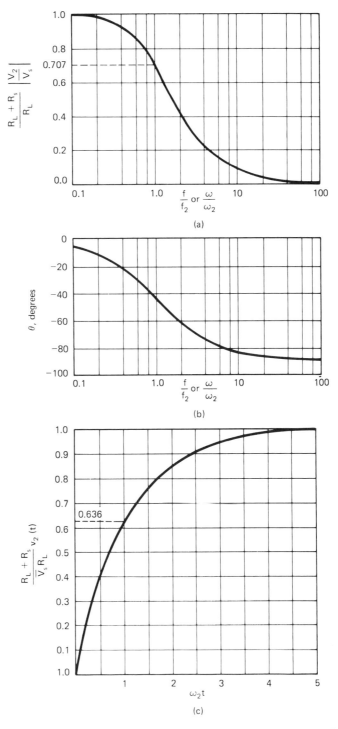

Figure 4.16 Frequency characteristics of single-pole low-pass networks. (a) Amplitude response. (b) Phase response. (c) Transient (step function) response.

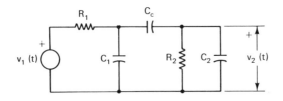

Figure 4.17 Combined low-pass, high-pass RC network.

$$\frac{V_2}{V_s} = \frac{R_2}{R_1 + R_2} \frac{1}{1 + (R_1C_1 + R_2C_2)\omega_1 + j[(\omega/\omega_2) - (\omega_1/\omega)]} \tag{4.88}$$

where

$$\omega_1 = \frac{1}{(R_1 + R_2)C_c} \tag{4.89}$$

$$\omega_2 = \frac{1}{(R_1R_2/R_1 + R_2)[C_1 + C_2 + (C_1C_2/C_c)]} \tag{4.90}$$

Also,

$$\omega_2 \cong \frac{1}{(R_1R_2/R_1 + R_2)(C_1 + C_2)} \tag{4.91}$$

for

$$C_c \gg \frac{C_1C_2}{C_1 + C_2}$$

Furthermore, if $\omega_1 \ll \omega_2$ and $(R_1C_1 + R_2C_2)\omega_1 \ll 1$, Eq. 4.88 may be written approximately as

$$\frac{V_2}{V_s} \cong \frac{R_2}{R_1 + R_2} \frac{1}{1 + j[(\omega/\omega_2) - (\omega_1/\omega)]} \tag{4.92}$$

This is simply a combination of Eqs. 4.74 and 4.81, and the overall frequency characteristic is the combined plots of Figs. 4.14 and 4.16. It should be observed that at low frequencies essentially full amplitude is reached for $f > 10f_1$, and at high frequencies for $f < 0.1f_2$. Thus, for a midband full-amplitude range to exist, it is required that $f_2 \geq 20f_1$.

4.12 FREQUENCY RESPONSE OF SINGLE-DEVICE INVERTING AMPLIFIERS

The circuit shown in Fig. 4.18 is representative of a single device, either BJT or FET, in its inverting mode connected to a source and load used as a small-signal amplifier. The basic device models used are the hybrid-π model for the BJT taken from Figs. 2.24(a) and 2.22 and the common-source model for the FET taken from Fig. 3.14(a). The correspondence to actual device parameters may be noted in Table 4.1.

For either device, C_F is a "feedback" capacitance from output to input, which has

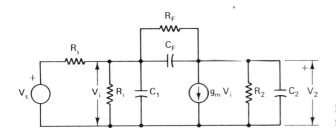

Figure 4.18 High-frequency circuit model for single-device inverting amplifier.

TABLE 4.1

Quantity	FET	BJT
V_s	V_s	V_s
R_s	R_s	R'_s (external) $+ r'_b$
R_i	$r_g \parallel R_g$ external	$r_{b'e} \cong r_d$
C_1	$C_{gs} + C$ external	$C_{b'e} + C$ (ext)
C_F	C_{gd}	$C_{cb'}$
R_F	R (external)	$r_{cb'}$ (usually very large)
V_i	V_{gs}	$V_{b'e}$
g_m	g_m	$g_m = \beta/r_d$
R_2	$r_d \parallel R_L$	$r_{ce} \parallel R_L$
C_2	C (structural) $+ C_L$ (load)	C_{ce} (structural) $+ C_L$ (load)

a marked effect on the input admittance of the amplifier. The amplifier will first be described in terms of its terminal-to-terminal gain

$$A = \frac{V_2}{V_i} \tag{4.93}$$

and its effective input admittance

$$Y_i = \frac{I_1}{V_i} \tag{4.94}$$

It is convenient to use the alternative circuit model as indicated in Fig. 4.19 with R_F assumed to be infinite, fully appropriate for the FET and usually good for an approximate solution for the BJT.

$$V_1 = \frac{R_i}{R_s + R_i} V_s$$

$$R_1 = \frac{R_s R_i}{R_s + R_i}$$

$$R_2 = \frac{r_o R_L}{r_o + R_L}; \quad \begin{array}{l} r_o = r_d \text{ for FET} \\ r_o = r_{ce} \text{ for BJT} \end{array}$$

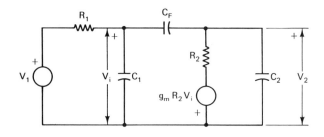

Figure 4.19 Simplified model for single-device inverting amplifier.

The two equations that can be written directly,

$$V_i = \frac{V_1 + j\omega R_1 C_F V_2}{1 + j\omega R_1 (C_1 + C_F)} \tag{4.95}$$

$$\frac{V_2}{V_i} = A = \frac{-g_m R_2 [1 - j\omega)C_F/g_m)]}{1 + j\omega R_2 (C_2 + C_F)} \tag{4.96}$$

describe the voltage at the input node in terms of source and load voltages and the output voltage in terms of the voltage at the input node.

Amplifier Gain

Equation 4.96 for the gain A is usually *almost* that of a single-pole response inasmuch as, when $g_m R_2$ is selected to provide substantial low-frequency voltage gain,

$$C_d g_m << R_2 (C_2 + C_F)$$

It is convenient to write Eq. 4.96 as

$$A = \frac{-g_m R_2 [1 - jk(\omega/\omega_2)]}{1 + j(\omega/\omega_2)} \tag{4.97}$$

where

$$k = \frac{C_F}{C_2 + C_F} \frac{1}{g_m R_2} \tag{4.98}$$

$$\omega_2 = \frac{1}{R_2 (C_2 + C_F)} \tag{4.99}$$

The magnitude of the gain is

$$|A| = \frac{g_m R_2 \{1 + [k(\omega/\omega_2)]^2\}^{1/2}}{\{1 + [(\omega/\omega_2)]^2\}^{1/2}} \tag{4.100}$$

The phase response (relative to the 180° equivalent of the polarity reversal) is

$$\theta_{rel} = -\tan^{-1} k \frac{\omega}{\omega_2} - \tan^{-1} \frac{\omega}{\omega_2} \tag{4.101}$$

Thus the frequency-dependent term in the numerator of Eq. 4.97 introduces excessive phase shift compared to the simple single-pole response, which, however, is negligible if $g_m R_2$ is sufficiently large. The situation is somewhat different if the device is used as an inverting voltage follower, (i.e., $g_m R_L \cong 1$).

Then, if $C_F \gg C_2$, $k \to 1$ and $|A|$ becomes essentially independent of frequency but with a phase characteristic of approximately

$$\theta \simeq 2 \tan^{-1} \frac{\omega}{\omega_2}$$

There are instances in which the inverting amplifier is used in this inverting voltage-follower mode.

Input Admittance

The input admittance Y_i of the inverting amplifier can be obtained using the circuit of Fig. 4.20 from Fig. 4.18. This input admittance consists of two components, direct and feedback,

$$Y_i = Y_{iD} + Y_{iF} \tag{4.102}$$

given by

$$Y_{iD} = j\omega C_1 + \frac{1}{R_i} \tag{4.103}$$

and

$$Y_{iF} = j\omega C_F (1 - A) \tag{4.104}$$

Using Eq. 4.97,

$$Y_{iF} = j\omega C_F \left\{ 1 + \frac{g_m R_2 [1 - jk(\omega/\omega_2)]}{1 + j(\omega/\omega_2)} \right\} \tag{4.105}$$

This can be expressed in terms of a frequency-dependent conductance and susceptance component as

$$Y_{iF} = \frac{\omega C_F (\omega/\omega_2) g_m R_2 (1 + k)}{1 + (\omega/\omega_2)^2} + j\omega C_F \frac{1 + g_m R_2 + (1 - g_m R_2 k)(\omega/\omega_2)^2}{1 + (\omega/\omega_2)^2} \tag{4.106}$$

and the total input admittance Y_i are these components in parallel with R_i and C_1 of Fig. 4.20.

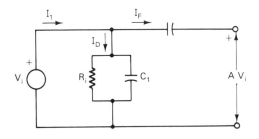

Figure 4.20 Circuit for determination of input admittance of inverting amplifier.

At low frequencies, $\omega \ll \omega_2$, the conductance term is negligible and

$$Y_{iF} \simeq j\omega C_F(1 + g_m R_2) \qquad (4.107)$$

Since, at low frequencies $g_m R_2 = -A_{0M} = |A_{0M}|$, where A_{0M} is the gain V_2/V_i at low frequencies,

$$Y_{iF} \simeq j\omega C_F(1 - A_{0M}) \simeq j\omega C_F(1 + |A_{0M}|) \qquad (4.108)$$

Thus, at low frequencies, there is effective in shunt with the direct input capacitance a magnified capacitance of value $C_F (1 + |A_{0M}|)$ as a result of the feedback capacitance. This has historically been referred to as the *Miller capacitance* as a result of the Miller effect of the increased effective input admittance of the feedback effect of the inverting amplifier.

As the frequency is increased, the importance of the conductance component increases, and at $\omega/\omega_2 = 1$ it is given by

$$[Y_{iF}]_{\omega = \omega_2} = \frac{\omega_2 C_F g_m R_2(1 + k)}{2} + \frac{j\omega_2 C_F[(1 + g_m R_2) + (1 - g_m R_2 k)]}{2} \qquad (4.109)$$

For sufficiently large values or gain, $g_m R_2 \gg 1$, the conductance and susceptance components are almost equal at $\omega = \omega_2$.

At much higher frequencies, $\omega/\omega_2 \gg 1$,

$$Y_{iF} = \omega_2 C_F g_m R_2(1 + k) + j\omega C_F(1 - g_m R_2 k) + j\frac{\omega_2^2}{\omega}(1 + g_m R_2)C_F \qquad (4.110)$$

which, with the last term negligible, is approximately

$$Y_{iF} \simeq \omega_2 C_F g_m R_2(1 + k) + j\omega C_F(1 - g_m R_2 k) \qquad (4.111)$$

which can be written as

$$Y_{iF} = \frac{C_F}{C_2 + C_F} g_m (1 + k) + j\omega C_F\frac{C_2}{C_2 + C_F} \qquad (4.112)$$

The conductance component at large values of gain approaches a limiting value of

$$g_m \frac{C_F}{C_2 + C_F}$$

while the susceptance component increasing as

$$\omega\frac{C_F C_2}{C_F + C_2}$$

Input Circuit Response

The response of the input to the amplifier when connected to a source can be determined from Fig. 4.21. This response can be expressed as

$$\frac{V_i}{V_1} = \frac{1}{1 + R_1(j\omega C_1 + Y_{iF})} \qquad (4.113)$$

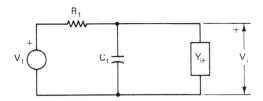

Figure 4.21 Input circuit of single-device inverting amplifier.

Using Eq. 4.105 for Y_{iF}, Eq. 4.113 can be expressed as

$$\frac{V_i}{V_1} = \frac{1 + j(\omega/\omega_2)}{1 - \dfrac{\omega}{\omega_2}\dfrac{\omega}{\omega_1}\left(1 - kg_mR_2\dfrac{\omega_1}{\omega_3}\right) + j\dfrac{\omega}{\omega_2}\left(1 + \dfrac{\omega_2}{\omega_1} + g_mR_2\dfrac{\omega_2}{\omega_3}\right)} \quad (4.114)$$

where

$$k = \frac{C_F}{C_2 + C_F}\frac{1}{g_mR_2}, \qquad \omega_2 = \frac{1}{R_2(C_2 + C_F)},$$

$$\omega_1 = \frac{1}{R_1(C_1 + C_F)}, \qquad \omega_3 = \frac{1}{R_1C_F}$$

Also, it is useful to note that

$$kg_mR_2\frac{\omega_1}{\omega_3} = \frac{C_F}{C_2 + C_F}\frac{C_F}{C_1 + C_F}$$

Equation (4.114) may be written as

$$\frac{V_i}{V_1} = \frac{1 + j(\omega/\omega_2)}{1 - m(\omega/\omega_1)(\omega/\omega_2) + jK(\omega/\omega_2)} \quad (4.115)$$

where

$$m = 1 - kg_mR_2\frac{\omega_1}{\omega_3} = 1 - \frac{C_F}{C_2 + C_F}\frac{C_F}{C_1 + C_F} \quad (4.116)$$

and

$$K = 1 + \frac{\omega_2}{\omega_1} + g_mR_2\frac{\omega_2}{\omega_3} = 1 + \frac{R_1(C_1 + C_F)}{R_2(C_2 + C_F)} + g_mR_2\frac{R_1C_F}{R_2(C_2 + C_F)} \quad (4.117)$$

$$= 1 + \frac{R_1}{R_2}\frac{C_1 + C_F}{C_2 + C_F} + (g_mR_2)^2 k\frac{R_2}{R_1}$$

The general nature of the responses V_i/V_1 and V_2/V_i and the overall response will be illustrated by the following example: $g_mR_2 \gg 1$, $\omega_1 = \omega_2$, $C_1 = C_2$, $R_1 = R_2$, $K = 10, m = 0.8$. Then from Eqs. 4.116 and 4.117 $C_1 = C_2 = 1.24C_F$, $g_mR_2 = 17.9$, and $k = 0.0249$.

The relative input response in decibels of $20\log_{10}(V_i/V_1)$ is plotted as curve A in Fig. 4.22. As a reference, curve B is a single-pole response whose -3-dB point is at ω_2. This would be the response $(V_2/V_i)\mid_{dB}$ if the term $k(\omega/\omega_2)$ in the numerator of Eq.

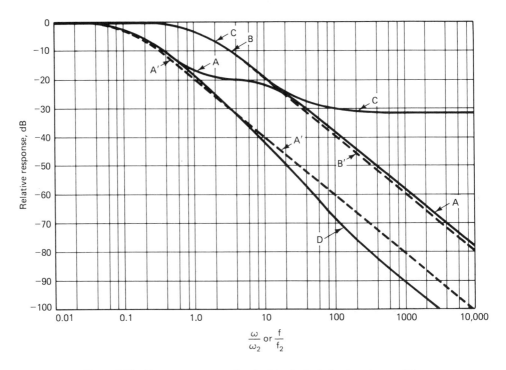

Figure 4.22 Frequency responses involved in single-device inverting amplifier.

4.100 can be neglected. It might be observed that at frequencies $\omega \ll 0.8\omega_2$ the actual response is similar to the single-pole response, where Eq. 4.115 is approximately

$$\left.\frac{V_i}{V_1}\right]_{\omega < 0.8\omega_2} \simeq \frac{1}{1 + jK(\omega/\omega_2)} \tag{4.118}$$

If this response were to continue, the result would be the dashed curve A' reaching an asymptotic value of -6 dB/octave (or 20 dB/decade) paralleling curve B with a separation of K. Instead, the input response because of the complex nature of Y_{iF}, which is part of Y_i, follows the curve shown, and at very high frequencies, $\omega > \omega_2$, reaches an asymptotic response of -6 dB/octave almost identical to that determined by ω_2 alone. This correlates with the fact that at such high frequencies Y_{iF} as given by Eq. 4.111 eventually reaches a value given by an equivalent parallel RC, with a value of R much lower than R_1 alone and a value of C slightly less than C_F.

The magnitude of the relative response of the amplifier itself given by Eq. 4.100 is plotted as curve C, which reaches a limiting value at very high frequencies of 20 log $k = -32$ dB. The overall response in decibels for $(V_i/V_1)(V_2/V_1)$ is the sum of curve A and curve C plotted as curve D, which reaches an asymptotic value of -6 dB/octave at very high frequencies after falling at a slightly greater rate at intermediate frequencies [e.g., approximately -8 dB/octave in the range $10 < (\omega/\omega_2) < 100$].

Interest is usually primarily in the range of frequencies over which substantial amplification is obtained, with that outside the range being of secondary interest. In this particular example, the response of the input circuit V_i/V_1 very closely follows that of the single-pole response out to $(\omega/\omega_2) \simeq 0.8$, where the response is -16 dB corresponding to $V_i/V_1 \doteq 0.15$. The amplifier response V_2/V_i is approximately that of a single-pole response whose 0.707 response point is approximately at ω_2. Furthermore, adding curves A and C shows that the single-pole response dominated by the effective input time constant K/ω_2, as given in Eq. 4.118, is a very good approximation of the overall response out to $\omega \simeq 10\omega_2$; that is,

$$\frac{V_2}{V_1} \simeq \frac{-g_m R_2}{1 + jK(\omega/\omega_2)}, \qquad \text{for } \frac{\omega}{\omega_2} < 10 \tag{4.119}$$

Overall Transient Response

The overall response V_2/V_1 is the product of Eqs. 4.115 and 4.97, which may be written in terms of $S = j(\omega/\omega_2)$ as

$$\frac{V_2(s)}{V_1(s)} = \frac{g_m R_2 k[s - 1/k]}{m(\omega_2/\omega_1) \, [s^2 + (K/m)(\omega_1/\omega_2) \, s + (1/m) \, (\omega_1/\omega_2)]} \tag{4.120}$$

This, when a step function V_1/s is applied, is of the form

$$F(s) = \frac{s - a_o}{s(s + \alpha)(s + \gamma)}$$

where

$$a_o = \frac{1}{k}$$

$$\gamma = \frac{K}{2m} \frac{\omega_1}{\omega_2} \left[1 + \left(1 - \frac{4m}{K^2} \frac{\omega_2}{\omega_1} \right)^{1/2} \right]$$

$$\alpha = \frac{K}{2m} \frac{\omega_1}{\omega_2} \left[1 - \left(1 - \frac{4m}{K^2} \frac{\omega_2}{\omega_1} \right)^{1/2} \right]$$

Its normalized transient response is

$$-\frac{\gamma\alpha}{a_o} f(t) = [1 - \frac{\gamma(a_o + \alpha)}{a_o(\gamma - \alpha)} e^{-\alpha t} + \frac{\alpha(a_o + \gamma)}{a_o(\gamma - \alpha)} e^{-\gamma t}] \tag{4.121}$$

The general form of the components of this response for $a_o \gg \alpha$ and $\gamma \gg \alpha$, together with the resultant response, is shown in Fig. 4.23. After a slight time delay due to the right half-plane zero and the presence of the pole far removed from the origin, the response is closely that of a single-pole response.

For the previous example, for $K = 10$, $m = 0.8$, $k = 0.025$, and $\omega_1 = \omega_2$, from Eq. 4.120,

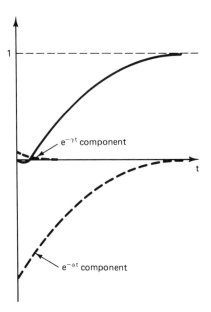

e^{-\gamma t} component

e^{-\alpha t} component

Figure 4.23 Transient response of inverting amplifier.

$$\alpha = 0.101, \qquad \gamma = 12.298, \qquad a_o = \frac{1}{k} = 40$$

$$-\frac{v_2(t)}{V_1} = 1 - 1.0108e^{-0.101\omega_2 t} + 0.0108e^{-12.298\omega_2 t}$$

In this case, with α and γ so far apart and a_o so large, the response is approximately

$$-\frac{v_2(t)}{V_1} \cong 1 - e^{-\alpha\omega_2 t}$$

This result confirms the results indicated in Fig. 4.22, where the overall frequency response is almost that controlled by a single time constant, with an $\omega_2' \cong 0.1\omega_2$ since the rise time is decreased by approximately the same factor.

Inverting Voltage Follower

The general conclusions drawn from the preceding example will differ somewhat when the amplifier is used in a low-gain mode, that is, an inverting voltage follower for $g_m R_2 \cong 1$, which will be discussed briefly as follows: The amplifier in its low-gain mode $g_m R_2 \cong 1$ is given by Eq. 4.97 as before, but using

$$k' = \frac{C_F}{C_2 + C_F}$$

and a new reference frequency

$$\omega_2' = \frac{1}{R_2'(C_2 + C_F)}$$

If A_0 is the magnitude of gain of the amplifier whose analysis was just concluded, and it is to be reduced to unity gain, then $R_2' = R_2/A_0$ and $\omega_2' = A_0\omega_2$, indicating a constant gain–bandwidth product.

Thus in Eq. 4.97, with $k' = g_m R_2 k$ and $\omega_2' = g_m R_2 \omega_2$, the form of the amplifier response is identical with $g_m R_2 = 1$ and the frequency scale shifted by the factor $g_m R_2$. Curve C of Fig. 4.20 represents the amplifier response where the frequency scale is given by ω/ω_2'. However, if $g_m R_2$ is further reduced such that $k = 1$, that is, $g_m R_2 = C_F/(C_2 + C_F)$, the magnitude of the gain A in Eq. 4.97 would be independent of frequency with a rate of change of phase twice that of the pure single-pole response.

The input admittance given by Eq. 4.106 for $g_m R_2 = 1$ can be written as

$$Y_{iF} = \frac{\omega C_F \left(\dfrac{\omega}{\omega_2'}\right)\left(1 + \dfrac{C_F}{C_2 + C_F}\right)}{1 + (\omega/\omega_2')^2}$$
$$+ j\omega C_F \frac{\left[2 + \left(1 - \dfrac{C_F}{C_2 + C_F}\right)\left(\dfrac{\omega}{\omega_2'}\right)^2\right]}{1 + (\omega/\omega_2')^2} \qquad (4.122)$$

The conductance component given by the first term is negligible out to a much higher frequency than in the previous example, because the multiplier $g_m R_2$ is missing and because ω_2' is greater than ω_2 by the same factor.

The second term can be represented by a pure capacitance of value $j\omega 2C_F$ to a good approximation out to a substantial fraction of ω_2' (at least to ω_2 with values used in the previous example).

4.13 FREQUENCY RESPONSE OF NONINVERTING AMPLIFIERS

Examples of noninverting single-device amplifiers capable of substantial voltage gain are the BJT common-base stage and the FET common-gate stage. Examples of noninverting voltage followers are the common-collector stage (emitter follower) and the common-drain stage (source follower).

Common-Gate FET Amplifier

The common-gate amplifier shown in Fig. 4.24(a) can be represented by the high-frequency model of Fig. 4.24(b), where R_F, C_F, C_1, and C_2 are device parameters combined with external circuit values.

The equation for amplifier gain, V_2/V_i, is

$$A = \frac{V_2}{V_1} = \frac{g_m R_2[1 + 1/(g_m R_F)]\{1 + j(\omega C_F/g_m)\,[1/(1 + 1/(g_m R_F))]\}}{[1 + j\omega(C_2 + C_F)R_2]} \qquad (4.123)$$

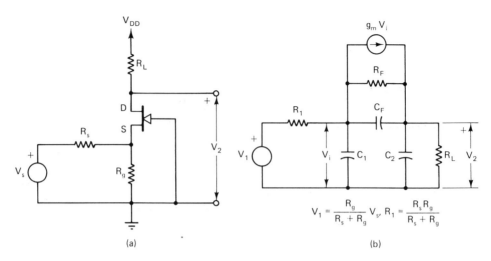

Figure 4.24 Common-gate FET amplifier. (a) Actual circuit. (b) High-frequency circuit model.

and the feedback component of input admittance is

$$Y_{iF} = g_m \left(1 + \frac{1}{g_m R_F} + j\omega \frac{C_F}{g_m} \right) \left[1 - \frac{R_2}{R_F} \frac{1 + j\omega R_F C_F}{1 + j\omega (C_2 + C_F) R_2} \right] \qquad (4.124)$$

where

$$R_2 = \frac{R_F R_L}{R_F + R_L}$$

The amplifier differs in important respects from the noninverting amplifier described in detail in Sec. 4.10. First, although the magnitude of the gain, $g_m R_2$, is the same, it is noninverting with respect to the input, which minimizes drastically the effect of C_F on the input admittance. Second, the net rate of change of phase is less, rather than greater, than that of the single-pole response. Also, the input admittance is low, rather than high, at low frequencies, given approximately by g_m, but its value changes much less rapidly with frequency than that of the inverting amplifier. Furthermore, the presence of R_F, if inherently present or added externally, permits some minor design flexibility to control the input admittance. For example, referring to Eq. 4.124, if the condition is met that if

$$\frac{R_2}{R_F} = \frac{C_F}{C_2 + C_F} = \frac{R_L}{R_L + R_F}$$

then

$$Y_{iF} = g_m \left(1 + \frac{1}{R_F g_m} \right) \left(\frac{R_F}{R_L + R_F} \right) + j\omega C_F \left(\frac{R_F}{R_L + R_F} \right) \qquad (4.125)$$

which is that of a parallel *RC* circuit. This is valid as an approximation except at very high frequencies, even if the condition is not closely met.

However, since the input admittance is high, V_i/V_s is small unless the amplifier is driven by a very low impedance source, which reduces the overall gain, V_2/V_s.

Common-Base BJT Amplifier

The common-base amplifier in Fig. 4.25(a) has the circuit model of Fig. 4.25(b) based on the hybrid-π model for the transistor. The complexity of the frequency response is increased because of the presence of r_b' in the common lead. However, if r_b' is small, the circuit model reverts to that of Fig. 4.25 and the general properties discussed with reference to Eqs. 4.123 and 4.124 are valid. Thus the general behavior of the common-base stage is similar to that of the common-gate stage, although the exact form of the response is somewhat more complicated.

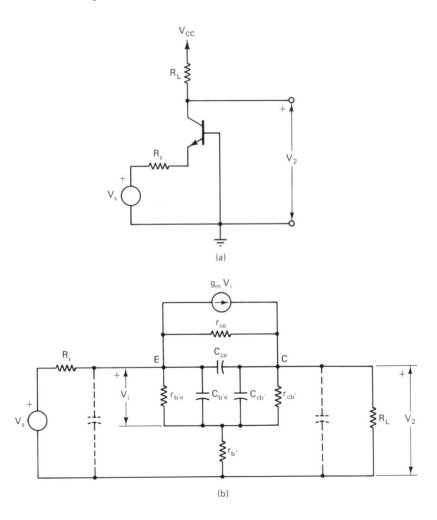

Figure 4.25 Common-base BJT amplifier. (a) Actual circuit. (b) High-frequency circuit model.

Voltage Follower

The BJT common-collector amplifier (emitter follower) and the common-drain FET (source follower) are examples of amplifiers whose voltage gain is less than, but approaches, unity if the load resistance is sufficiently large, and hence they are referred to as *voltage followers*.

The common-drain FET amplifier shown in Fig. 4.26(a) has the high-frequency circuit model in Fig. 4.26(b).

The amplifier gain is

$$
A = \frac{V_2}{V_i}
$$

$$
= \frac{g_m \left(1 + \dfrac{1}{g_m R_F}\right)\left(1 + j\omega \dfrac{C_F}{g_m} \dfrac{1}{1 + (1/g_m R_F)}\right)}{g_m \left[1 + \dfrac{1}{g_m}\left(\dfrac{1}{R_F} + \dfrac{1}{R_2}\right)\right]\left[1 + j\omega \dfrac{C_2 + C_F}{g_m} \dfrac{1}{1 + \dfrac{1}{g_m}\left(\dfrac{1}{R_F} + \dfrac{1}{R_2}\right)}\right]} \tag{4.126}
$$

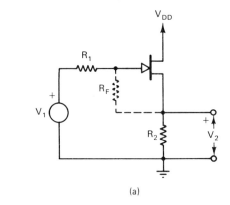

(a)

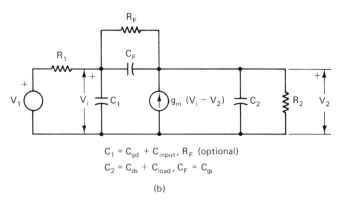

$C_1 = C_{gd} + C_{input}$, R_F (optional)
$C_2 = C_{ds} + C_{load}$, $C_F = C_{gs}$

(b)

Figure 4.26 FET source follower. (a) Actual circuit. (b) High-frequency circuit model.

and the feedback component of input admittance is

$$Y_{iF} = \left(\frac{1}{R_F} + j\omega C_F\right)(1 - A) \tag{4.127}$$

The gain A is noninverting and can be made exactly independent of frequency by equating the frequency-dependent terms in Eq. 4.126. The condition to be met is

$$R_F = \frac{R_2}{(C_F/C_2) - g_m R_2} \tag{4.128}$$

If R_F is very large (i.e. $R_F \to \infty$), the condition is approximately

$$\frac{C_F}{C_2} = g_m R_2 \tag{4.129}$$

The feedback component of input admittance

$$Y_{iF} = \left(\frac{1}{R_F} + j\omega C_F\right)(1 - A) \tag{4.130}$$

becomes that of a simple parallel RC circuit if the frequency independence of A is established; then

$$Y_{iF} = \frac{1 - A}{R_F} + j\omega C_F(1 - A) \tag{4.131}$$

The emitter follower in Fig. 4.27(a) has the approximate high-frequency circuit model of Fig. 4.27(b) based on the hybrid-π model for the transistor. With the appropriate substitutions as indicated, this model fits the model for the FET, and the same basic equations are reasonably valid.

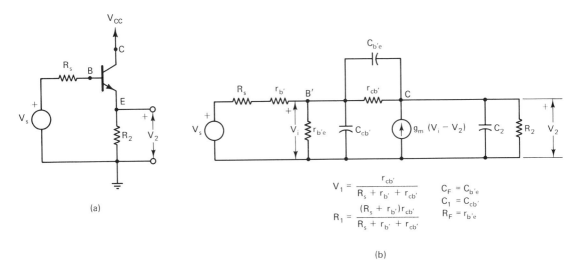

(a)

(b)

Figure 4.27 BJT emitter follower. (a) Actual circuit. (b) High-frequency circuit model.

4.14 NOISE MODELS FOR SINGLE-DEVICE AMPLIFIERS

In the discussions in previous chapters of noise sources in circuit components and active devices, noise models were developed for both the BJT and the FET, which included various representations of internal noise sources. The overall noise output of such devices used as amplifiers in various circuit configurations depends on how the particular noise sources enter into the complete circuit. To illustrate how noise determinations are made, the simplified h-parameter model of the BJT with source and load connected and including the noise sources described in Sec. 2.12 is shown in Fig. 4.28.

The output noise currents resulting from the separate internal noise sources are determined to be

$$\overline{i^2_{2rb'}} = \frac{\beta^2 \overline{v^2_{rb'}}}{(R_s + r'_b + r_d)^2} \tag{4.132}$$

$$\overline{i^2_{2b'c}} = \beta^2 \left(\frac{R_s + r'_b}{R_s + r_{b'} + r_d}\right)^2 \overline{i^2_{b'e}} \tag{4.133}$$

$$\overline{i^2_{2c}} = \overline{i^2_c} \tag{4.134}$$

$$\overline{i^2_{2cb'}} = \left(1 + \beta \frac{R_s + r'_b}{R_s + r_{b'} + r_d}\right)^2 \overline{i^2_{cb'}} \tag{4.135}$$

The mean-square output noise current is the sum of these components and would also include the thermal noise of R_s and R_L, which will be considered separately.

Each noise current source, because of its placement in the circuit, appears at the output in a slightly different way. It is convenient to compare them by expressing each in terms of an equivalent input mean-square noise voltage, which appears as an addition to $\overline{v^2_{rb'}}$, as follows:

The mean-square output current due to any input voltage v_s is

$$\overline{i^2_{2s}} = \beta^2 \left(\frac{v_s}{R_s + r'_b + r_d}\right)^2 \tag{4.136}$$

Each of the preceding noise current sources is equated in turn to Eq. 4.136 and solved for the corresponding $\overline{v^2_s}$. The results are

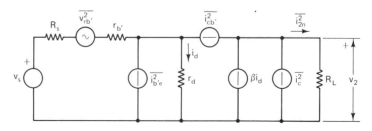

Figure 4.28 Low-frequency noise model for BJT amplifiers.

$$\overline{v_{b'es}^2} = (R_s + r_b')^2 \, \overline{i_{b'e}^2} \tag{4.137}$$

$$\overline{v_{cs}^2} = \left(\frac{R_s + r_b' + r_d}{\beta} \right)^2 \overline{i_c^2} \tag{4.138}$$

$$\overline{v_{cb's}^2} = \left(1 + \beta \, \frac{R_s + r_b'}{R_s + r_b' + r_d} \right)^2 \left(\frac{R_s + r_b' + r_d}{\beta} \right)^2 \overline{i_{cb'}^2} \tag{4.139}$$

which reduces to

$$\overline{v_{cb's}^2} \cong (R_s + r_b')^2 \, \overline{i_{cb'}^2} \tag{4.140}$$

for

$$\beta \, \frac{R_s + r_b'}{R_s + r_b' + r_d} >> 1$$

The total equivalent input mean-square noise voltage, including $\overline{v_{rb'}^2}$ and using the approximate Eq. 4.140 for $\overline{v_{cb's}^2}$, is

$$\overline{v_{ns}^2} = 4kTr_b' \, \Delta f + (R_s + r_b')^2 \, (\overline{i_{b'e}^2} + \overline{i_{cb'}^2}) + \frac{R_s + r_b' + r_d)^2}{\beta^2} \, \overline{i_c^2} \tag{4.141}$$

The equivalent noise model of Fig. 4.28 becomes that shown in Fig. 4.29.

The equivalent input voltage noise as indicated by Eq. 4.141 is often expressed as the noise of an equivalent resistance generating a noise voltage $\overline{v_n^2} = 4kTR_{eq} \, \Delta f$. Such an equivalent noise resistance from Eq. 4.141 is

$$R_{eq} = r_b' + \frac{(R_s + r_b')^2}{4kT \, \Delta f} \, (\overline{i_{b'e}^2} + \overline{i_{cb'}^2}) + \frac{(R_s + r_d' + r_d)^2}{4kT \, \Delta f} \, \frac{\overline{i_c^2}}{\beta^2} \tag{4.142}$$

All the noise sources that have been discussed have been sources within the device itself and have not included the noise of the source and load resistance.

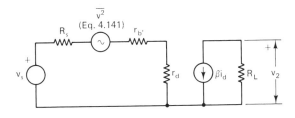

Figure 4.29 Simplified equivalent noise model for BJT amplifiers.

Noise Figure

The noise generated within a transistor used as an amplifier, no matter how expressed, is of importance only in its magnitude relative to that of the signal that is being amplified or, more important, the magnitude of the contribution of all noise sources including those within the signal itself compared with the amplitude of the signal being amplified.

One commonly used measure of the relative noise contribution of the device is the *noise figure*, which is a measure of the noise at the output of the device compared with that which would be present if the device itself were noiseless.

If the input to the device is a voltage source with a source resistance R_s and the output applied to a load resistance R_L, the *noise figure* of the device and its associated circuitry may be defined as

$$\text{NF} = \frac{P_{no}}{P_{nso}} \qquad (4.143)$$

where P_{no} is the total output noise power and P_{nso} is the output power if the source itself were the only source of noise.

The total noise model for the input portion of Fig. 4.29 in terms of R_{eq} is shown in Fig. 4.30.

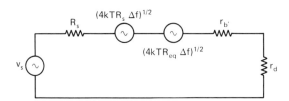

Figure 4.30 Model for determination of noise figure.

Since both noise sources are amplified in the same manner, the noise figure is

$$\text{NF} = 1 + \frac{R_{eq}}{R_s} \qquad (4.144)$$

Thus a noise figure of $\text{NF} \cong 1$ would mean that the device itself is a negligible noise contributor.

In the discussions so far, the contribution of R_L to the total noise output has been neglected. If it needs to be included, an additional output collector current mean-square noise current component of

$$\overline{i_{cL}^2} = \frac{4kT\,\Delta f}{R_L}$$

may be applied in the same manner as $\overline{i_C^2}$ itself to any of the preceding results.

Frequency-dependent Noise Models

The noise sources suggested in Fig. 4.28 may be applied to any of the frequency-dependent circuit models developed in Chapter 2 to illustrate how the output noise contribution may be frequency dependent. The simplified *h*-parameter model shown in Fig. 4.31 is chosen here for such an illustration. In solving such circuits, it should be pointed out that any available techniques of ac circuit analysis may be used for both noise and signal sources at a single frequency, but the end result can only be expressed in terms of magnitude, with the response for each source solved separately.

For the circuit shown, the frequency-dependent equivalent noise resistance as derived in Eq. 4.142 may be obtained by substituting Z_d for r_d and h_{fe} for β with the result

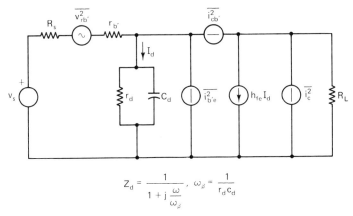

$$Z_d = \cfrac{1}{1 + j \cfrac{\omega}{\omega_\beta}} \;,\quad \omega_\beta = \frac{1}{r_d C_d}$$

Figure 4.31 Frequency-dependent noise model for BJT amplifiers.

$$R_{eqf} = r_b' + \frac{(R_s + r_b')^2}{4kT\,\Delta f}\,(\overline{i_{b'e}^2} + \overline{i_{cb'}^2}) + \frac{|R_s + r_b' + Z_d|^2}{4kT\,\Delta f}\,\frac{\overline{i_c^2}}{|h_{FE}^2|} \qquad (4.145)$$

The corresponding noise figure from Eq. 4.144 at a specific frequency at which Z_d and h_{fe} are determined is termed the *narrow-band* or *spot* noise figure at that frequency. For this particular circuit, the spot noise figure is an increasing function of frequency due to the decrease in h_{fe} with frequency unless $(R_s + r_b') << |Z_d|$, even though the noise sources themselves may not be frequency dependent. The *signal-to-noise* ratio will be degraded still more as the frequency is increased owing to the decrease in gain which is due to the decreasing h_{fe}.

The output noise components over a bandwidth f_1 to f_2 from solutions using Fig. 4.31 are

$$\overline{i_{2rb'}^2} = \frac{4kTr_b'\beta^2}{(R_s + r_b' + r_d)^2}\int_{f_1}^{f_2} \frac{1}{1 + \left(\dfrac{\omega}{\omega_\beta}\right)^2\left(\dfrac{R_s + r_d'}{R_s + r_b' + r_d}\right)^2}\,df \qquad (4.146)$$

$$\overline{i_{2b'e}^2} = \left[\frac{\beta(R_s + r_d')}{R_s + r_b' + r_d}\right]^2\int_{f_1}^{f_2} \frac{1}{1 + \left(\dfrac{\omega}{\omega_\beta}\right)^2\left(\dfrac{R_s' + r_b'}{R_s + r_b' + r_d}\right)^2}\,\overline{i_{b'e}^2}\,df \qquad (4.147)$$

$$\overline{i_{2c}^2} = \int_{f_1}^{f_2} \overline{i_c^2}\,df \qquad (4.148)$$

$$\overline{i_{2cb'}^2} = \left[\frac{\beta(R_s + r_b')}{R_s + r_b' + r_d}\right]^2\int_{f_1}^{f_2} \frac{1 + \left(\dfrac{\omega}{\omega_\beta}\dfrac{1}{\beta}\right)^2}{1 + \left(\dfrac{\omega}{\omega_\beta}\right)^2\left(\dfrac{R_s + r_d'}{R_s + r_b' + r_b'}\right)^2}\,\overline{i_{cb'}^2}\,df \qquad (4.149)$$

The terms $\overline{i_{b'e}^2}$, $\overline{i_c^2}$, and $\overline{i_{cb'}^2}$ represent the respective mean-square noise currents per unit frequency (power spectral density) at the points at which they originate and may or may not themselves be functions of frequency depending on the presence of excess noise components. Over a broad frequency spectrum, the integrated noise over the band is most important, whereas in a narrow-band region for a frequency-dependent circuit the spot noise in the neighborhood of the frequency of interest is most important.

Similar procedures may be followed for FET amplifiers using the device model of Fig. 3.25 with source and load connected in a common-source configuration as shown in Fig. 4.32(a).

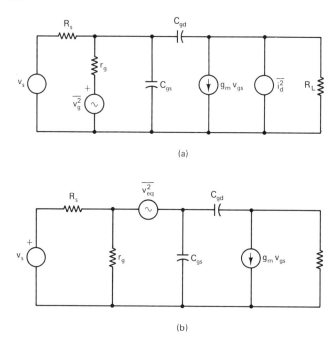

(a)

(b)

Figure 4.32 Noise models for FET amplifiers.

The effects of $\overline{v_g^2}$ and $\overline{i_d^2}$ can be combined into an equivalent noise voltage at the source as indicated in Fig. 4.32(b), which is the noise voltage of an equivalent noise resistance determined at very low frequencies at which C_{gs} and C_{ds} are negligible and can be expressed as

$$R_{eq} = \frac{\overline{v_{eq}^2}}{4kT\,\Delta f} \qquad (4.150)$$

Spot noise figures and wide-band noise components can be determined in a manner similar to what was done for bipolar transistors.

Other Circuit Configurations

Noise calculations, including spot noise figure and broad-band noise components, can be made for other than common-emitter and common-source amplifiers using the same device noise models. The results will be different for other circuit configurations.

Usually, the other configurations are used in composite or compound structures involving two or more devices, and therefore amplifier noise models using these other configurations will be deferred to Chapter 5, where such composite amplifiers are discussed and analyzed in detail.

PROBLEMS

4.1 The transistor shown in Fig. 4.4 has voltage supplies and external resistances $Z_S = R_B$, $Z_T = R_E$, and $Z_L = R_L$, such that the dc emitter current is $I_E = 2.0$ mA. At this value of current, $\beta = 100$, $r_{bes} = 1.5$ kΩ, $r_{ceo} = 200$ kΩ, and $\mu_{re} = 1.5 \times 10^{-4}$. The external resistances are $R_B = 10$ kΩ, $R_L = 5$ kΩ, and $R_E = 0.1$ kΩ.

(a) Determine the values of overall voltage gain and input and output resistances using the approximate relationships of Eqs. 4.15, 4.16, and 4.17.

(b) If R_L is changed to $R_L = 50$ kΩ while maintaining the same transistor operating point, is the approximate $Z_L \ll 1/h_{oe}$ still valid? If not, determine the gain from Eq. 4.11 retaining whatever approximations *are* still valid.

4.2 The transistor in the amplifier circuit shown is expected to have the dc operating point $I_B = 0.02$ mA, $I_E = 2.0$ mA, $V_{BE} = 0.8$ V, and $V_{CE} = 3.0$ V. At the specified operating point, $r_{bes} = 1.5$ kΩ, $r_{ceo} = 100$ kΩ, and $\beta = 100$.

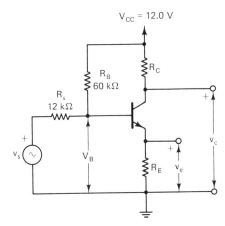

Prob. 4.2

(a) For $v_S = 0$, determine V_B, and then h_{FE}, R_E, and R_C.

(b) Make an estimate of μ_{re} at the prescribed operating point.

(c) Determine the incremental voltage gains, v_c/v_s and v_e/v_s. *Note:* It is suggested that the input circuit be replaced by its Thevenin equivalent.

(d) What is the incremental input resistance, R_i?

(e) Determine R_{oc} (looking from R_C back into the collector terminal) and R_{oe} (looking from R_E back into the emitter terminal).

4.3 It is desired that the *pnp* emitter follower shown be biased for zero offset voltage between input and output, as shown (i.e., $V_O = 0$ when $v_s = 0$). Operating conditions are specified as $|I_B| = 0.02$ mA, $|I_E| = 1.0$ mA, and $V_{BE} = -0.7$ V.

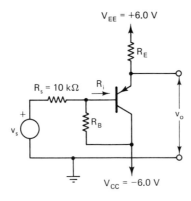

Prob. 4.3

(a) Determine R_B and R_E.

(b) In the absence of any experimental measurement, assume $\beta \cong h_{FE}$ at the operating point, that r_d can be estimated for the specified emitter current, and that r_b' is estimated as $r_b' \cong 0.2r_d$. Find approximately the voltage gain v_o/v_s and the small signal input and output resistances R_i and R_o.

(c) Assuming that the collector–base junction is not allowed to become forward biased, that $\beta \cong h_{FE}$ is constant throughout the active region, and that the minimum collector current permitted is 0.1 mA, what are the most positive and negative input and output small-signal voltage excursions?

4.4 Consider the simplified h-parameter circuit model for the common-emitter amplifier shown (R_L sufficiently small that the $\mu_{re}\beta$ term can be neglected). It is found that $\beta \cong 100$ and $r_{ceo} \cong 100$ kΩ over a reasonably wide operating range.

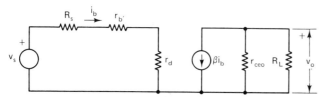

Prob. 4.4

(a) Draw a new equivalent circuit using the hybrid-π model for the transistor and estimate and tabulate the hybrid-π parameters at emitter current levels of $I_E = 0.1, 1.0,$ and 10 mA.

(b) Discuss the validity of neglecting $r_{cb'}$ at the various current levels, particularly for $R_L \ll r_{ceo}$.

4.5 The transistor in the circuit shown is found to be reasonably linear through most of its normal active region, and its parameters may be approximated over a wide range by $h_{FE} \cong \beta \cong 100$

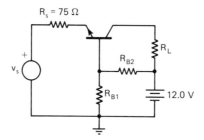

Prob. 4.5

and $V_{BE} \cong 0.7$ V. Also, at the operating point, $r_{bes} \cong 1.5$ kΩ for $R_{B1} = 6.6$ kΩ, $R_{B2} = 66$ kΩ, and $R_L = 1.5$ kΩ.

(a) Determine approximately the incremental input resistances R_i and the overall small-signal voltage gain v_o/v_s.

(b) Find approximately all dc voltages and currents in the circuit.

4.6 The emitter follower shown is designed to have an output resistance to match the input resistance of a properly terminated 75-Ω almost lossless coaxial transmission line at a dc output voltage level of zero volts. The transistor is operated at an emitter current of $I_E = 3$ mA and, at this value, β = 150 and $r_{bes} = 1.5$ kΩ.

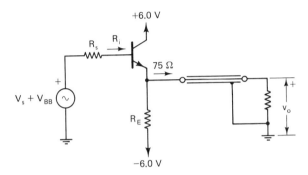

Prob. 4.6

(a) Determine R_S and R_E.

(b) What is the small-signal voltage gain, v_o/v_s?

(c) What is the input resistance, R_i?

4.7 The two amplifiers shown are to be compared when biased at identical operating points at which $r_{bes} = 2.0$ kΩ and β = 150. For both amplifiers, $R_B = 2.0$ kΩ, $R_E = 0.2$ kΩ, $R_L = 2.0$ kΩ and $r_{ceo} = 200$kΩ.

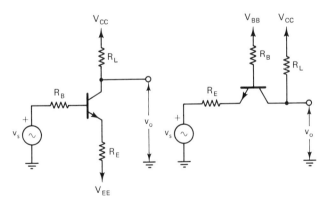

Prob. 4.7

(a) Calculate the approximate overall voltage gain v_o/v_s and R_i and R_o for both amplifiers.

(b) Repeat part (a) for the common-base amplifier with the values for R_E and R_B interchanged (i.e., making the source resistance and the common leg resistance the same as those for the common-emitter amplifier).

4.8 A junction FET having characteristics similar to those shown in Fig. 3.10 is used in the circuit shown. The operating point is chosen to be at $V_{DS} = 10.0$ V, $I_D = 2.5$ mA.

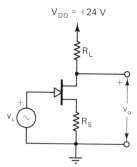

Prob. 4.8

(a) What value of R_S is required to make $V_{GS} = -1.0$ V, and what value of R_L is required for $V_{DS} = 10.0$ V?

(b) Estimate g_m and r_d at the operating point and calculate the small-signal voltage gain v_o/v_s.

(c) What are the incremental output resistances looking back into the drain and source terminals?

4.9 A simple amplifier using a p-channel MOSFET having the characteristics of Fig. 3.23 is shown.

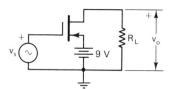

Prob. 4.9

(a) Find the value of R_L to estimate the operating point $V_{GS} = -9$ V and $V_{DS} = -6$ V.

(b) Make estimates of g_m, r_d, and μ_A at the selected operating point.

(c) What is the voltage gain, v_o/v_s?

4.10 An enhancement-mode IGFET is connected as a source follower as shown with an intentional 4.0 V offset from input to output (i.e., $V_G = +4$ for $V_O = 0$). The FET is to be designed to have $g_m = 5$ mS for $V_{GS} = 4.0$ V and a threshold voltage of $V_T = 2.0$ V. It will be assumed that $r_d \to \infty$ in the saturation region.

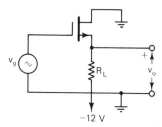

Prob. 4.10

(a) Determine the required value for $K = \mu_n C_o/L^2$.

(b) How much drain current will flow?

(c) What is the required value for R_L?

(d) What is the incremental resistance looking back into the source?

(e) What would be the effect of connecting the drain to a voltage $V_{DD} = +12$ particularly with respect to the permissible signal voltage swing?

4.11 A p-channel depletion-mode MOSFET is used in the circuit shown. At a drain current of 2 mA, $g_m = 4$ mS and $r_d = 22$ kΩ. The drain current i_D reduces to zero for $V_{GS} \cong +4$ V.

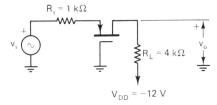

R$_s$ = 1 kΩ

R$_L$ = 4 kΩ

V$_{DD}$ = -12 V   **Prob. 4.11**

(a) Determine the overall gain v_o/v_s, R_i, and R_o at the operating point.

(b) Make an estimate for μ_A, λ, and K in the approximate saturation model in the neighborhood of the operating point.

4.12 Important high-frequency parameters of a particular transistor are to be approximated from a few external measurements and computations. At an operating point $I_C = 2.0$ mA at $V_{CE} = 5.0$ V and $V_{BE} = 0.7$ V, $\beta = 120$, $r_{bes} = 2.3$ kΩ, $r_{ceo} = 240$ kΩ, and $f_\beta = 4.8$ MHz. Assuming graded junctions at both emitter–base and collector–base junctions, external capacitance measurements and calculations of the type suggested in Sec. 1.6 lead to a determination at junction capacitances at the operating point of $C_{je} \cong 1.2$ pF and $C'_{cb} = 0.10$ pF.

(a) As completely as you can from the given data, determine all the circuit elements in both the h-parameter circuit model and the hybrid-π circuit model.

(b) Assuming that r'_{cb} is sufficiently large that the circuit of Fig. 4.19 is valid for a transistor amplifier, and that $R_L = 4.0$ kΩ and $C_2 = 5$ pF, determine the effective input capacitance at low frequencies ($\omega \ll \omega_\beta$).

(c) If the amplifier is driven from a voltage source with $R'_s = 10$ kΩ, write an approximate equation for V_2/V_1 using Eq. 4.119 as an approximation.

4.13 The transistor amplifier shown is to be approximated by the equivalent circuit of Fig. 4.19 with the following parameters predetermined: $R'_S = 9.8$ kΩ, $R_L = 2.5$ k$\Omega \approx R_2$, $C_1 = 10$ pF, $C_F = 0.1$ pF, $r_d = 2.0$ kΩ, $r'_b = 0.2$ kΩ, $\beta = 150$, and $C_2 = 10$ pF.

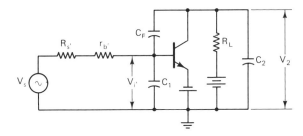

R$_s'$ r$_{b'}$ C$_F$ R$_L$ C$_2$ V$_2$

V$_s$ V$_i'$ C$_1$

Prob. 4.13

(a) What is the effective input capacitance at frequencies $\omega \ll \omega_2$?

(b) Determine the complex input admittance at $\omega = \omega_2$.

(c) Plot the magnitude of the relative response at the input $|V'_i/V_s|$ in decibels relative to the zero frequency value on the scale of Fig. 4.22.

(d) Plot $|V_2/V'_i|$ in decibels relative to its zero frequency value on the same scale as in part (c).

(e) Plot the overall response V_2/V_s by combining parts (c) and (d) and compare it with a plot of the magnitude of Eq. 4.119.

4.14 The FET amplifier shown has equal source and load resistances, $R_1 = R_2 = 4.5$ kΩ. At the operating point chosen, $g_m \cong 4.0$ mS and the total effective input, output, and feedback capacitances are $C_1 \cong C_2 \cong 2.5$ pF and $C_F \cong 2.0$ pF.

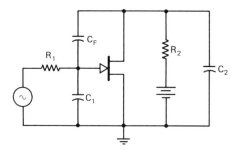

Prob. 4.14

(a) What is the low-frequency gain and the effective input capacitance at low frequencies?
(b) For the generalized response equations, determine the constants k, K, and m and compare these with the example used in Sec. 4.12.
(c) Determine approximately the overall -3-dB response point.

4.15 The simple common-gate enhancement-mode MOSFET amplifier shown has parameters at a selected operating point as follows: $I_D = 1$ mA, $g_m = 5$ mS, $r_d = 50$ kΩ, $C_{gs} = 1.2$ pF, $C_{gd} = 0.2$ pF, $C_2 \cong 2.0$ pF, and $R_S = R_L = 5.0$ kΩ.

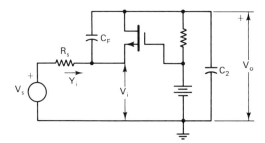

Prob. 4.15

(a) Find a value for C_F that will make the input admittance Y_i equivalent to that of a parallel RC circuit. What are R and C in the parallel RC combination?
(b) What is the overall low-frequency gain V_o/V_s written as $(V_i/V_s)(V_o/V_i)$?
(c) Determine approximately the overall bandwidth and indicate the validity of whatever approximations you make.

4.16 The FET of Problem 4.15 is used as a source follower as indicated; that is, $C_{gs} = 1.2$ pF, $C_{gd} = 0.2$ pF, and $g_m = 5$ mS.

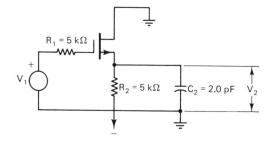

Prob. 4.16

(a) Determine the low-frequency gain and the effective input capacitance at low frequencies.

(b) Since $R_F \to \infty$, write a simplifed equation for the voltage gain V_2/V_i (from Eq. 4.126) and, assuming that R_2 can be changed, indicate what the approximations are for the gain to be written as

$$\frac{V_2}{V_i} \cong \frac{g_m R_2}{1 + g_m R_2} \frac{1 + j\omega \, (C_F/g_m)}{1 + j\omega \, [(C_2 + C_F)/g_m]}$$

Using such an approximation, determine the bandwidth for the circuit. Is the approximation fairly good for $R_2 = 5 \text{ k}\Omega$?

(c) Assuming that A is constant with frequency for $R_2 = 5 \text{ k}\Omega$, determine the bandwidth of the input circuit (i.e., for V_i/V_1), and compare it with that of the amplifier itself. Is there sufficient difference to indicate that the overall frequency response is approximately that of the amplifier alone (i.e., that of V_2/V_i)?

4.17 A voltage follower is sometimes used to drive a common-gate stage; that is, the effective R_2 is that of the input resistance of the common-gate stage.

(a) Using the FET of Problem 4.16 as a source follower, determine the required value of R_2 to make the voltage gain V_2/V_i independent of frequency. What is the resultant gain?

(b) What is the bandwidth of the overall circuit (i.e., for V_2/V_1)?

REFERENCES

GHAUSI, M. S. *Electronic Devices and Circuits,* Holt, Rinehart and Winston, New York, 1985.

GLASFORD, G. M. *Linear Analysis of Electronic Circuits,* Addison-Wesley, Reading, Mass., 1965.

GRAY, P. R., AND R. G. MEYER. *Analog Integrated Circuits,* Wiley, New York, 1984.

MILLMAN, J. *Microelectronics,* McGraw-Hill, New York, 1979.

SEDRA, A. S., AND H. C. SMITH. *Microelectronic Circuits,* Holt, Rinehart and Winston, New York, 1982.

Chapter 5

Composite Amplifiers: Structure, Performance, and Biasing

INTRODUCTION

A composite discrete amplifier structure may be described as an amplifier comprising a combination of two or more discrete devices interconnected in such a way as to impart desirable characteristics to the combination that cannot be realized readily by an amplifier using a single device, or to realize the desirable characteristics of a single-device amplifier in an improved manner.

The composite structure then assumes a set of terminal characteristics that can be treated analytically as though one discrete device having the desired characteristics were used. However, such composite amplifiers are not usually realized as "wired connections" of individual devices, but more often through a fabrication process that forms the individual devices and their connections as an overall series of diffusions or ion implantations, with the result being what might be termed a VSSI (very small scale integrated circuit).

These second-order building blocks can then be further combined to yield a complete amplifier, such as the high-gain operational amplifier in widespread use, which in turn is the building blocks of most analog and some digital systems. In analyzing these composite devices, the general device equations developed in Chapter 4 may be used.

5.1 APPROXIMATE EQUATIONS FOR SINGLE-DEVICE AMPLIFIERS

The approximate low-frequency equations for single-device amplifier structures developed in Chapter 4 may be simplified further by eliminating some of the internal feedback terms. These modified equations can then be used in a simplified analysis of various composite structures. These simplified equations for the amplifiers shown in Fig. 5.1 for the BJT are as follows:

1. *Common-emitter amplifier:*

$$R_i = r_{bes} + (1 + \beta)R_E - \mu_{re}\beta R_L \tag{5.1}$$

$$A_v = \frac{-\beta R_L}{(R_s + r_{bes})\left[1 + R_E\dfrac{(1 + \beta)}{R_{S_s} + r_{bes}} - \dfrac{\mu_{re}\beta R_L}{R_{S_s} + r_{bes}}\right]} \tag{5.2}$$

$$R_o = r_{ceo}\left[1 + \frac{R_E(1 + \beta)}{R_s + r_{bes}}\right] \tag{5.3}$$

2. *Common-collector amplifier:*

$$R_i = r_{bes} + (1 + \beta)R_L - \mu_{re}\beta R_C \tag{5.4}$$

$$A_v = \frac{1}{1 + [(R_s + r_{bes})/(1 + \beta)R_L]} \tag{5.5}$$

$$R_o = \frac{R_s + r_{bes}}{1 + \beta} \tag{5.6}$$

3. *Common-base amplifier:*

$$R_i = \frac{r_{bes} + R_B}{1 + \beta} + \frac{\mu_{rb}\beta R_L}{1 + \beta} \tag{5.7}$$

$$A_v = \frac{[\beta/(1 + \beta)]R_L}{R_s + [(r_{bes} + R_B)/(1 + \beta)]} \tag{5.8}$$

$$R_o = \frac{r_{ceo}(1 + \beta)}{1 + (1 + \beta)R_B/[r_{bes} + R_E(1 + \beta)]} \tag{5.9}$$

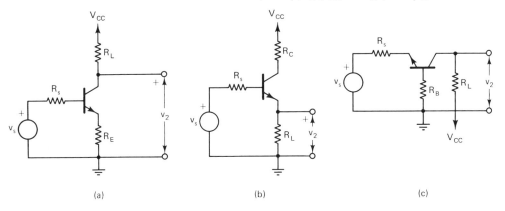

(a) (b) (c)

Figure 5.1 Single-transistor amplifiers. (a) Common emitter. (b) Common collector. (c) Common base.

The comparable low-frequency equations for the FET amplifier in its common-source and common-drain connections, as discussed in Sec. 4.7, are somewhat simpler than those for the BJT common-emitter and common-collector voltage amplifiers, both because the input resistance is very high and because there is no comparable internal feedback term.

The common-gate amplifier discussed in Sec. 4.8, comparable to the common-base BJT amplifier, does exhibit a low input resistance as a result of internal feedback resulting from the particular configuration, as indicated in Fig. 4.10(c), where the input resistance is

$$R_i = \frac{R_D + R_L}{\mu + 1}$$

or

$$R_i \simeq \frac{1}{g_m}$$

where $r_d \gg R_L$.

At the same time, the output resistance is given by

$$R_o = r_d + R_s(\mu + 1)$$

5.2 CASCODE AMPLIFIER

The two-device structure consisting of a common-emitter stage driving a common-base stage as indicated in Fig. 5.2(a) is one form of a circuit known as a *cascode amplifier*. The particular circuit shown includes resistive networks in order to bias the transistors appropriately.

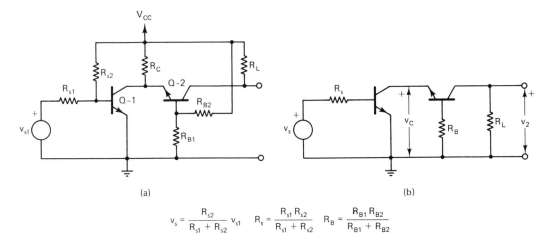

$$v_s = \frac{R_{s2}}{R_{s1} + R_{s2}} v_{s1} \qquad R_s = \frac{R_{s1} R_{s2}}{R_{s1} + R_{s2}} \qquad R_B = \frac{R_{B1} R_{B2}}{R_{B1} + R_{B2}}$$

Figure 5.2 Cascode amplifier circuits. (a) With bias network. (b) Simplified circuit for small-signal analysis.

For purposes of small-signal amplifier analysis, the circuit reduces to that of Fig. 5.7(b) for

$$R_C \gg \frac{r_{bes2} + R_B}{1 + \beta_2}$$

The overall gain is

$$A_v = \frac{v_c}{v_s}\frac{v_2}{v_c} = \frac{v_2}{v_s} \qquad (5.10)$$

for $R_{S2} \gg R_S$, which from Eqs. 5.2 and 5.7, along with Eq. 5.8, is

$$A_v = \frac{-\beta_1[(r_{bes2} + R_B)/(1 + \beta_2)]}{R_s + r_{bes1}} \cdot \frac{[\beta_2/(1 + \beta_2)]R_L}{[(r_{bes2} + R_B)/(1 + \beta_2)]} \qquad (5.11)$$

or

$$A_v = \frac{-\beta_1[\beta_2/(1 + \beta_2)]R_L}{R_s + r_{bes1}} \qquad (5.12)$$

Also,

$$R_i \simeq r_{bes1} \quad \text{and} \quad R_o \simeq (1 + \beta_2)r_{ceo2} \quad \text{for } R_B \ll r_{ceo1}$$

The actual overall gain taking the input network into account is

$$\frac{v_2}{v_{s1}} = \frac{R_{s1} + R_{s2}}{R_{s2}} A_v$$

Equation 5.12 shows that the gain of the cascode stage is essentially that of the common-emitter amplifier since the current gain of the common-base stage is near unity. Equation 5.11 shows that the overall gain is independent of the common resistance in the base lead. However, the fraction of the overall gain attributable to the common-emitter stage increases with R_B because $(r_{bes2} + R_b)/(1 + \beta_2)$ is the load resistance of that stage.

One important use of the cascode stage is to reduce the effective capacitance loading at the input of Q-1 as a result of its collector-base capacitance (Miller effect). Reduction of the voltage gain of Q-1 decreases this capacitance; hence R_B should be as low as possible for this consideration.

The circuit of Fig. 5.2(a) causes a dc level shift upward from input to output. This "offset" can be eliminated by using the *npn–pnp* combination shown in Fig. 5.3(a) to produce either a negative or zero offset output voltage. The modification shown in Fig. 5.3(b) can be used at very small signal levels, with the transistors operating in the vicinity of $v_{CB} = 0$.

An enhancement-mode IGFET in a common-source connection may be combined with a common-base BJT to form a cascode stage with the features of very high input impedance, very low feedback capacitance loading of the driving source, and both input and output referenced to the same level (zero offset). Such a circuit is shown in Fig. 5.4(a) using a *p*-channel FET with a *pnp* BJT. The overall voltage gain for

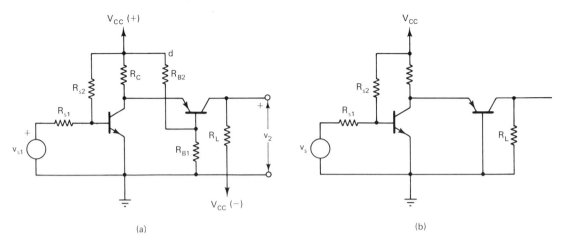

(a)

(b)

Figure 5.3 Cascode amplifier using *npn–pnp* transistor. (a) For dc level shifting. (b) For low-voltage operation.

$$r_{d1} >> \frac{r_{bes2} + R_B}{1 + \beta_2}$$

is

$$A_v = -g_{m1} \frac{r_{bes2} + R_B}{1 + \beta_2} \frac{[\beta_2/(1 + \beta_2)]R_L}{[(r_{bes2} + R_B)/(1 + \beta_2)]} \tag{5.13}$$

or

$$A_v = -g_{m1} \frac{\beta_2}{1 + \beta_2} R_L$$

Cascode stages may also be composed entirely of FETs. A combination of *p*- and *n*-channel devices permits design for specific offset requirements as indicated in Fig.

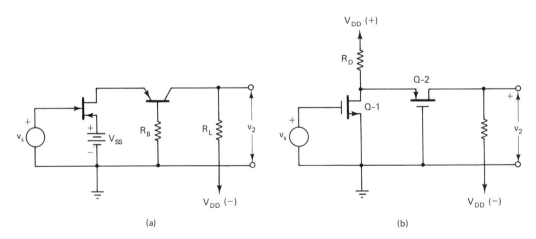

(a)

(b)

Figure 5.4 Cascode amplifier using FETs. (a) FET–BJT pair. (b) Two-MOSFET amplifier.

5.4(b), which combines an n-channel enhancement–depletion mode device with a p-channel enhancement-mode device to permit zero offset between input and output. For such a circuit, the overall voltage gain is

$$\frac{v_2}{v_s} = \frac{v_{d1}}{v_s} \frac{v_2}{v_{d1}}$$

Regardless of the particular configuration and assuming only that

$$R_D >> \frac{r_{d2} + R_L}{\mu_2 + 1}$$

the overall gain is

$$\frac{v_2}{v_s} = \frac{-\mu_1[(r_{d2} + R_L)/(\mu_2 + 1)]}{R_{d1} + (r_{d2} + 1)/(\mu_2 + 1)} \cdot \frac{(\mu_2 + 1)R_L}{r_{d2} + R_L} \tag{5.14}$$

$$= \frac{-\mu_1 R_L}{r_{d1} + \dfrac{r_{d2} + R_L}{\mu_2 + 1}}$$

or

$$\frac{v_2}{v_s} \simeq -g_{m1}R_L \tag{5.15}$$

for

$$r_{d1} >> \frac{r_{d2} + R_L}{\mu_2 + 1}$$

Frequency Response of Cascode Amplifiers

The general nature of the frequency response of the cascode amplifier may be deduced from the general equations of the response of inverting and noninverting amplifiers discussed in Secs. 4.12 and 4.13. For example, consider the FET cascode amplifier of Fig. 5.4 with all internal and external capacitances as indicated in Fig. 5.5.

For this circuit, Q-1 operates in its approximate inverting voltage follower mode since its load consists principally of the input impedance of the common-gate stage, which at low frequencies is approximately $R_{i2} \simeq 1/g_{m2}$; hence, the low-frequency gain of Q-1, $A_1 = V_x/V_i \simeq g_{m1}/g_{m2} \simeq 1$ for $g_{m1} = g_{m2}$. Therefore, if the cascode amplifier is to have substantial gain, it is provided by the gain of Q-2 with sufficiently large R_L such that $A = A_1 A_2 \simeq g_{m1} R_L'$, even for $g_{m1} \neq g_{m2}$. It is in the context of these assumptions that the overall frequency response is considered.

Although the input admittance of Q-1 is a complex function of frequency as discussed in Sec. 4.12, at relatively low frequencies it can be shown to be essentially capacitive, with its feedback component given approximately, from Eq. 4.107, by

$$C_{i1F} \simeq C_{gd1}\left(1 + \frac{g_{m1}}{g_{m2}}\right) \tag{5.16}$$

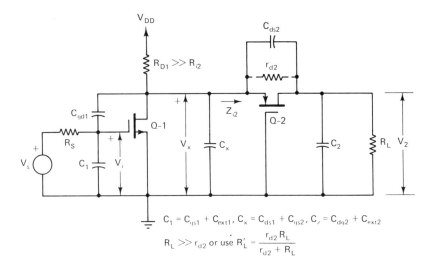

Figure 5.5 FET cascode amplifier including capacitive elements.

This equation assumes that the conductive component (the first term of Eq. 4.122) is low relative to $1/R_s$, and that ω_2' is high relative to other frequency dependencies because $R_2' \simeq \dfrac{1}{g_{m2}}$ is very low. Under these conditions, the response of the input circuit is given approximately by

$$\frac{V_i}{V_s} \simeq \frac{1}{1 + j\omega R_s[C_1 + C_{gd1}(1 + g_{m1}/g_{m2})]} \tag{5.17}$$

The frequency response of Q-1 given by $A_1 = V_x/V_i$, modeled by an equation of the form of Eq. 4.96, depends on the input admittance of Q-2, which can be shown to be that of a parallel RC circuit if the condition suggested by Eq. 4.125 is met. Even if the condition is not met precisely, but R_F is sufficiently high, the feedback component of input admittance of Q-2 is given approximately by

$$Y_{iF2} \simeq g_{m2} + j\omega C_{ds2} \tag{5.18}$$

which is an approximate form of Eq. 4.125.

Then the frequency response of Q-1 is approximately

$$A_1 = \frac{V_x}{V_i} \simeq -\frac{g_{m1}}{g_{m2}} \frac{1 - j\omega(C_{gd1}/g_{m1})}{1 + j\omega[(C_x + C_{ds2})/g_{m2}]} \tag{5.19}$$

The magnitude of this gain is relatively independent of frequency and would be exactly independent of frequency by making

$$\frac{g_{m1}}{g_{m2}} = \frac{C_{gd1}}{C_x + C_{ds2}}$$

However, the amplifier exhibits excess phase shift as discussed in Sec. 4.12, the importance of which can best be determined after considering the response of Q-2, which is $A_2 = V_2/V_x$, given by Eq. 4.123 for large R_F written as

$$A_2 = \frac{V_2}{V_x} \simeq \frac{g_{m2}R_L[1 + j\omega(C_{ds2}/g_{m2})]}{1 + j\omega(C_2 + C_{ds2})R_L} \tag{5.20}$$

The overall response, V_2/V_s, is the product of Eqs. 5.17, 5.19, and 5.20:

$$\frac{V_2}{V_s} = \frac{V_i}{V_s}\frac{V_x}{V_i}\frac{V_2}{V_x} \tag{5.21}$$

To illustrate the relative importance of individual pole and zero locations, this product can be rearranged in the form

$$\frac{V_2}{V_s} \simeq \frac{-g_{m1}R_L}{\left\{1 + j\omega R_s\left[C_1 + C_{gd1}\left(1 + \frac{g_{m1}}{g_{m2}}\right)\right]\right\}[1 + j\omega(C_2 + C_{ds2})R_L]}$$

$$\times \frac{\left(1 - j\omega\frac{C_{gd1}}{g_{m1}}\right)\left(1 + j\omega\frac{C_{ds2}}{g_{m2}}\right)}{1 + j\omega[(C_x + C_{ds2})/g_{m2}]} \tag{5.22}$$

Where substantial gain is to be provided and where the input source and output loading time constants are somewhat comparable, it should be noted that

$$(C_2 + C_{ds2})R_L >> \left(\frac{C_{gd1}}{g_{m1}}, \frac{C_{ds2}}{g_{m2}}, \text{ or } \frac{C_x + C_{ds2}}{g_{m2}}\right)$$

and the response is essentially that of the input circuit cascaded with the output. That is,

$$\frac{V_2}{V_s} \simeq \frac{-g_{m1}R_L}{\left\{1 + j\omega R_s\left[C_1 + C_{gd1}\left(1 + \frac{g_{m1}}{g_{m2}}\right)\right]\right\}[1 + j\omega(C_2 + C_{ds2})R_L]} \tag{5.23}$$

Thus, in comparing the cascode stage with an overall gain comparable to that of the single inverting amplifier illustrated by the example of Sec. 4.12, the overall bandwidth will be found to be substantially greater because it is not limited by the excessively large frequency-dependent input admittance resulting from the feedback of a high-gain inverting amplifier, inasmuch as the inverting amplifier is operating at very low gain.

Therefore, a major advantage of the cascode stage over a single inverting amplifier is the improvement in overall gain–bandwidth product by eliminating excessive input loading due to the high-gain inverting amplifier.

For a more accurate solution, all terms of Eq. 5.22 must be considered, in particular the excess phase term $[1 - j\omega(C_{gd1}/g_{m1})]$. However, through integrated-circuit process techniques, control of relative values of g_{m1} and g_{m2}, and the various capacitances, the response can be simplified almost to that of the single-pole response, particularly if the amplifier is driven from a source having a very low value of R_s.

5.3 COMPOSITE VOLTAGE FOLLOWERS

The basic action of an emitter follower can be improved by the addition of another transistor, as shown in Fig. 5.6. Such an improved circuit is often simply referred to as a voltage follower. Two alternative viewpoints can be used to describe the performance of this circuit:

1. The emitter follower, Q-1, also provides an inverted output from its collector, a fraction of which is applied to another inverter that provides additional output of the correct polarity to reinforce the output of the emitter follower alone.
2. Transistors Q-1 and Q-2 constitute a two-stage common-emitter amplifier with negative feedback applied from the output to the emitter of the input. Such feedback is negative voltage feedback, which decreases the output impedance and reduces the gain compared to that with no feedback. It has more overall gain than the emitter follower, but a lower output impedance because more of the increased gain is fed back to the input.

If Q-2 in the circuit of Fig. 5.6 is replaced by a *pnp* transistor, the circuit may be coupled directly and still be properly biased, as shown in Fig. 5.7(a). As a practical matter, there is some difference in the behavior of the two circuits. In Fig. 5.6, the current in one transistor is increasing, while that in the other is decreasing. This improves the linearity somewhat, because the nonlinearities of the two devices tend to cancel each other. In the circuit of Fig. 5.7(a), the currents in the two devices increase or decrease together and there is no self-compensation.

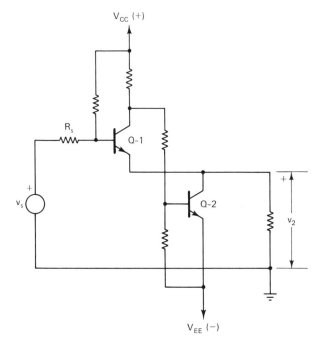

Figure 5.6 Dual BJT voltage follower.

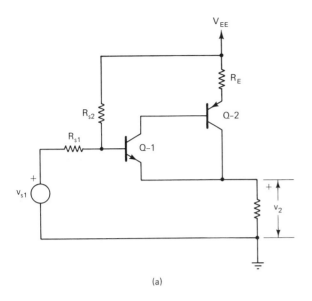

(a)

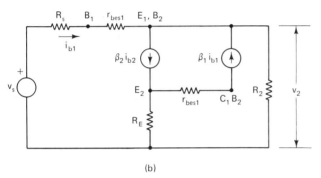

(b)

Figure 5.7 Direct-coupled *npn–pnp* pair voltage follower. (a) Actual circuit. (b) Equivalent circuit.

Because of the compound interconnection of the two devices, the composite voltage follower cannot be readily analyzed using the simple equations of input impedance, voltage gain, and output impedance. Instead, the simplified *h*-parameter models for each device are inserted, as indicated in Fig. 5.7(b), and the complete circuit analyzed. The results are

$$R_i = r_{bes1} + [1 + \beta_1(1 + \beta_2)]R_L \tag{5.24}$$

$$A_v = \frac{1}{1 + (R_s + r_{bes1})/R_L\,[1 + \beta_1(1 + \beta_2)]} \tag{5.25}$$

$$R_o = \frac{R_s + r_{bes1}}{1 + \beta_1(1 + \beta_2)} \tag{5.26}$$

The compound voltage follower has the exact characteristics of the single emitter follower with an effective current gain of

$$\beta_{\text{eff}} = \beta_1(1 + \beta_2)$$

With this substitution, the emitter-follower equations hold.

These equations appear to be independent of R_E. This arises because the approximate model neglecting r_{ceo1} was used. This would not hold unless

$$r_{ceo1} \gg r_{bes2} + R_E(1 + \beta_2)$$

It should also be observed that in Fig. 5.7(a) the base current for Q-2 is the collector current for Q-1. Hence Q-1 operates at a collector current less than that of Q-2 by the factor β_2. This might result in a reduced β for Q-1. However, because of this, the input resistance, r_{bes1}, is relatively high, and the composite structure becomes more that of a natural voltage amplifier rather than a current amplifier for $R_s \ll r_{bes1}$.

A compound voltage follower using IGFETs is shown in Fig. 5.8(a). With the simple biasing circuit shown, Q-1 would be a depletion-mode and Q-2 an enhancement-mode device.

Using the simple current generator model for $r_{d1} \gg R_D$ and $r_{d2} \gg R_L$, as indicated in Fig. 5.8(b), the overall voltage gain may be expressed as

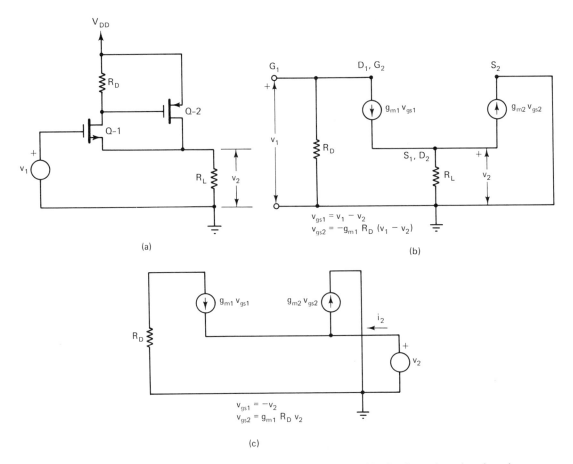

Figure 5.8 Compound voltage follower using FETs. (a) Circuit using n-channel, p-channel combination. (b) Equivalent circuit. (c) Equivalent circuit for output impedance.

$$\frac{v_2}{v_1} = \frac{1}{1 + 1/[g_{m1} R_L(1 + g_{m2} R_D)]} \tag{5.27}$$

This compares with a value of $1/[1 + (1/g_m R_2)]$ obtained from Eq. 4.51 for the simple source follower, indicating that the additional current gain provided by Q-2 improves the voltage follower action.

The output resistance from the circuit of Fig. 5.8(c) is

$$R_o = \frac{1}{g_{m1}[1 + g_{m2}R_D]} \tag{5.28}$$

which is less than $1/g_m$ for the source follower.

5.4 DARLINGTON AMPLIFIER

The compound structure in Fig. 5.9(a) is known as the Darlington connection and can be used as a composite common-emitter, common-collector, or common-base amplifier. If it is used as a common-collector amplifier with the collectors connected to a fixed supply voltage, as shown in Fig. 5.9(b), it is simply two cascaded emitter followers, and Eqs. 5.4, 5.5, and 5.6 may be used to establish its external characteristics almost by inspection. The results are

$$R_i = r_{bes1} + (1 + \beta_1)[r_{bes2} + (1 + \beta_2)R_L] \tag{5.29}$$

$$R_o = \frac{r_{bes2} + [(R_s + r_{bes1})/(1 + \beta_1)]}{1 + \beta_2} \tag{5.30}$$

$$A_v = \frac{v_e}{v_s}\frac{v_2}{v_e} = \frac{1}{1 + \dfrac{R_s + r_{bes1}}{(1 + \beta_1)[r_{bes2} + (1 + \beta_2)R_L]}} \cdot \frac{1}{1 + \dfrac{r_{bes2}}{(1 + \beta_2)R_L}} \tag{5.31}$$

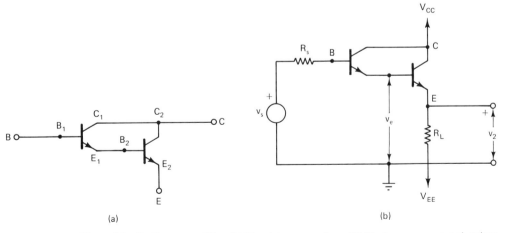

(a)

(b)

Figure 5.9 Darlington amplifier. (a) Transistor connections. (b) Biasing as a compound emitter follower.

A circuit of this type is used primarily to increase the input impedance and decrease the output impedance to a greater extent than for a single emitter follower. The equations are similar, but not identical, to those of Sec. 5.3 for the composite voltage follower.

When the Darlington configuration is used as an emitter follower but with some resistance in the collector circuit, or when it is used in the common-emitter or common-base connection, the separate device (cascade) analysis becomes very cumbersome, and it is more useful to work with the composite common-emitter h parameters derived from the h parameters of each device, as indicated in the equivalent circuit of Fig. 5.10.

The solutions for the composite h parameters are

$$h_{ie}^d = h_{ie1} + (1 - h_{re1}) \frac{h_{ie2}(1 + h_{ie2})}{1 + h_{ie2}h_{oe1}} \tag{5.32}$$

$$h_{fe}^d = \frac{h_{fe1} + h_{fe2}(1 + h_{fe1}) - h_{ie2}h_{oe1}}{1 + h_{ie2}h_{oe1}} \tag{5.33}$$

$$h_{oe}^d = h_{oe2} + \frac{h_{oe2}(1 + h_{fe2})(1 - h_{re2})}{1 + h_{ie2}h_{oe1}} \tag{5.34}$$

$$h_{re}^d = \frac{h_{re1} + h_{re2} + h_{ie2}h_{oe1}}{1 + h_{ie2}h_{oe1}} \tag{5.35}$$

In all cases, $h_{re} \ll 1$ and $h_{oe}h_{ie} \ll 1$ for each device, which allows considerable simplification of the preceding equations. In particular, the low-frequency parameters reduce approximately to

$$r_{bes}^d \simeq r_{bes1} + (1 + \beta_1)r_{bes2} \tag{5.36}$$

$$\beta^d \simeq \beta_1 + \beta_2 + \beta_1\beta_2 \tag{5.37}$$

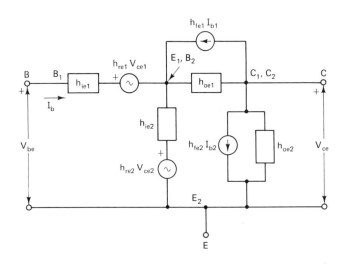

Figure 5.10 Complete h-parameter circuit model for Darlington amplifier.

$$r_{ceo}^d \simeq \frac{r_{ceo1}}{1 + \beta_2} \tag{5.38}$$

$$\mu_{re}^d \simeq \mu_{re1} + \mu_{re2} + \frac{r_{bes2}}{r_{ceo1}} \simeq \frac{r_{bes2}}{r_{ceo1}} \tag{5.39}$$

Examination of these parameters is a reminder that the input resistance of the Darlington common-emitter connection is greater by at least the factor $(1 + \beta_1)$ than that of the single device; its output resistance is lower by approximately the same amount, and its μ_{re} is usually higher because of the nature of Eq. 5.39. With these differences in mind, the Darlington composite common-emitter amplifier can usually be used to replace the single device in each of its usual circuit configurations, usually with improved performance, except that more care must be taken to be sure that the composite μ_{re} can be neglected in the analysis.

Comments on DC Operating Conditions and Biasing

If a Darlington amplifier is fabricated with two similar devices and in such a way that only terminals B, C, and E are accessible, the input section is forced to operate at a much lower collector current level than that of the output section. Then, for the output operating at relatively low current levels, the input stage may be at such a low level that its current gain is relatively low. However, this situation can be improved if the internal point is available by using the resistor R_{B2} as shown in Fig. 5.11. The bias current for Q-1 can then be increased markedly while bypassing very little signal current.

On the other hand, with very careful internal device design, β_1 can be held high even at quite low currents, while the reduced current allows even higher input resistances. Thus the unequal-current Darlington pair can be made to more closely resemble a voltage-controlled device with very high input impedance and low output impedance.

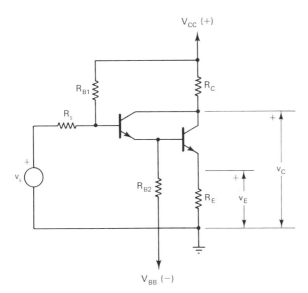

Figure 5.11 Darlington amplifier with biasing network.

5.5 SINGLE-INPUT, SINGLE-OUTPUT EMITTER-COUPLED AMPLIFIER

The emitter follower driving a common-base amplifier, as shown in Fig. 5.12, is usually referred to as an emitter-coupled amplifier. The overall voltage gain can be expressed as

$$A_v = \frac{v_2}{v_s} = \frac{v_e}{v_s}\frac{v_2}{v_e}$$

which is, by combining Eqs. 5.5 and 5.8,

$$A_v = \frac{1}{1 + (R_s + r_{bes1})/[(1 + \beta_1)R'_E]} \cdot \frac{[\beta_2/(1 + \beta_2)]R_L}{(r_{bes2} + R_{B2})/(1 + \beta_2)} \tag{5.40}$$

where

$$R'_E = \frac{R_E[(r_{bes2} + R_{B2})/(1 + \beta_2)]}{R_E + [(r_{bes2} + R_{B2})/(1 + \beta_2)]}$$

If the emitter resistance R_E is very large compared with the input resistance of the common-base stage, that is, for $R_E \gg (r_{bes2} + R_{B2})/(1 + \beta_2)$,

$$A_v \simeq \frac{1}{1 + (R_s + r_{bes1})\Big/\left[(1 + \beta_1)\dfrac{R_{B2} + r_{bes2}}{1 + \beta_2}\right]} \cdot \frac{\beta_2 R_L}{r_{bes2} + R_{B2}} \tag{5.41}$$

For matched transistors,

$$\beta = \beta_1 = \beta_2 \text{ and } r_{bes} = r_{bes1} + r_{bes2}, \tag{5.42}$$

$$A_v = \frac{\beta R_L}{2r_{bes} + R_s + R_{B2}} \tag{5.43}$$

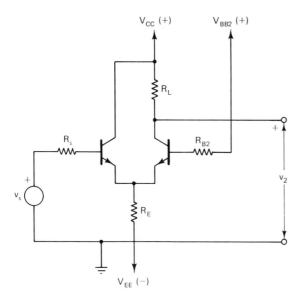

Figure 5.12 Basic emitter-coupled amplifier.

The gain is maximized for $R_{B2} = 0$; then for $R_{B2} = R_s$

$$A_v = \frac{\beta R_L}{2(r_{bes} + R_s)}$$

5.6 EMITTER-COUPLED DIFFERENTIAL AMPLIFIER PAIR

The emitter-coupled differential amplifier is a modification of the emitter-coupled amplifier, as shown in Fig. 5.13, where input signals are applied to either or both base inputs and outputs are taken from either or both collectors. Using the output from Q-2, the amplifier functions as an emitter-coupled amplifier for signal v_{s1} and as a common-emitter amplifier with emitter degeneration for a signal applied at v_{s2}. For an output from Q-1, the roles of the input signals are reversed.

The various gain functions are defined as follows:

$$A_{21} = \left.\frac{v_{o2}}{v_{s1}}\right]_{v_{s2}=0} \tag{5.44}$$

$$A_{22} = \left.\frac{v_{o2}}{v_{s2}}\right]_{v_{s1}=0} \tag{5.45}$$

$$A_{12} = \left.\frac{v_{o1}}{v_{s2}}\right]_{v_{s1}=0} \tag{5.46}$$

$$A_{11} = \left.\frac{v_{o1}}{v_{s1}}\right]_{v_{s2}=0} \tag{5.47}$$

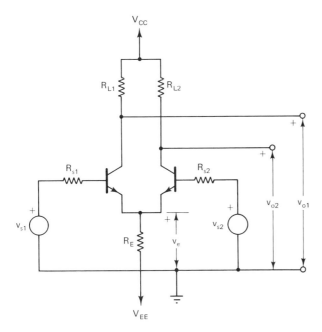

Figure 5.13 Emitter-coupled differential amplifier with two inputs and two outputs.

Ideally, the amplifier would have complete symmetry; that is, it would be desirable for many applications for $A_{21} = A_{12} = -A_{11} = -A_{22}$.

However, to assess the basic properties of the differential amplifier, it is convenient first to assume load conditions such that the approximate gain equations neglecting internal feedback can be used; then

$$A_{21} = \frac{v_e}{v_{s1}} \frac{v_{o2}}{v_e} = \frac{1}{1 + \dfrac{(R_{s1} + r_{bes1})}{(1 + \beta_1)R'_{E2}}} \cdot \frac{[\beta_2/(1 + \beta_2)] R_{L2}}{(r_{bes2} + R_{s2})/(1 + \beta_2)} \tag{5.48}$$

$$A_{12} = \frac{v_e}{v_{s2}} \frac{v_{o1}}{v_e} = \frac{1}{1 + \{(R_{s2} + r_{bes2})/[(1 + \beta_2)R'_{E1}]\}} \cdot \frac{[\beta_1/(1 + \beta_1)] R_{L1}}{(r_{bes1} + R_{s1})/(1 + \beta_1)} \tag{5.49}$$

$$A_{22} = -\frac{1}{1 + [R'_{E1}(1 + \beta_2)/(R_{s2} + r_{bes2})]} \cdot \frac{\beta_2 R_{L2}}{R_{s2} + r_{bes2}} \tag{5.50}$$

$$A_{11} = -\frac{1}{1 + [R'_{E2}(1 + \beta_1)](R_{s1} + r_{bes1})} \cdot \frac{\beta_1 R_{L1}}{R_{s1} + r_{bes1}} \tag{5.51}$$

where

$$R'_{E2} = \frac{R_E [(r_{bes2} + R_{s2})/(1 + \beta_2)]}{R_E + [(r_{bes2} + R_{s2})/(1 + \beta_2)]}$$

$$R'_{E1} = \frac{R_E [(r_{bes1} + R_{s1})/(1 + \beta_1)]}{R_E + [(r_{bes1} + R_{s1})/(1 + \beta_1)]}$$

The amplifier would be a true difference amplifier if the conditions

$$v_{o2} = K_1 (v_{s1} - v_{s2})$$

$$v_{o1} = K_2 (v_{s2} - v_{s1})$$

are met. Then also the differential output is a true difference; that is,

$$v_{o2} - v_{o1} = (K_1 + K_2)(v_{s1} - v_{s2})$$

However, if the amplifier is to be perfectly balanced; that is, $v_{o2} = v_{o2}$ for $(v_{s1} - v_{s2}) = (v_{s2} - v_{s1})$, then

$$K = K_1 = K_2$$

Examination of the gain equations indicates what degree of balance or other conditions are required for any or all of the preceding conditions to hold. For example, looking at v_{o2} using Eqs. 5.48 and 5.50, the amplifier is a true difference amplifier only if

$$\frac{R_{s1} + r_{bes1}}{(1 + \beta_1)R'_{E2}} = \frac{R'_{E1} (1 + \beta_2)}{R_{s2} + r_{bes2}}$$

This requires that

$$R'_{E2} = \frac{R_{s2} + r_{bes2}}{1 + \beta_2}$$

and

$$R'_{E1} = \frac{R_{s1} + r_{bes1}}{1 + \beta_1}$$

which in turn requires that $R_E \to \infty$.

Thus *the amplifier is a true difference amplifier for a single output only if R_E in the emitter circuit is very large compared with the effective input resistance of each common-base stage.* It does not otherwise depend upon matched conditions. However, for a differential output,

$$v_{o2} - v_{o1} = (A_{21} - A_{11})v_{s1} - (A_{12} - A_{22})v_{s2}$$

Examination of the gain equations shows that the differential output identifies a true difference amplifier only if R_E is very large *or* if the two sides are perfectly matched with respect to R_s, r_{bes}, β, and R_L.

5.7 EFFECTS OF INTERNAL DEVICE FEEDBACK ON BALANCE

The balance conditions just stated were determined under the assumption that internal device feedback through the $\mu\beta$ terms could be neglected. It remains to determine what residual unbalance may exist in cases where that assumption is not valid. This can be done starting with the approximate characteristic amplifier equations from Chapter 4, with some further minor modifications as follows:

1. *Common-emitter amplifier:*

$$R_i = r_{bes} + (1 + \beta)R_E - \mu_{re}\beta R_L, \qquad \text{from Eq. 4.15} \qquad (5.52)$$

$$A_v = \frac{-\beta R_L}{R_s + r_{bes} + (1 + \beta)R_E - \mu_{re}\beta R_L}, \qquad \text{from Eq. 4.17} \qquad (5.53)$$

$$R_o = r_{ceo}\left(1 + \frac{R_E(1 + \beta)}{R_s + r_{bes}}\right), \qquad \text{Eq. 4.16 modified} \qquad (5.54)$$

The $\mu_{re}\beta$ term is relatively unimportant in Eq. 4.16.

2. *Common-collector amplifier:*

$$R_1 = r_{bes} + (1 + \beta)R_L - \mu_{re}\beta R_c, \qquad \text{from Eq. 4.30} \qquad (5.55)$$

$$A_v = \frac{1}{1 + [(R_s + r_{bes})/(1 + \beta)R_L]}, \qquad \text{Eq. 4.32 modified} \qquad (5.56)$$

The $\mu_{re}\beta$ term is relatively unimportant in Eq. 4.32.

$$R_o = \frac{R_s + r_{bes}}{1 + \beta}, \qquad \text{Eq. 4.31 modified} \qquad (5.57)$$

The internal feedback path through the collector is relatively unimportant.

3. *Common-base amplifier:*

$$R_i = \frac{r_{bes} + R_B + \mu_{rb}\beta R_L}{1 + \beta}, \qquad \text{from Eq. 4.39} \tag{5.58}$$

$$A_v = \frac{[(\beta/(1 + \beta)]R_L}{R_s + [(r_{bes} + R_B)/(1 + \beta)] + \mu_{rb}[\beta/(1 + \beta)]R_L}, \qquad \text{from Eq. 4.41} \tag{5.59}$$

$$R_o = (1 + \beta)r_{ceo}, \qquad \text{from Eq. 4.40} \tag{5.60}$$

and neglecting the term containing $\mu_{rb}\beta$.

Retaining the approximation of $R_E \rightarrow \infty$, but using the preceding single-device expressions, the differential amplifier gain equations can be expressed as

$$A_{21} = \frac{\beta_2 R_{L2}}{r_{bes2} + R_{s2}} \frac{1}{1 + \dfrac{1 + \beta_2}{1 + \beta_1} \dfrac{R_{s1} + r_{bes1}}{R_{s2} + r_{bes2} + \mu_{rb2}\beta_2 R_{L2}}} \frac{1}{1 + \dfrac{\mu_{rb2}\beta_2 R_{L2}}{r_{bes2} + R_{s2}}} \tag{5.61}$$

$$A_{22} = \frac{-\beta_2 R_{L2}}{r_{bes2} + R_{s2}} \frac{1}{1 + \dfrac{(1 + \beta_2)(R_{s1} + r_{bes1} + \mu_{rb1}\beta_1 R_{L1} - \mu_{re2}\beta_2 R_{L2})}{(1 + \beta_1)(R_{s2} + r_{bes2})}} \tag{5.62}$$

Inasmuch as $v_{o2} = A_{21}v_{s1} + A_{22}v_{s2}$ and from the preceding equations $A_{21} \neq A_{22}$ because of the internal feedback terms, the amplifier for a single output is not a perfect differential amplifier even for matched devices and components. However, examination of the μ_{re} terms shows that the unbalance may be quite small. By symmetry, equations similar to Eqs. 5.61 and 5.62 may be written for A_{12} and A_{11} simply by subscript interchanges.

Since the differential output is

$$v_{o2} - v_{o1} = (A_{21} - A_{11})v_{s1} - (A_{12} - A_{22})v_{s2}$$

it would be a true difference amplifier if perfectly matched devices and components were used (i.e., if $A_{21} = A_{12}$ and $A_{11} = A_{22}$). Otherwise, there would be a slight unbalance even for the differential output.

5.8 HIGH-IMPEDANCE (CURRENT) SOURCE FOR COMMON EMITTERS

The high impedance in the common-emitter lead ($R_E \rightarrow \infty$) required to achieve balance in the differential amplifier without at the same time requiring high emitter bias voltage supplies (because of the large voltage drop across a large R_E) can be achieved by utilizing the collector–emitter circuit of another transistor, as indicated in Fig. 5.14(a).

The effective incremental resistance as seen from the common emitters is given by Eq. 4.17, which can be made to be very large even for very small values of R_E. This can be achieved with a voltage drop across Q-3 on the order of $v_{CE} = v_{BE}$. Improved current sources using Q-3 in conjunction with other devices will be discussed in a later section. Any such circuit which permits the ideal high-impedance (current source) approximations is generally represented by a current source as shown in Fig. 5.14(b).

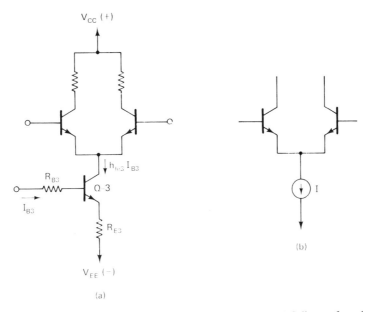

Figure 5.14 Differential amplifier with current source biasing. (a) Collector of transistor as current source. (b) Symbolic representation of current source.

5.9 DIFFERENTIAL-CASCODE AMPLIFIER

If a signal is applied to the base–emitter circuit of the transistor in the common-emitter circuit of the differential amplifier, as indicated in Fig. 5.15, the circuit becomes a cascode stage with respect to the Q-3 input, with Q-3 being loaded by both common-base stages with outputs being available from either collector of the differential pair, Q-1 and Q-2.

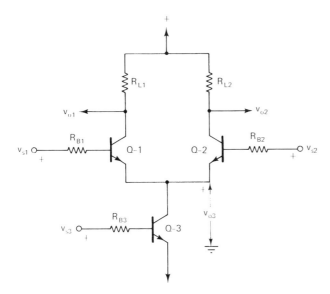

Figure 5.15 Composite differential-cascode amplifier.

If Q-1 and Q-2 are perfectly balanced with $r_{bes} = r_{bes1} = r_{bes2}$, $R_B = R_{B1} = R_{B2}$, and $\beta = \beta_1 = \beta_2$, then the overall gain is given approximately by

$$\frac{v_{o2}}{v_{s3}} = \frac{v_{o3}}{v_{s3}} \frac{v_{o2}}{v_{o3}} = \frac{-\beta_3 \, [(r_{bes} + R_B)/2(1 + \beta)]}{R_{B3} + r_{bes3}} \cdot \frac{[\beta/(1 + \beta)] \, R_{L2}}{R_B + r_{bes}}$$

or   (5.63)

$$\frac{v_{o2}}{v_{s3}} = \frac{-\beta_3 R_L}{2(r_{bes3} + R_{B3})} \cdot \frac{\beta}{1 + \beta}$$ (5.64)

If $R_{B3} = R_B$ and $r_{bes3} = r_{bes}$,

$$\frac{v_{o2}}{v_{s3}} = \frac{-\beta \, [\beta/(1 + \beta)]R_L}{2(R_B + r_{bes})}$$ (5.65)

It may be observed that the magnitude of this gain is exactly that of the emitter-coupled section except for the factor $\beta/(1 + \beta)$.

5.10 COMPOSITE-COMPOUND BJT AMPLIFIERS AT LOW CURRENT LEVELS: ANOTHER VIEWPOINT

In this chapter and the preceding, BJT amplifiers have been analyzed on the basis of current gain mode operation because of their inherently low input impedance and the relatively linear current gain function. However, at increasingly lower collector current levels, it becomes practical to consider characterization of the device by its transconductance, g_m, rather than its current gain, β, because of its increasingly large input impedance, whose diffusion component is given approximately by

$$r_d \simeq \frac{kT}{q} \frac{\beta}{I_C}$$ (5.66)

taken from Eq. 2.87. For example, for $\beta = 100$, $I_C = 0.1$ mA, and $kT/q = 0.026$ V, $r_d \simeq 26 \, k\Omega$, which is sometimes large compared to a proposed driving source impedance.

At the same time, from Eq. 2.89,

$$g_m \simeq \frac{I_C}{kT/q}$$ (5.67)

Under such conditions, the other parameters in the hybrid-π model for which g_m is derived, r_b', and r_{cb}', may often be ignored and the transistor characterized by its g_m and r_{ce} alone.

In particular, the differential amplifier often is used as the input stage of a multistage amplifier and is operated at very low current levels in order to achieve a very high input impedance. Or the input resistance of the Darlington amplifier, which is high normally, is made even higher because its input section operates at very low levels.

The voltage-amplifier mode of differential BJT amplifiers becomes analogous to the g_m models to be used for FETs as discussed in the next section.

5.11 DIFFERENTIAL PAIR FET AMPLIFIER

A differential (source-coupled) amplifier using IGFETs with resistance coupling is shown in Fig. 5.16(a). The coupling resistance, R_s, is replaced by the drain circuit of another FET in Fig. 5.16(b) to make it more like current source biasing, similar to that suggested for the BJT amplifier, without being forced to an abnormally high bias voltage supply.

In either case, for the present analysis it will be assumed that

$$R_s >> \frac{r_d + R_L}{\mu + 1}$$

in order to simplify the overall equations. Then the overall gain of the source-coupled stage

$$\frac{v_{o2}}{v_1} = \frac{v_s}{v_1} \cdot \frac{v_{o2}}{v_s}$$

using Eqs. 4.58 and 4.65 along with Eq. 4.61, for the source follower is

$$\frac{v_{o2}}{v_1} = \frac{\mu_1 \left[(r_{D2} + R_{L2})/(\mu_2 + 1) \right]}{R_{d1} + R_{L1} + (\mu_1 + 1) \left[(r_{d2} + R_{L2})/(\mu_2 + 1) \right]} \cdot \frac{(\mu_2 + 1)R_{L2}}{r_{d2} + R_{L2}} \quad (5.68)$$

or

$$\frac{v_{o2}}{v_1} = \frac{\mu_1 R_{L2}}{r_{d1} + R_{L1} + \left[(\mu_1 + 1)/(\mu_2 + 1) \right] (r_{d2} + R_{L2})} \quad (5.69)$$

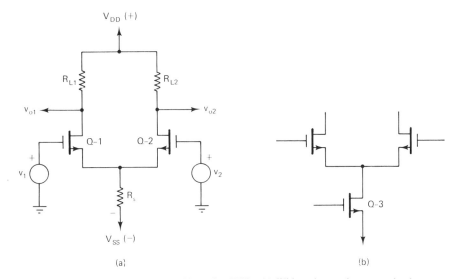

Figure 5.16 Differential amplifier using FETs. (a) With resistance in source circuit. (b) Current biasing with FET.

For matched devices and loads, Eq. 5.69 reduces to

$$\frac{v_{o2}}{v_1} = \frac{\mu R_L}{2(r_d + R_L)} \tag{5.70}$$

Then, using the $r_d \gg R_L$ approximation,

$$\frac{v_{o2}}{v_1} \simeq \frac{g_m R_L}{2} \tag{5.71}$$

There is a shortcut to the preceding result, which can be made by using the $r_d \gg R_L$ approximation from the very beginning, noting that the input resistance of the common-source stage, which is the load of the source follower, is approximately $1/g_{m2}$; hence, from Eq. 5.68,

$$\frac{v_{o2}}{v_1} \simeq \frac{g_{m1}/g_{m2}}{1 + (g_{m1}/g_{m2})} \cdot g_{m2} R_L \tag{5.72}$$

Then, for $g_{m1} \simeq g_{m2}$, the gain of the source follower is approximately $\frac{1}{2}$, and

$$\frac{v_{o2}}{v_1} \simeq +\frac{g_m R_L}{2}$$

The effect of unbalances due to an insufficient value for R_s, as well as due to unmatched devices, may be determined by using more exact forms for the device equations, as was done for the bipolar junction transistor.

Cascode-Differential-Pair FET Amplifier

If an input signal v_3 is applied to the gate of Q-3 in Fig. 5.16(b), the overall gain v_{o2}/v_3 or v_{o1}/v_3 is that of a cascode stage with the common-source stage (Q-3) loaded by the parallel inputs of two common-gate stages.

The overall gain for $\mu = \mu_1 = \mu_2$, $r_d = r_{d1} = r_{d2}$, and $g_m = g_{m1} = g_{m2}$ is

$$\frac{v_{o2}}{v_3} = \frac{v_{o1}}{v_3} = \frac{v_s}{v_3}\frac{v_{o2}}{v_s} = \frac{v_s}{v_3}\frac{v_{o1}}{v_s}$$

which is

$$\frac{v_o}{v_s} = \frac{v_{o2}}{v_s} = \frac{v_{o1}}{v_s} = \frac{-\mu_3\,[(r_d + R_L)/2(\mu + 1)]}{r_{d3} + [(r_d + R_L)/2(\mu + 1)]} \cdot \frac{(\mu + 1)R_L}{r_d + R_L}$$

or

$$\frac{v_o}{v_s} = \frac{-\mu_3 R_L}{2[r_{d3} + (r_d + R_L)/2(\mu + 1)]} \tag{5.73}$$

or in terms of g_m

$$\frac{v_o}{v_s} = \frac{-g_{m3} R_L}{2\{1 + (1/r_{d3})\,[(r_d + R_L)/2(\mu + 1)]\}} \tag{5.74}$$

Then in terms of the $r_d \gg R_L$ approximation

$$\frac{v_o}{v_s} \simeq \frac{-g_{m3}R_L}{2[1 + (1/2g_m r_{d3})]} \simeq \frac{-g_{m3}R_L}{2} \tag{5.75}$$

This is the same magnitude as that of the single output differential amplifier itself.

5.12 FREQUENCY RESPONSE OF VOLTAGE FOLLOWER DRIVING A NONINVERTING GAIN STAGE

The emitter-coupled amplifier described in Sec. 5.5 and the source-coupled amplifier described in Sec. 5.11 with $R_{L1} = 0$ and $v_2 = 0$ in Fig. 5.15 are examples of a composite amplifier, which, like the cascode amplifier, offers an alternative to the single-device inverting gain stage as a means of achieving an improved gain–bandwidth product through a reduction of the effective frequency-dependent input admittance. The circuit of Fig. 5.17 forms the basis of an approximate frequency-response analysis of such a stage.

Using the equations for the source follower of Fig. 4.26 and for the common-gate stage of Fig. 4.24 as reference, the following analysis is made:

The feedback component of the input admittance of Q-2, which, in parallel with R_{s2} constitutes the load for the source follower, can, as deduced from Eqs. 4.124 and 4.125, for large R_F be shown to be approximately

$$Y_{iF2} \simeq g_{m2} + j\omega C_{ds2} \tag{5.76}$$

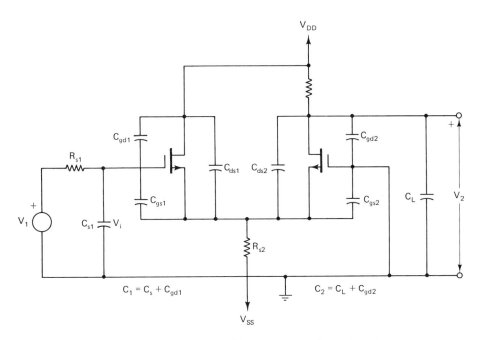

Figure 5.17 Differential amplifier showing capacitance elements.

with the total capacitive load for Q-1 being

$$C = C_{ds1} + C_{gs2} + C_{ds2}$$

From Eq. 4.126, the gain of the source follower $A_1 = V_x/V_i$ is independent of frequency if the condition of Eq. 4.129 is met; that is,

$$\frac{C_{gs1}}{C_2} = g_{m1}R_2$$

where $C_2 \simeq (C_{gs2} + C_{ds1} + C_{ds2})$.

$$R_2 = \frac{R_S/g_{m2}}{R_s + 1/g_{m2}}$$

The required value for R_{s2} from the preceding condition is

$$R_{s2} = \frac{C_{gs1}/C_2}{g_{m1}[1 - (g_{m1}/g_{m2})(C_{gs1}/C_2]}$$

Inasmuch as normally $C_{gs} \gg C_{ds}$, under most conditions R_s will be very large and in any case can be shown not to be very critical. Under conditions approximating those of the preceding analysis, the gain of Q-1 from Eq. 4.126 may be written as

$$A_1 \simeq \frac{V_x}{V_i} = \frac{1}{1 + (g_{m1}/g_{m2})} \tag{5.77}$$

which for matched devices (i.e., $g_{m1} = g_{m2}$)

$$A_1 \simeq \frac{1}{2} \tag{5.78}$$

Under the condition leading to Eq. 5.77 for A_1, the total input admittance is

$$Y_{i1} = j\omega C_1 + j\omega C_{gs1}\left[1 - \frac{1}{1 + (g_{m1}/g_{m2})}\right] \tag{5.79}$$

which for $g_{m1} = g_{m2}$ is

$$Y_i = j\omega C_1 + j\omega \frac{C_{gs1}}{2} \tag{5.80}$$

Thus the input circuit frequency response is approximately

$$\frac{V_i}{V_s} = \frac{1}{1 + j\omega R_s[C_1 + (C_{gs}/2)]} \tag{5.81}$$

The gain of Q-2, V_2/V_x of the common-gate stage, is that previously given by Eq. 5.20, which is

$$A_2 = \frac{g_{m2}R_L[1 + j\omega(C_{ds2}/g_{m2})]}{1 + j\omega(C_2 + C_{ds2})R_L} \tag{5.82}$$

for $R_L \ll r_{d2}$. Then the overall response for matched devices ($g_{m1} = g_{m2}$) is

$$\frac{V_2}{V_s} = \frac{g_{m2}R_L}{2} \frac{1 + j\omega(C_{ds2}/g_{m2})}{\{1 + j\omega R_s\ [C_1 + (C_{gs}/2)]\}\ [1 + j\omega R_L(C_2 + C_{ds2})]} \tag{5.83}$$

which can be reduced to the simple two-pole response

$$\frac{V_2}{V_s} = \frac{g_m R_L/2}{\{1 + j\omega R_s\ [C_1 + (C_{gs}/2)]\}\ [1 + j\omega(C_2 + C_{ds2})R_L]} \tag{5.84}$$

for the usual condition of

$$R_L(C_2 + C_{ds2}) \gg \frac{C_{ds2}}{g_{m2}}$$

for substantial gain to be realized.

Frequency Response Compared with That of Cascode Stage

The preceding equation for the optimized source-coupled amplifier is very similar to Eq. 5.23 for the approximation of the cascode stage, taking into account the slight difference in capacitance values. The effective input capacitance tends to be smaller since $C_{gs}/2$ in Eq. 5.84 $< C_{gd1}\ [1 + (g_{m1}/g_{m2})]$ in Eq. 5.23. However, the gain of the source-coupled stage is less by a value of approximately a factor of 2. Then, depending on the relative values of R_s and R_L, the overall gain–bandwidth product might be roughly comparable in the two cases.

However, if the more exact form of the response of the cascode stage shown in Eq. 5.22 is considered, it is seen that the excess phase term not present in the source-coupled amplifier might make it a preferable choice in some applications.

5.13 COMPLEMENTARY-PAIR TRANSISTOR AMPLIFIERS

A two-device amplifier is the complementary pair consisting of a *pnp–npn* bipolar transistor pair or complementary *n*-channel, *p*-channel FET pair operating either in a composite voltage-follower mode or in composite inverting-amplifier mode. Such an amplifier is the composite emitter follower shown conceptually in Fig. 5.18(a). The inputs are shown as two identical time-varying signals with their average (dc) values separated by the two base–emitter bias voltages if the bias is set to place Q-1 and Q-2 each in the center of its prescribed operating range; if they are not overdriven, both transistors function as emitter-follower amplifiers at all times. However, the current in one is decreasing while that in the other is increasing, and any device nonlinearities over a wide operating range tend to be self-compensating. This linearity compensation feature could be verified by graphically deriving a set of composite characteristics for the complementary pair. Such a mode of operation with both transistors conducting simultaneously is referred to as *class A* operation.

A slightly different mode of operation is shown in Fig. 5.18(b); in this case the

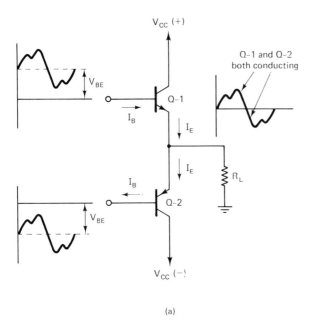

(a)

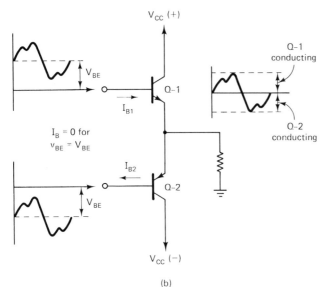

(b)

Figure 5.18 Complementary composite emitter follower. (a) With both transistors always conducting. (b) With biasing for class B operation.

biases are adjusted such that with no signal applied almost zero current flows in each transistor. Then *Q*-1 functions as an emitter follower during positive excursions of the input signal, with *Q*-2 nonconducting, and *Q*-2 functions during negative excursions with *Q*-1 off. Such operation is historically referred to as *class B* operation. As a practical matter, because of nonlinearity near zero current, pure class B operation would result in nonlinear distortion near the zero signal axis, referred to as *crossover distortion*. Such distortion can be minimized by biasing the transistors to both conduct slightly under zero signal conditions, a mode referred to as *class AB* operation.

As a practical matter, it is not necessary to provide separate offset input signals to each transistor, as indicated by the circuit of Fig. 5.19. The bias voltage, $2V_{BE}$ is provided by the divider network comprising R_1, R_2, and the two diodes with the current in the network, and hence with the voltage across the diodes adjusted for class A, class B, or class AB operation as desired. The signal is derived from a single output of a driving transistor, but appears almost equally at the two inputs because of the very small incremented resistance of the diodes, but with the desired dc voltage offset. There are many ways in which the input signal to the complementary pair can be derived. These will be discussed in much greater detail in connection with operational amplifiers in Chapter 7.

Complementary transistors can be used as composite common-emitter amplifiers as indicated in Fig. 5.20(a). In this case, the biasing is a bit more complicated because of the larger offsets required of the two input signals. One of many practical methods of providing such offsets is shown in Fig. 5.20(b), where a complementary transistor pair is used as the signal source. In this circuit the input transistor collectors provide the biasing current for the output transistors as required. The input transistors should be operated at relatively low current levels and in the class A mode.

Complementary JFETs or MOSFETs can be used as composite source followers or common-source amplifiers. A complementary almost zero threshold voltage MOSFET source-follower pair is shown in Fig. 5.21(a). With the two gates connected (no dc offset), class A, B, or AB operation can be achieved by structuring the FETs themselves for zero V_T or slightly depletion-mode operation.

For enhancement-mode FETs, where input offset voltages would be required for the two gates, the circuit of Fig. 5.21(b) is suggested. For this circuit, the signal is applied directly to the gate of Q-1 at the appropriate dc level and to the gate of Q-2 through the source follower having the required threshold voltage.

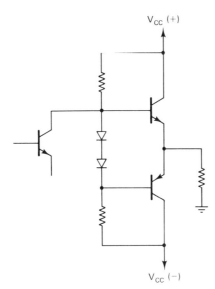

V_CC (+)

V_CC (−)

Figure 5.19 Single-input complementary emitter follower with diode input voltage biasing.

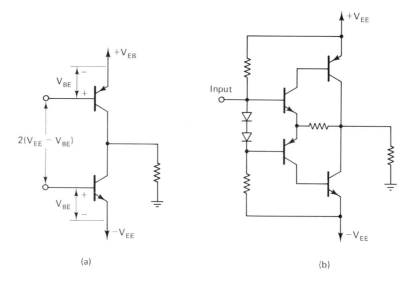

(a)   (b)

Figure 5.20 Complementary common-emitter amplifier. (a) Without input source indicated. (b) Biasing with complementary input source.

For the complementary common-source amplifier, the required offset voltage is much greater, and a circuit similar to that shown in Fig. 5.22 is suggested as one of many possibilities. For this circuit, all the FETs are enhancement-mode types, with the driven transistors operating as source followers with the desired offset voltage to drive the output transistors.

Complementary-pair transistors are most often used as power-output stages where considerable voltage and/or current variations are to be supplied to a load impedance, and therefore have signal swings through most of their dynamic range. For class A

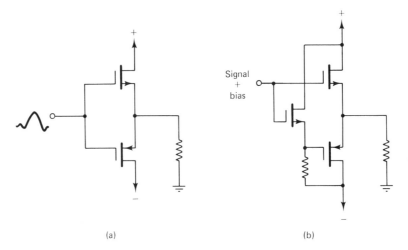

(a)   (b)

Figure 5.21 Complementary MOSFET source followers. (a) Simple structure for depletion-mode devices. (b) Biasing for enhancement-mode devices.

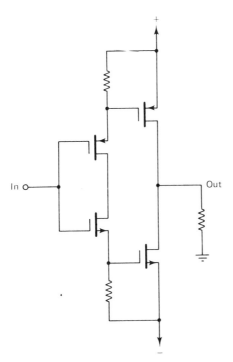

In o——

Out

Figure 5.22 Complementary common-source MOSFET amplifier with complementary input circuit.

operation, the principal merit is to provide a more stable dc output level than is possible with a single transistor and to improve linearity over a wide operating range. For class AB or class B operation, the added advantage is greater efficiency, since zero or only a very small dc current flows under zero signal conditions.

Various aspects of nonlinearity, particularly with respect to class B or class AB operation, will be discussed in Chapter 8, which treats nonlinear distortion in detail.

5.14 CURRENT SOURCES AND LOADS FOR BJTS

The concepts discussed in Secs. 5.8 and 5.11 of using the collector or drain circuits to provide simultaneously a dc voltage-controlled current bias source and a high incremental resistance at a low voltage drop have much wider applications than in the emitter- or source-coupled amplifier, and the performance of such sources in all important respects can be improved by using more elaborate composite circuits.

The basic current source that was discussed in connection with Fig. 5.14 is shown again in Fig. 5.23. In this circuit, V_{BB} and V_{EE} are dc bias voltages. The resistance R_E in the emitter circuit, if used at all, should be relatively low if the requirement at a low series voltage drop, V_C, is to be met.

The incremental output resistance R_o from Eq. 4.17, and using $r_e = \mu_{re}r_{ceo}$ from Eq. 2.76 is

$$R_o = \frac{r_{ceo} (1 + \{[R_E (1 + \beta)]/(R_B + r_{bes})\})}{1 - \{[r_e \beta/(R_B + r_{bes})] + [R_E/(R_B + r_{bes})]\}} \tag{5.85}$$

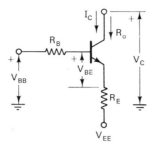

Figure 5.23 Basic BJT circuit for current source application.

With a small but finite value of $R_E \geq r_e\beta$, the output resistance is independent of internal feedback; then, for larger values of R_B with reasonably large β,

$$R_o \simeq r_{ceo} \left[1 + \frac{R_E (1 + \beta)}{R_B + r_{bes}} \right]$$

Thus R_o increases with increasing values of the R_E/R_B ratio. However, it is not desirable to make R_E large from the standpoint of required small voltage drop; nor is it desirable to make R_B small from the standpoint of operating point stabilization, as explained as follows:

The dc collector current from Fig. 5.23 can be expressed as

$$I_C = \frac{h_{FE} [(V_{BB} - V_{EE}) - V_{BE}]}{R_B + R_E (1 + h_{FE})} \tag{5.86}$$

This equation shows that making R_B small makes the collector current relatively independent of h_{FE}. However, it makes the current more dependent on v_{BE} variations, which are highly temperature dependent, and it is not desirable to make R_E too large, as explained previously. The seemingly incompatible requirements can be met by using composite circuits, as follows:

In Fig. 5.24(a), R_B of Q-1 is replaced by the diode-connected transistor and R_2 with $R_E = 0$. If the dc current gains are h_{FE1} and h_{FE2}, and the transistor emitter areas are proportioneed such that $I_{B2} = KI_{B1}$ and $(A_{E2} = KA_{E1})$, then

$$I_{C1} = \frac{h_{FE1}I_2}{1 + K(1 + h_{FE2})} \tag{5.87}$$

and in terms of V_2

$$I_{C1} = \frac{h_{FE1}}{1 + K(1 + h_{FE2})} \cdot \frac{V_{CC} - V_{BE}}{R_2} \tag{5.88}$$

For matched transistors,

$$I_{C1} = \frac{h_{FE}}{2 + h_{FE}} \frac{V_{CC} - V_{BE}}{R_2} \tag{5.89}$$

with I_C relatively independent of h_{FE}. Then if $V_{CC} >> V_{BE}$, the current is also relatively independent of V_{BE} temperature variations.

It is apparent from Eq. 5.88 that fabricating Q-2 with a larger emitter area than Q-1 (i.e., making $K > 1$) makes I_{C2} relatively larger than I_{C1} and allows a small value

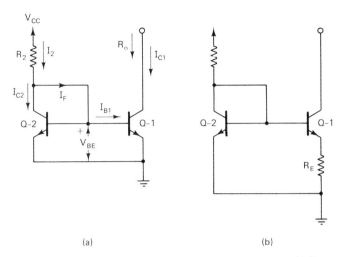

(a)   (b)

Figure 5.24 Basic current mirror circuits. (a) For balanced currents. (b) For unequal source and mirrored currents.

of R_L, which is desirable from the standpoint of integrated-circuit fabrication. Because of the approximate required inverse relationship between K and R_2, the I_{C1} dependence on h_{FE} is relatively independent of K at a specified operating current.

It is often desirable to control very small collector currents with small values of R_2 more than is possible by using devices with unequal emitter areas alone. Modification of the circuit to that of Fig. 5.24(b) accomplishes the desired result. The emitter feedback provided by R_E forces Q-1 to operate at a lower value of v_{BE} than Q-2 and thus further unequalizes the two currents, which allows further minimization of R_2.

The incremental output resistance R_o of Q-1 from Eq. 5.85 depends on the effective resistance R_B (eq) in the base load. In the case of Fig. 5.24, as opposed to Fig. 5.23, this is the effective resistance looking into Q-2 as indicated in Fig. 5.25(a), which can be determined by the equivalent circuit of Fig. 5.25(b). Neglecting the $\mu_{re} v_{be2}$ term,

$$R_i = \frac{R_2' \, r_{bes2}}{R_2' \, (1 + \beta) + r_{bes2}} \tag{5.90}$$

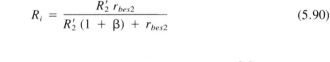

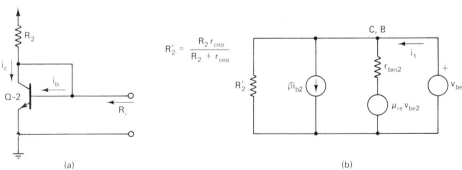

(a)   (b)

Figure 5.25 Circuit for output resistance of source side of current mirror. (a) Circuit. (b) Equivalent circuit for resistance determination.

Then for $R_2'(1 + \beta) \gg r_{bes}$,

$$R_i \simeq \frac{r_{bes2}}{1 + \beta} \tag{5.91}$$

which is the same as the input resistance of a common-base stage.

Circuits similar to that of Fig. 5.24 are frequently referred to as *current mirrors* because they force a desired current to be like or mirror another current, which can be controlled to a high degree of precision.

Various modifications of the simple current mirror offer improved independence of current on h_{FE}. For example, in the circuit of Fig. 5.26 the emitter follower Q-3 provides current gain to the base of Q-1. It can be shown that the desired collector current, I_{C1}, can be given by

$$I_{C1} = \frac{h_{FE1}(1 + h_{FE3})I_2}{1 + K + K(1 + h_{FE3})h_{FE2}} \tag{5.92}$$

where K is as previously defined. For $K = 1$ and $h_{FE} = h_{FE1} = h_{FE2} = h_{FE3}$,

$$I_{C1} = \frac{1}{1 + [2/(h_{FE} + h_{FE}^2)]} I_2 \tag{5.93}$$

These equations show a much greater independence of h_{FE} than the previous circuit.

However, a larger value of V_{CC} is necessary to provide the same independence from v_{BE} temperature variations because of the two v_{BE} voltage drops in series.

A slightly more elaborate circuit, which might be viewed as a modification of that of Fig. 5.24(b), is shown in Fig. 5.27. In this circuit the collector circuit of Q-4 provides a very high incremental emitter resistance for Q-1, which effectively increases R_o. The base of Q-4 is driven by Q-5 in a mirror arrangement to provide a very low effective R_B. At the same time, the collector circuit of Q-5, which provides the base current for $Q - 4$, provides a very low incremental emitter resistance for Q-2. From another point of view, Q-4 is a current-driven common-emitter stage driving a common-base stage Q-1, and for this reason the circuit is referred to as a *cascode current source*.

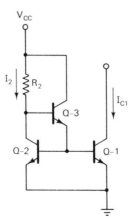

Figure 5.26 Emitter follower feedback current mirror.

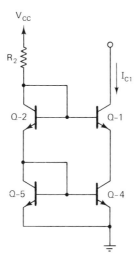

V_{CC}

R_2

$Q\text{-}2$

$Q\text{-}1$

I_{C1}

$Q\text{-}5$

$Q\text{-}4$

Figure 5.27 Cascode current mirror.

Still another version of a current mirror is shown in Fig. 5.28, which is something of a reverse twist to the foregoing circuits. The input–output current relationships of the three devices are indicated on the schematic.

The feedback current, I_F, from the circuit of Fig. 5.28 is

$$I_F = (1 + h_{FE1})(I_2 - h_{FE3}I_{B3}) - h_{FE2}I_{B2}$$

Now if Q-3 and Q-2 are matched, $I_{B3} = I_{B2} = I_B = I_F/2$, then

$$I_B = \frac{(1 + h_{FE1})I_2}{2 + h_{FE2} + h_{FE3}(1 + h_{FE1})}$$

and

$$I_{C1} = \frac{2h_{FE1} + h_{FE1}h_{FE2}}{2 + h_{FE2} + h_{FE3} + h_{FE1}h_{FE3}} I_2 \qquad (5.94)$$

or

$$I_{C1} = \frac{1}{1 + [2/(2h_{FE} + h_{FE}^2)]} I_2 \qquad (5.95)$$

for identical h_{FE}'s.

Sensitivity to Variations in Supply Voltage

The current sources (current mirrors) that have been discussed in this section have varying degrees of sensitivity to voltage-supply variation. For example, the simple and most commonly used current mirror of Fig. 5.24(a) yields a current source that is directly proportional to $V_{CC} - V_{BE}$, according to Eq. 5.89, with V_{BE} being almost constant over a wide range of currents.

A modified version of the circuit of Fig. 5.28, with Q-2 replaced by a resistor as

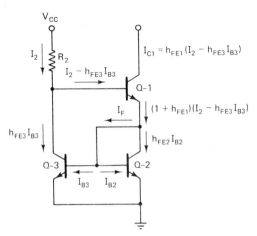

Figure 5.28 Compound current mirror.

shown in Fig. 5.29, offers improved voltage-supply sensitivity, with the current to be controlled relying on the "almost constant" V_{BE3} as a reference voltage. In this circuit, the controlled current is

$$I_{C1} = \frac{h_{FE1}(I_2 + h_{FE3}V_{BE3}/R_E)}{1 + h_{FE3}(1 + h_{FE1})} \tag{5.96}$$

which in terms of supply voltage is

$$I_{C1} = \frac{h_{FE1}\{[(V'_{CC} - (V_{BE1} + V_{BE3})]/R_2\} + (h_{FE3}V_{BE3}/R_E))}{1 + [h_{FE3}(1 + h_{FE1})]} \tag{5.97}$$

when $V'_{CC} = V_{CC} - V_{EE}$.

To illustrate relative supply-voltage dependency, a simplification assuming $h_{FE} = h_{FE1} = h_{FE3}$ and $V_{BE} = V_{BE1} = V_{BE3}$ will be considered. This reduces Eq. 5.97 to

$$I_{C1} = \frac{h_{FE}^2}{1 + h_{FE} + h_{FE}^2}\left(\frac{V_{BE}}{R_E} - \frac{2V_{BE}}{h_{FE}R_2} + \frac{V_{CC}}{h_{FE}R_2}\right) \tag{5.98}$$

Many design variations are possible, but usually the V_{BE}/R_E term dominates in the limit for large h_{FE}, with

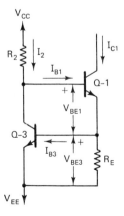

Figure 5.29 Current mirror to minimize supply voltage dependence.

$$I_{C1} \cong \frac{V_{BE}}{R_E}$$

being essentially independent of variations in V_{CC}.

Temperature Sensitivity

All the current sources are temperature dependent to varying degrees as they are to voltage, due primarily to the V_{BE} temperature dependence of the involved transistors. From Eq. 1.129 as applied to a base–emitter junction,

$$\frac{dv_{BE}}{dT}\bigg]_{v_{BE}=V_{BE}} = -\frac{1}{T_r}\left(V_{gO} - V_{BE} + \frac{nkT_r}{q}\right) - \frac{nk}{q}\ln\frac{T}{T_r} + \frac{k}{q}\ln\frac{I_E}{I_{Er}} \quad (5.99)$$

The last two terms are relatively small, resulting in the approximation

$$\frac{dv_{BE}}{dT} \cong -\frac{1}{T_r}\left(V_{gO} - V_{BE} + \frac{nkT_r}{q}\right) \quad (5.100)$$

demonstrating the only slight dependence on current level.

For the circuit of Fig. 5.24, using the approximate Eq. 5.98 for collector current and Eq. 5.100 for temperature variation of V_{BE}, $(di_{C1}/dT) = f(dv_{BE}/dT)$,

$$\frac{di_C}{dT} = -\frac{h_{FE}^2}{1 + h_{FE} + h_{FE}^2}\left(\frac{1}{R_E} - \frac{2}{h_{FE}R_2}\right)\frac{1}{T_r}\left(V_{gO} - V_{BE} + \frac{nkT_r}{q}\right) \quad (5.101)$$

or

$$\frac{di_C}{dT} \cong -\frac{1}{T_r}\frac{1}{R_E}(V_{gO} - V_{BE}) \quad (5.102)$$

for $h_{FE}/R_E \gg 2/R_L$ and $(V_{gO} - V_{BE}) \gg (nk/q) T_r$. This represents a very slight temperature dependence, which, however, is greater than that for the circuit of Fig. 5.19.

Temperature-Compensated and Controlled-Current Sources

It is possible and often desirable to establish conditions for more complete independence of a current source from temperature variations after an initial set of operating conditions has been established. On the other hand, it is sometimes desirable to force a variation of current with temperature in either the positive or negative direction to compensate for other factors. For example, as suggested in Sec. 5.10, at low current levels it is often useful to use transconductance, g_m, as the basic parameter for a low-current transistor amplifier. Since, from Eq. 5.67, $g_m \cong (q/k)(I_C/T)$, it would be desirable to maintain g_m constant by making I_C an increasing function of temperature to compensate for the otherwise inverse g_m versus T relationship. Bias networks for various degrees of temperature compensation are discussed in some detail as follows.

First, consider the network shown in Fig. 5.30 with a total supply voltage $V_H - V_L$ and the N diode-connected transistors. Using Eq. 2.126 and $i(R_1 + R_2) = V_H - V_L - Nv_{BE}$, the current can be written as

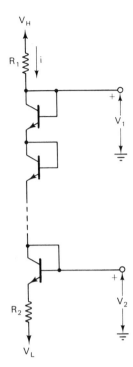

Figure 5.30 Current-mirror source temperature compensation.

$$\ln i + \frac{R_1 + R_2}{N} \frac{q}{kT} i = \frac{V_H - V_L}{N} \frac{q}{kT} + \ln A_o' + \ln T - \frac{qV_{gO}}{kT} \quad (5.103)$$

and

$$\left. \frac{di}{dT} \right]_{i=I} = \frac{\dfrac{R_1 + R_2}{N} \dfrac{q}{kT} \dfrac{I^2}{T} - \dfrac{V_H - V_L}{N} \dfrac{q}{kT} \dfrac{I}{T} + n \dfrac{I}{T} + \dfrac{qV_{gO}}{kT} \dfrac{I}{T}}{1 + [(R_1 + R_2)/N] (q/kT) I} \quad (5.104)$$

which is approximately

$$\frac{di}{dT} \cong \frac{I}{T} - \frac{V_H - V_L}{(R_1 + R_2)T} + \frac{n}{T} \frac{N}{R_1 + R_2} \frac{kT}{q} + \frac{V_{gO}}{T} \frac{N}{R_1 + R_2} \quad (5.105)$$

for

$$\frac{R_1 + R_2}{N} \frac{q}{kT} I \gg 1$$

For a known value of V_{BE} at the specified value for I, Eq. 5.105 is

$$\frac{di}{dT} = \frac{N}{T} \frac{1}{R_1 + R_2} \left(V_{gO} - V_{BE} + \frac{n}{T} \frac{kT}{q} \right) \quad (5.106)$$

The current through the network is always an increasing function of temperature since

$$\frac{n}{T}\frac{kT}{q} << (V_{gO} - V_{BE})$$

If an output voltage V_1 is taken from the collector of the top transistor, $V_1 = V_H - IR_1$ and $dV_1/dT = -Rdi/dt$ or, from Eq. 5.106,

$$\frac{dV_1}{dT} = -\frac{R_1}{R_1 + R_2}\frac{N}{T}\left(V_{gO} - V_{BE} + \frac{n}{T}\frac{kT}{q}\right) \tag{5.107}$$

Since $V_2 = V_L + IR_2$,

$$\frac{dv_2}{dT} = +\frac{R_2}{R_1 + R_2}\frac{N}{T}\left(V_{gO} - V_{BE} + \frac{n}{T}\frac{kT}{q}\right) \tag{5.108}$$

Therefore, using outputs V_1 or V_2, voltages can be obtained that are either decreasing or increasing functions of temperature. Outputs can be taken from other points in the divider for other temperature variations.

Now let us consider separately the temperature-dependent output current of a voltage-driven transistor as shown in Fig. 5.31. Again using Eq. 1.126 for i_E,

$$i_E = A_o'T^n e^{-q(V_{gO}/kT)}\, e^{q(v_{B'E}/kT)}$$

and

$$v_{B'E} = V_1 - \frac{i_E}{1 + h_{FE}}R_B' - i_E R_E$$

The emitter current can be expressed as

$$\ln i_E + \frac{q}{kT}\left(1 + \frac{R_B'}{R_E}\frac{1}{1 + h_{FE}}\right)i_E R_E = \ln A_o' + n \ln T + \frac{q}{kT}(V_1 - V_{gO}) \quad (5.109)$$

Considering T and V_1 both variable,

$$\frac{di_E}{dT}\bigg]_{i_E = I_E} =$$

$$\frac{\dfrac{n}{T}I_E + \left(1 + \dfrac{R_B'}{R_E}\dfrac{1}{1 + h_{FE}}\right)\dfrac{q}{kT}\dfrac{1}{T}(I_E R_E) - \dfrac{q}{kT}\dfrac{1}{T}(V_1 - V_{gO})I_E + \dfrac{q}{kT}I_E\dfrac{dv_1}{dT}}{1 + \{1 + (R_B'/R_E)/(1 + h_{FE})\}\dfrac{q}{kT}I_E R_E} \tag{5.110}$$

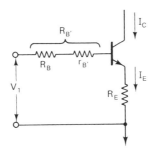

Figure 5.31 Temperature-dependent output current of current mirror.

which can be rewritten as

$$\frac{dv_1}{dT} = -\frac{kT}{q}\frac{n}{T} - \left(1 + \frac{R_B'}{R_E}\frac{1}{1 + h_{FE}}\right)\frac{1}{T}I_ER_E + \frac{1}{T}(V_1 - V_{gO}) \quad (5.111)$$

$$+ \frac{kT}{q}\frac{1}{I_E}\left[1 + \left(\frac{R_B'}{R_E}\frac{1}{1 + h_{FE}}\right)\frac{q}{kT}I_ER_E\right]\frac{di_E}{dT}$$

The condition required to make i_E temperature independent (i.e., $di_E/dt = 0$) is

$$\frac{dv_1}{dT}\bigg]_{i_E = I_E} = \frac{1}{T}(V_1 - V_{gO}) - \frac{kT}{q}\frac{n}{T} - \left(R_E + \frac{R_B'}{1 + h_{FE}}\right)\frac{I_E}{T} \quad (5.112)$$

This can be written approximately as

$$\frac{dV_1}{dT} = -\frac{1}{T}\left(V_{gO} - V_{BE} + \frac{kT}{q}\frac{n}{T}\right) \quad (5.113)$$

for

$$\frac{R_B'}{1 + h_{FE}} << R_E \qquad \text{and} \qquad V_{BE} = V_1 - I_ER_E$$

at the prescribed operating point.

The circuits of Fig. 5.30 and Fig. 5.31 can be combined to yield a set of conditions to make I_E temperature independent in the neighborhood of a specified operating point or to make it increase or decrease in a prescribed manner.

As an example, to make I_E independent of temperature, Eq. 5.107 may be equated to Eq. 5.112 or 5.113 as an approximation. Using Eq. 5.113,

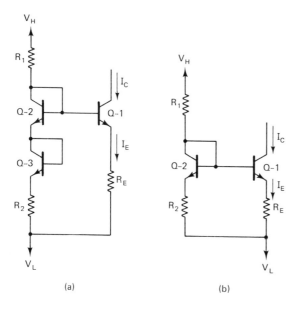

(a)

(b)

Figure 5.32 Divider network in current mirror for temperature-dependence control. (a) Transistor in divider. (b) No transistor in divider.

$$\frac{R_1}{R_1 + R_2} \cong \frac{1}{N} \frac{[V_{gO} - V_{BE1} + (kT/q)(n/T)]}{[V_{gO} - V_{BE2} + (kT/q)(n/T)]}$$

where V_{BE1} is the base–emitter voltage for the transistor in Fig. 5.31 and V_{BE2} is that for each transistor in the divider network of Fig. 5.30. This analysis is slightly in error because it neglects the effect of the small base current supplied to the transistor of Fig. 5.31 in the determination of V_1 from Fig. 5.30. For the circuit of Fig. 5.32(a), using only two transistors in the divider, $N = 2$, and if all transistors are identical, $R_1/(R_1 + R_2) = \frac{1}{2}$. The collector current

$$I_C = \frac{h_{FE}}{1 + h_{FE}} I_E$$

is specified, and R_E is chosen to achieve the required operating point. For the circuit of Fig. 5.32(b), $N = 1$ and for all transistors identical R_2 and R_E should be zero. For unequal currents (i.e., for $V_{BE1} < V_{BE2}$), R_2 should be small but finite, and R_E is chosen to meet the required operating point conditions.

Combined Source and Load for Differential Amplifier

Current mirrors can be used, among many other things, as an aid in optimizing differential amplifiers to meet balance conditions as discussed in Secs. 5.6 and 5.7. For example, the current mirror Q-3, Q-4 in the common-emitter circuit of Fig. 5.33 provides the constant current source ($R_s \to \infty$) necessary for balance conditions for the V_1 and V_2 inputs. In the collector circuits, the current of the amplifier, Q-1, provides the current

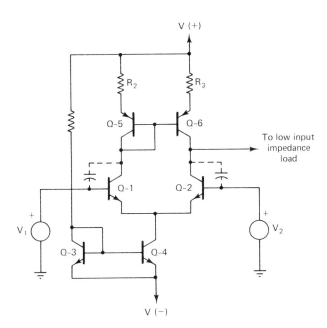

Figure 5.33 Current mirrors for source and loads of transistor differential amplifier.

source for the upper current mirror, Q-5, Q-6, which in turn provides the balanced collector current for the other side of the differential amplifier, Q-2.

The input of the upper current mirror, Q-5, acts as a low impedance load (voltage source) for the collector of Q-1, appearing essentially as a diode-connected transistor and hence, because of the low voltage gain of Q-1, does not present substantial Miller capacitance loading for the input signal voltage, V_1. However, since Q-6 provides a high-impedance (current) source load for the collector of Q-2, there would be substantial Miller capacitance loading for V_2 unless the load to be driven were a low impedance, which is the usual case if it is the input to a common-base or common-emitter transistor. Unequal R's in Q-5 and Q-6 can force precise balance conditions, or for integrated circuits unequal emitter areas for Q-5 and Q-6 can achieve the same result.

5.15 FETS AS SOURCES AND LOADS

The drain circuit of an FET can be used to provide a high-impedance (current) source and may be controlled in the current-mirror arrangement analogous to those using BJTs, as illustrated in Fig. 5.34(a). With the gate connected to the drain, the enhancement-mode IGFET, Q-1, is forced to operate in its saturation region, and since the current to be controlled is $I = I_1$ for identical FETs, the current source transistor, Q-S, is also in its saturation region, with the current being given by

$$I = \frac{V_{DD} - V_{EE} + V_{GS}}{R}$$

The resistor R can be eliminated by replacing it with another transistor, Q-2, as indicated in Fig. 5.34(b), also in saturation.

In the design and fabrication of an integrated-circuit structure, the current I determined by its gate control voltage V_{GS1} can be controlled by the parameters K and V_T, as defined in Sec. 3.8, which permit V_{GS1} and V_{GS2} to be unequal for a design value for I_1.

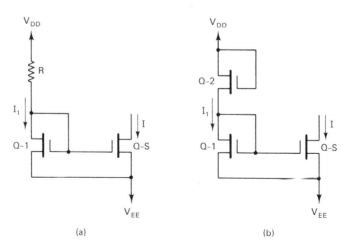

(a)   (b)

Figure 5.34 FETs as current mirrors. (a) Basic circuit. (b) With R replaced by an FET.

Assuming the use of the basic saturation equations with no channel shortening effects or other factors that lead to other than a very high incremental drain resistance in saturation, the first-order equation

$$i_{dsat} = \frac{K}{2}(V_{GS} - V_T)^2 \qquad (5.114)$$

is applicable for Q-1 and Q-2. Since the currents in Q-1 and Q-2 are equal,

$$\frac{V_{GS1} - V_{T1}}{V_{GS2} - V_{T2}} = \left(\frac{K_2}{K_1}\right)^{1/2} \qquad (5.115)$$

where, from Eqs. 3.53 and 3.54,

$$K = \frac{\mu_n \varepsilon_o \varepsilon_i W}{dL}$$

Hence

$$\frac{V_{GS1} - V_{T1}}{V_{GS2} - V_{T2}} = \left(\frac{W_2}{W_1}\frac{L_1}{L_2}\right)^{1/2} \qquad (5.116)$$

For two FETS having the same threshold voltage, $V_{GS} = V_{GS2}$ can be controlled by the ratio of channel widths, channel lengths, or a combination of both. Further control of the current I can be effected by varying V_{GS1} to make it a fraction of V_{DS1}, as indicated in Fig. 5.35.

For $V_{GS2} < V_{DS2}$, I_1 is reduced, V_{DS1} is increased, and V_{GS2} is decreased, which decreases I to match the decrease in I_1.

A variety of combinations of n- and p-channel transistors can be used for the Q-1, Q-2 combination depending on the polarity desired, required voltage level shift, or

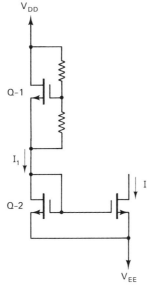

Figure 5.35 Current variation in current mirror using gate divider network.

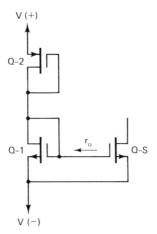

Figure 5.36 Current mirror using complementary FETs.

other design factors. For example, complementary devices can be used for Q-1 and Q-2 as indicated in Fig. 5.36.

It is useful to determine the incremental resistance r_o looking back from the gate of Q-S into the source, which can be shown to be

$$r_o \simeq \frac{1}{g_{m1} + g_{m2}} \tag{5.117}$$

This applies for the circuit of either Fig. 5.34 or 5.36 inasmuch as the resistance looking back into Q-2 is $1/g_{m2}$ for $V_{GS} = V_{DS}$ for both circuits.

This low resistance value shows that a combination of Q-1 and Q-2 is a voltage divider and functions as a low-impedance voltage source useful for applications other than simply the control voltage of a current mirror, such as level shifting circuits in general.

In the circuits of Figs. 5.34, 5.35, and 5.36, Q-S is not quite an ideal current source because it does have a finite incremental drain resistance, r_d. Various combinations can improve this, just as various modifications improved the current mirrors using bipolar transistors. An example of an improved circuit is the cascode-connected current source

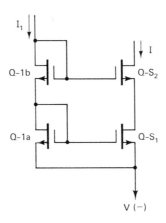

Figure 5.37 Cascode current mirror using FETs.

shown in Fig. 5.37, which is similar in general form to the circuit of Fig. 5.27 using bipolar transistors. With Q-1a and Q-1b structured to supply the correct gate voltages to Q-S_1 and Q-S_2 to control the current I, the circuit functions essentially the same as the previous circuits. However, r_o is the output resistance of a cascode stage given approximately by

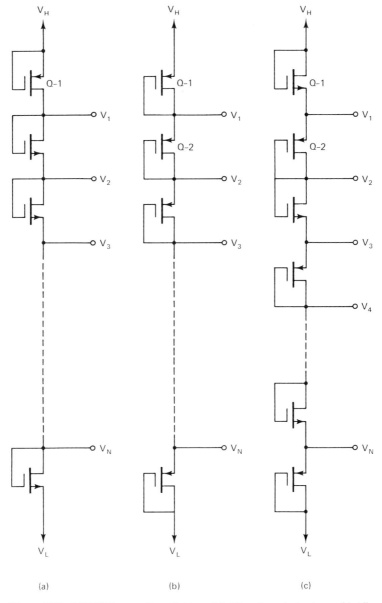

Figure 5.38 MOSFETs as voltage dividers. (a) All *n*-channel structures. (b) All *p*-channel structures. (c) Complementary pair structures.

$$r_o = r_{ds1} + r_{ds2}(1 + g_{ms2} r_{ds1}) \qquad (5.118)$$

This is shown to offer a greatly improved current source for whatever application it is used for.

FETs as Voltage Dividers

If low-impedance voltage sources at different voltage levels are needed, arrays of FETs can be employed in a voltage-divider network to provide such sources. Some examples are shown in Fig. 5.38. The divider in Fig. 5.38(a) uses all *n*-channel enhancement-mode devices, the one in Fig. 5.38(b) is all *p*-channel, and that of Fig. 5.38(c) uses an array of complementary devices. In any case, the voltages can be proportioned by appropriate dimensional factors, as indicated by Eq. 5.116. A choice of a particular array might be based on the positioning of other MOS elements in the circuit, as influenced by fabrication processes and the combination of other *n*- and *p*-channel devices.

FET Differential Amplifier Biasing

The circuit of Fig. 5.39 is basically an integrated MOSFET version of the BJT circuit of Fig. 5.33 with constant current biasing. Here, however, MOSFET voltage dividers can be used to eliminate the need for passive resistors. The divider network may be any one of the three versions illustrated in Fig. 5.38.

If the load draws no appreciable current, that is, the gate of another transistor, the currents in *Q*-1 and *Q*-2 are essentially balanced for *Q*-1 and *Q*-2 identical and for *Q*-3

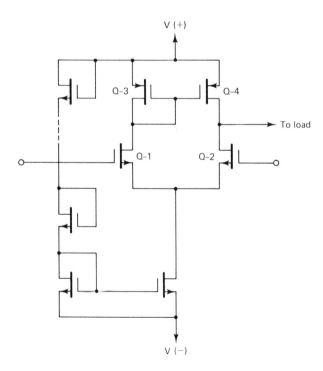

Figure 5.39 FET differential amplifier with constant current biasing.

and Q-4 identical. However, if the load draws appreciable current as for the input to a common gate stage, the currents in Q-1 and Q-2 will be unbalanced. This can be corrected by using different K-factors for Q-3 and Q-4.

PROBLEMS

5.1 A two-stage transistor amplifier is connected as shown in Fig. 5.2(a) with R_c not considered. Assume approximate transistor characteristics of $V_{BE} \cong 0.8$ V and $\beta \cong h_{FE} \cong 100$ for both transistors over a quite wide operating range.

   (a) For $R_{S1} = 10$ kΩ, $R_{B1} = 10$ kΩ, $R_L = 2$ kΩ, and $I_{C2} = 2.0$ mA, and for $V_{CE1} = V_{CE2}$, determine the required values of R_{S2} and R_{B2} to bias the transistors as specified.

   (b) Assuming that for each transistor $r_{bes} \cong 1.2\ r_d$, what is the resistance R_{i2} looking into the emitter circuit of Q-2? What would it be if V_{B2} were obtained from a fixed voltage rather than from the divider network?

   (c) What are the voltage gains of Q-1 and Q-2 separately, that is, v_{C1}/v_{S1} and v_{C2}/v_{C1}, and the overall gain for both cases of part (b)?

   (d) What is the overall gain using the composite relationship of Eq. 5.12?

5.2 For the circuit shown, there is to be zero input–output offset voltage (i.e., for $V_{S1} = 0$, $V_2 = 0$) with operating conditions specified as follows: For Q-1, $\beta_1 \cong h_{FE1} = 100$ at $I_C = 1.0$ mA and $V_{CE} = 6.0$ V. For Q-2, $\beta_2 \cong h_{FE2} \cong 75$ at $I_C = 2.0$ mA and $V_{CE} = 6.0$ V. The base–emitter voltage for each transistor is estimated at $V_{BE} \cong 0.7$ V. The breakdown voltage of the zener diode is approximately $V_B = 5.3$ V at $I = 2.0$ mA.

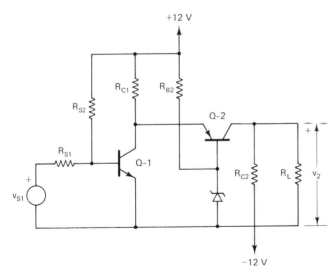

Prob. 5.2

   (a) Determine R_{S1} and R_{S2} to make the effective $R_S = 2.0$ kΩ and find values for R_{C1}, R_{B2}, and R_{C2} under the specified conditions.

   (b) Determine R_L to make the effective load of the collector of Q-2 be $R_L' = 2.0$ kΩ.

   (c) Assuming that for each transistor $r_{bes} \cong 1.2 r_d$, what is the effective input resistance R_{i2} of Q-2?

   (d) What is the approximate overall gain of v_2/v_{S1}?

5.3 Two exactly complementary MOSFETs are used in the circuit shown operating at the current and voltage levels of $I_{D1} = I_{D2} = 2.0$ mA at $V_{GS1} = -V_{GS2} = 6.0$ V for $V_S = V_2 = 0$ V. At these levels, $g_{m1} = g_{m2} = 4.0$ ms.

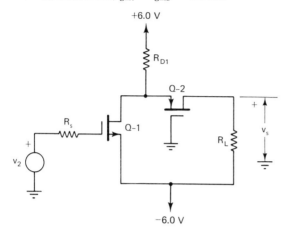

Prob. 5.3

(a) Determine R_{D1}, R_L, and the overall gain v_2/v_s at the operating point. What is the input resistance of Q-2?

(b) Assuming ideal square-law-type FETs, determine the K coefficient and threshold voltage for each transistor.

5.4 A compound voltage follower has approximately the same dc input and output levels established by the extra diode-connected transistor adapted from the basic circuit of Fig. 5.6. At the operating level $V_2 = V_1 = 0$ determined by the ± 6-V supplies and the resistance $R_E \cong 2.5$ kΩ,

$$\beta_1 \cong h_{FE1} \cong 100$$

$$\beta_2 \cong h_{FE2} \cong 75$$

$$V_{BE1} \cong V_{BE2} \cong V_{BE3} \cong 0.7 \text{ V}$$

Prob. 5.4

(a) Determine the base, collector, and emitter currents for transistors Q-1 and Q-2.

(b) What is approximately the incremental input resistance R_i, assuming $r_{b'1} \ll r_{d1}$?

(c) What is the output resistance R_o (assuming that the diode resistance is negligible)? At this value, is it reasonable to assume negligible diode resistance?

(d) What is the overall voltage gain v_2/v_1?

5.5 The circuit shown uses FETs similar to these given by Figs. 3.21 and 3.23. The circuit is to be designed at an operating level $V_{GS1} = 0$, $V_{DS1} = 9$ V, and from the curves of Fig. 3.21, $I_{D1} = 3$ mA.

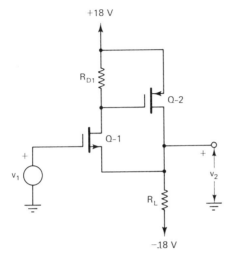

Prob. 5.5

(a) Determine the operating point for Q-2 and find values for R_{D1} and R_L.

(b) Make estimates for r_{d1}, r_{d2}, g_{m1}, and g_{m2}.

(c) Calculate the output resistance, R_o, and the small-signal voltage gain.

5.6 The circuit shown is suggested as an impedance-matching circuit capable of providing appreciable insertion voltage gain. Under impedance-matching conditions for $R_L = R_S$, it is desired to make $R_i = R_S$ and $R_O = R_L$.

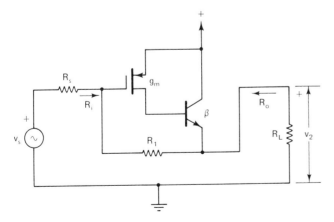

Prob. 5.6

(a) State the condition for which the FET can be characterized by its g_m only and the BJT by its β only.

(b) Derive approximate equations for R_i and R_o.

(c) Obtain an equation for R_1 to meet the impedance-matching conditions stated.

(d) Under impedance-matching conditions as stated, write an equation for insertion voltage gain.

5.7 The characteristics of the transistor evaluated in Sec. 2.5 illustrate the possibility of achieving large current gains over extremely wide ranges of collector current and thus are quite applicable to the Darlington circuit of Fig. 5.9(a). Consider such a circuit with an output current of 10 mA for $V_{CE2} \cong 14$ V.

(a) From Fig. 2.16, find values for h_{FE2}, β_2, and r_{ceo2}, and estimate r_{bes2} assuming that at such a current level $r_b' \cong 0.5\, r_d$.

(b) Find a value for I_{E1}, estimate β_1 assuming that at low currents $\beta_1 \cong \beta_{N1}$, and estimate r_{bes1}.

(c) After an additional measurement, it is determined that $r_{ceo1} \cong 300$ kΩ; determine the composite Darlington h-parameters at the operating levels specified.

5.8 An emitter-coupled amplifier as shown in Fig. 5.12 has supply voltages $V_{CC} = +12$ V and $V_{EE} = -12$ V, with $V_{BB2} = 0$ and $R_{B2} = R_S = 10$ kΩ. Also, $R_E = 2.8$ kΩ and $R_L = 4$ kΩ. At the operating point $\beta_1 = \beta_2 = 100$, and $V_{BE} \cong 0.6$ V. Also, $r_{bes1} = r_{bes2} = 1.2r_d$.

(a) What is the effective input resistance, R_{i1}?

(b) Determine the overall gain in two segments; that is, $A_v = (v_e/v_s)(v_2/v_e)$.

(c) Repeat parts (a) and (b) for $R_{B2} = 0$.

5.9 The circuit of Fig. 5.13 is like that of Fig. 5.12 except that R_{L1} is added.

(a) Using the parameters and results of Problem 5.8 for $A_{21} = A_{12}$, calculate values for A_{11} and A_{22}, and write the value for A_{22}/A_{21} and A_{11}/A_{12}.

(b) Compute the incremental input resistance looking from R_s into the amplifier.

(c) Compute the "impedance level" at the two emitters.

(d) Repeat part (a) if R_E is replaced by a current source as indicated in Fig. 5.14.

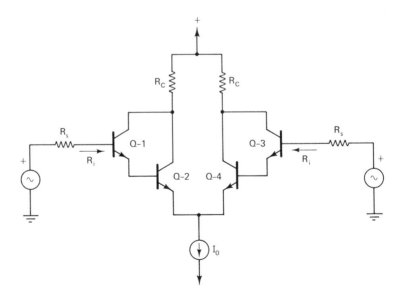

Prob. 5.10

5.10 One purpose of using a Darlington amplifier in the differential amplifier shown is to increase the input resistance R_i in order that the amplifier might better function as a voltage amplifier. Assume that R_C is sufficiently small that internal feedback terms can be neglected.

(a) Assuming that $\beta_1 = \beta_3$, $\beta_2 = \beta_4$, $r_{bes1} = r_{bes3}$, and $r_{bes2} = r_{bes4}$, write an approximate expression for R_i involving only r_{bes1}, r_{bes2}, β_1, and R_s.

(b) For $R_s = 2$ kΩ, $r_{bes} = 1.5$ kΩ at $I_E = 2$ mA, $h_{FE} \cong \beta = 100$ for all transistors, and $I_0 = 4$ mA, determine R_i at 300 K.

(c) Using the appropriate composite h parameters, determine A_{21} and A_{22} using numerical values with everything except R_C.

5.11 The curves shown are those for an actual low-current *pnp* transistor. Five of these are used in the complete low-voltage, low-current differential amplifier shown.

(a) At the operating current $I_0 \cong 28$ μA, estimate β, r_{ceo}, r_{bes}, and g_m for the *pnp* transistors (assume $V_{BE} \cong -0.7$ V).

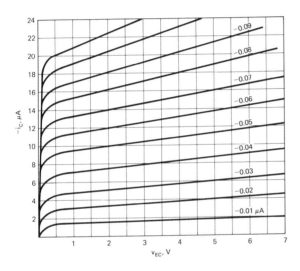

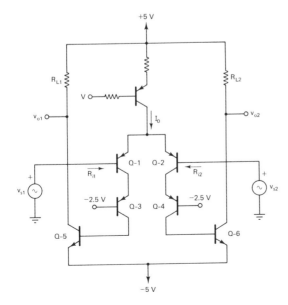

Prob. 5.11

(b) Calculate approximately R_{i1} or R_{i2} at the prescribed operating point.

(c) For Q-5 and Q-6 with $h_{FE} \cong \beta \cong 100$, calculate the value for R_{L1} and R_{L2} to make V_{o2} and $V_{o1} = 0$ for $v_{s1} = 0$ and $v_{s2} = 0$.

(d) Calculate the overall small-signal voltage gain v_{o2}/v_{s1} and v_{o1}/v_{s1}, using $i_{c1} = g_m v_{s1}$ for the input stage Q-1, Q-2 and the β's for succeeding stages.

5.12 The MOSFETs in the accompanying circuit are structured such that the voltage and/or current levels are as indicated and $g_m = 4\text{mS}$.

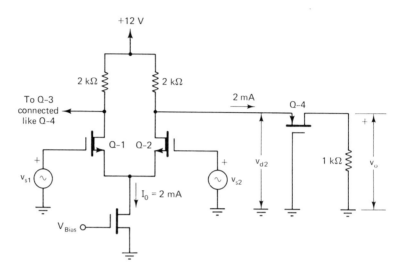

Prob. 5.12

(a) What, approximately, is the incremental input resistance of Q-4?

(b) Determine the small-signal voltage gains for v_{d2}/v_{s1}, v_o/v_{d2}, and v_o/v_{s1}, and also, for $v_{s1} = 0$, v_{d2}/v_{s2}.

5.13 A two-stage MOSFET amplifier of which the circuit of Prob. 5.3 might be a typical example has capacitances that are combinations of circuit and channel capacitances, as indicated in Fig. 5.5. For the parameters of Prob. 5.3, $R_{D1} = R_L = 1.5\text{ k}\Omega$; also $R_s = 1.5\text{ k}\Omega$ and also $g_{m1} = g_{m2} = 4\text{ mS}$ at $I_D = 2.0\text{ mA}$. The effective capacitances are $C_1 = 2.1\text{ pF}$, $C_{gd1} = 0.4$ pF, $C_x = 2.4\text{ pF}$, $C_2 = 3.0\text{ pF}$, and $C_{ds2} = 0.2\text{ pF}$.

(a) What value of r_{d2} would make Z_{i2} be exactly a parallel RC combination, and what would be the values for R and C? (Note how close these values are to $1/g_{m2}$ and C_{ds2}.)

(b) What, approximately, is the total effective shunt capacitance at the input of Q-1?

(c) Write approximate equations for $|V_i/V_s|$, $|V_x/V_i|$, and $|V_2/V_x|$, determine the -3-dB response frequency of each, and compare them. Which response is dominant? Would Eq. 5.23 be a good approximation of the overall response?

5.14 An integrated source-coupled amplifier of the general form illustrated in Fig. 5.17, but with R_{s2} being replaced by a current source, is assumed to have the following parameters: $g_{m1} = g_{m2} = 5\text{ mS}$, $C_{gs1} = C_{gs2} = 2.0\text{ pF}$, $C_{gd1} = C_{gd2} = 0.2\text{ pF}$, $C_{s1} = 0.4\text{ pF}$, $C_L = 1.4$ pF, $C_{ds1} = C_{ds2} \to 0$, and $R_s = R_L = 10.0\text{ k}\Omega$.

(a) Determine the effective input admittance and gain of the source follower alone (i.e., V_{s2}/V_i). Is the condition met that the input admittance is that of a pure capacitance and that the gain is independent of frequency? Explain.

(b) Write equations for the magnitude of the input circuit response, V_i/V_1, and for the response of the common-gate stage. What is the -3-dB response frequency for each?

(c) What is the overall -3-dB response frequency for $R_S \ll R_L$?

5.15 The resistance R_E in the bias circuit of Fig. 5.24(b) for identical Q-1 and Q-2 forces the current of Q-1 to be less than the control current I_2. Suppose the opposite is desired, that is, $I_2 < I_{C1}$ with R_E switched to the emitter of Q-2.

(a) Show for identical transistors that

$$R_E = \frac{kT}{q} \frac{h_{FE1}}{h_{FE1} I_2 - I_{C1}} \ln \frac{(1 + h_{FE1})I_{C1}}{h_{FE1}I_2 - I_{C1}}$$

Hint: Make use of the fact that

$$\frac{I_{E1}}{I_{E2}} \cong \frac{e^{q(V_{BE1}/kT)}}{e^{q(V_{BE2}/kT)}}$$

(b) Show that for $R_E = 0$ the equation reduces to Eq. 5.89.

(c) For transistors with $h_{FE} = 100$, it is desired to produce a current $I_{C1} = 2$ mA for a control current of $I_2 = 0.5$ mA; determine R_E.

5.16 Biasing circuits are to be specified for the complete differential amplifier shown in Fig. 5.28 for supply voltage $V(+) = +12$ V and $V(-) = -12$ V. Assume that, for Q-3 and Q-4, $h_{FE} \cong \beta = 100$, and $V_{BE} \cong 0.8$ V. The emitter currents for Q-1 and Q-2 are to be $I_{E1} = I_{E2} = 0.5$ mA, assuming appropriate loading conditions.

(a) What is the value required for R_1?

(b) What are the emitter currents for Q-1 and Q-2 if the supply voltages are reduced to ± 6.0 V?

(c) Suppose the collector of Q-2 is to drive the input of a common-base *pnp* transistor with $I_E = 2.0$ mA with $h_{FE} = 50$. Find a value for R_2 or R_3 with the other being zero so that the currents in Q-1 and Q-2 remain balanced. *Hint:* Make use of the results of Problem 5.15(a) if you wish.

5.17 For the circuit of Fig. 5.14(a), the bias circuit is replaced by that of Fig. 5.24(b) and designed for $I_{C1} = 1.0$ mA. The supply voltages are $V_{CC} = +12$ V and $V_{EE} = -12$ V. The transistors Q-1 and Q-3 are identical with $V_{BE} \cong 0.7$ V and $h_{FE} \cong 100$.

(a) For $R_E = 0.7$ kΩ, determine I_2 and R_2.

(b) If the supply voltages are changed to $V_{CC} = +6.0$ V and $V_{EE} = -6.0$ V determine I_{C1}.

(c) Without permitting the base–collector voltage to become forward biased for the modified circuit, what approximately are the minimum possible values for V_{CC} and V_{EE}?

(d) For part (a), determine di_{C1}/dT at the reference temperature of 300 K (assume $n = 1.5$ in the T^n relationship).

5.18 The bias circuit shown is to supply current for a differential amplifier with the collector

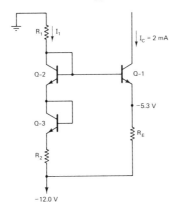

Prob. 5.18

current $I_C = 2.0$ mA to be essentially temperature independent. Transistors are identical with $h_{FE} = 100$ and $V_{BE} = 0.7$ V.

(a) Determine R_1 and R_2 for $I_1 = 1$ mA for approximate temperature independence.

(b) What, approximately, is R_E for the specified current?

REFERENCES

GHAUSI, M. S. *Electronic Devices and Circuits,* Holt, Rinehart and Winston, New York, 1985.

GLASFORD, G. M. *Linear Analysis of Electronic Circuits,* Addison-Wesley, Reading, Mass., 1965.

GRAY, P. R., AND R. G. MEYER. *Analog Integrated Circuits,* Wiley, New York, 1984.

MILLMAN, J. *Microelectronics,* McGraw-Hill, New York, 1979.

SEDRA, A. S., AND H. C. SMITH. *Microelectronic Circuits,* Holt, Rinehart and Winston, New York, 1982.

Chapter 6

Properties of Amplifiers with Feedback

INTRODUCTION

The concept of feedback in amplifiers was introduced in Chapter 4 when the characteristics of single-device amplifiers were discussed. Internal feedback in the bipolar junction transistor was assumed to exist because of the effect of the terms involving h_r, which were proportional to output voltage, in turn modifying the effective input to the device terminals. External feedback was also found to exist because of the impedance Z_T, which was introduced into the circuit common to the input and output. It was pointed out that both external and internal feedback affected gain and input and output impedance in different ways depending on the particular circuit configuration and that these properties were analogous to those exhibited by specific classifications of feedback, which will be discussed in this chapter.

Feedback is frequently deliberately introduced into an amplifier by making one or more external connections between the output and input through some sort of impedance connected to the output and input in some manner. Such feedback can drastically modify the characteristics of an amplifier in both desirable and undesirable ways. It is the purpose of this chapter to develop a formal way of classifying feedback connections and to develop general methods of analyzing the effects of feedback on amplifier performance.

6.1 AMPLIFIER REPRESENTED AS A GAIN BLOCK; SOME MATTERS OF TERMINOLOGY

A single-input linear amplifier without internal feedback is generally represented as a gain block with voltage gain A such that

$$A = \frac{v_o}{v_i} \qquad (6.1)$$

as indicated in Fig. 6.1(a) or (b). The + or − sign indicates whether the amplifier is noninverting or inverting (i.e., whether the value of A is positive or negative).

For a two-input differential amplifier input stage, such as the differential amplifier discussed in Chapter 5, if the magnitude of the gain is the same for both inverting and noninverting inputs, as shown in Fig. 6.1(c), the input–output relationship is

$$v_o = |A| (v_{iN} - v_{iI}) = |A|v_i \qquad (6.2)$$

A two-input differential amplifier may of course be used as a single-input amplifier with one input common, as indicated in Fig. 6.1(d).

If the amplifier is frequency dependent (as all amplifiers are), as the input frequency is increased, we are reminded of the fact by using $A = A(s)$, where $s = j\omega$, as indicated in Fig. 6.2(a). Also, the fact that the amplifier has a finite input and output impedance is sometimes indicated as shown in Fig. 6.2(b). When input and output impedances are considered, the amplifier may be represented by its Thevenin equivalent, as shown in Fig. 6.2(c) for the noninverting amplifier and Fig. 6.2(d) for the inverting amplifier, where A_o is the magnitude of the open circuit voltage gain and Z_i and Z_o are input and output impedance, respectively.

It should be noted that a frequency-dependent amplifier nominally designated as inverting or noninverting at its designated operating frequency may exhibit as much as 180° phase shift (i.e., again invert at same other frequency).

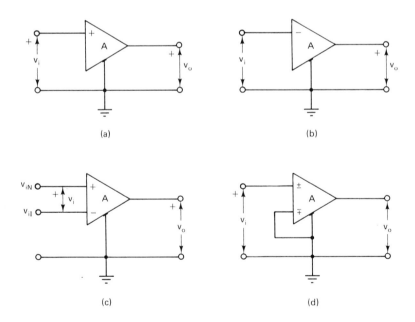

Figure 6.1 Block representation of voltage amplifier. (a) Noninverting. (b) Inverting. (c) Differential amplifier. (d) Differential amplifier with one input fixed.

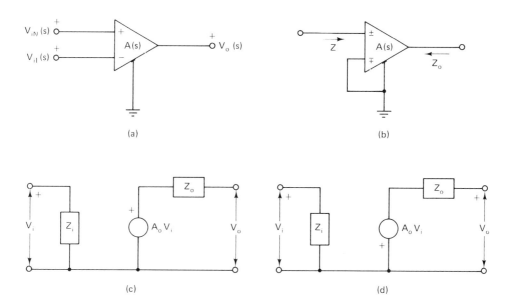

Figure 6.2 Frequency-dependent amplifier representation. (a) Input–output voltage relationships. (b) Impedance relationships. (c) Circuit model of noninverting amplifier. (d) Circuit model of inverting amplifier.

6.2 GAIN EQUATIONS FOR AMPLIFIERS WITH FEEDBACK

A single-input amplifier, which may be either inverting or noninverting, is shown in Fig. 6.3 in which there is an input–output connection through networks N_1 and N_2, where

$$A(s) \equiv \frac{V_o(s)}{V_i(s)} \tag{6.3}$$

with everything connected. For a linear amplifier with linear interconnecting networks, superposition applies, and the voltage at the amplifier input may be written as

$$V_i = K_1 V_s + B V_o \tag{6.4}$$

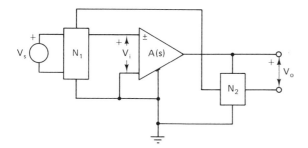

Figure 6.3 Generalized amplifier feedback model.

where

$$K_1 = \left.\frac{V_i}{V_s}\right]_{V_o=0}, \qquad B = \left.\frac{V_i}{V_o}\right]_{V_s=0} \tag{6.5}$$

Then, by combining Eqs. 6.3 and 6.4, the overall gain is

$$\frac{V_o}{V_s} = \frac{K_1 A}{1 - AB} \tag{6.6}$$

This is a general feedback equation that exhibits a variety of properties depending on the nature of the $1 - AB$ term. For example, if A is inverting but not frequency dependent, given by $A = -|A|$,

$$\frac{V_o}{V_s} = \frac{-K_1|A|}{1 + |A|B} \tag{6.7}$$

the output is still inverting, but its magnitude is reduced because of the feedback connection. This reduction is termed *negative feedback* because the signal feedback is used to reduce the effective input voltage at the terminals. On the other hand, if A is noninverting (i.e., $A = |A|$),

$$\frac{V_o}{V_s} = \frac{K_1|A|}{1 - |A|B} \tag{6.8}$$

which indicates that the term AB increases the overall gain up to the point $|A|B \to 1$, where a condition of instability is approached. This property is termed *positive feedback*.

In both Eqs. 6.6 and 6.7, the term AB in the denominator exists as a unit, and in general either A or B can be positive or negative or may be frequency dependent. Then negative or positive feedback is determined by the magnitude of $(1 - AB)$ as a unit.

Open-Loop Gain

If the input impedance is included as part of the network N_1, V_S is short-circuited, the feedback path opened, and a signal applied at the input terminals, all as indicated in Fig. 6.4, the gain, $A = V_o/V_i$, is the same as before and the ratio is the same as B as previously defined. The combined term

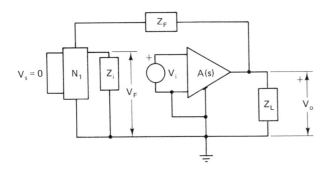

Figure 6.4 Open-loop gain of feedback amplifier.

$$\frac{V_f}{V_i} = AB \qquad (6.9)$$

can be determined as a unit by opening the loop, applying V_i, and determining V_f. The result is termed the *open-loop gain* and is identical to the term AB as previously derived. Sometimes the factor

$$\frac{V_f}{V_i} = -T_r$$

where T_r is called the *return ratio* and the term $1 - AB = 1 + T_r$ is called the *return difference*.

6.3 PROPERTIES OF VOLTAGE FEEDBACK: SPECIFIC EXAMPLES

When the feedback signal is derived as a sampling of the output voltage (in shunt with the terminals of the load impedance) it is generally termed *voltage feedback* or *shunt feedback*. Two examples that differ only in the manner in which the input connection is made will be discussed in this section.

Shunt–Shunt Feedback

A very commonly used feedback connection is that shown in Fig. 6.5, where shunt or voltage feedback is delivered to the input via the voltage-divider network, that is, in shunt with the input; hence the designation of *shunt–shunt feedback*.

Solving this circuit directly or using superposition as defined by Eq. 6.4, K_1 and B, can be expressed as

$$K_1 = \frac{Z_i}{Z_i + Z_S} \frac{Z_F}{Z_1 + Z_F} \qquad (6.10)$$

$$B = \frac{Z_1}{Z_1 + Z_F} \qquad (6.11)$$

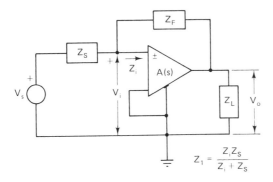

$$Z_1 = \frac{Z_i Z_S}{Z_i + Z_S}$$

Figure 6.5 Shunt–shunt feedback connection.

where

$$Z_1 = \frac{Z_i Z_S}{Z_i + Z_S}$$

The overall gain is written as

$$\frac{V_o}{V_s} = \frac{Z_i}{Z_i + Z_S} \frac{Z_F}{Z_1 + Z_F} \frac{A(s)}{1 - A(s)[Z_1/(Z_1 + Z_F)]} \tag{6.12}$$

If the amplifier is inverting and not frequency dependent, $A = -|A|$, and if Z_S, Z_F, and Z_L are resistances, the overall gain is still that of an inverting amplifier, the gain is reduced, and the feedback is classified as negative.

The effect of the load and output impedances on gain may be included in Eq. 6.12 by solving the circuit of Fig. 6.6 to express A in terms of A_o and Z_o as

$$A = \frac{A_o + (Z_o/Z_F)}{1 + Z_o/Z_L[1 + (Z_L/Z_F)]} \tag{6.13}$$

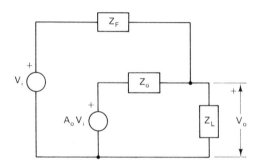

Figure 6.6 Terminal gain in terms of open circuit gain and output impedance.

This equation used in the $1 - AB$ term in Eq. 6.12 allows it to be expressed as

$$1 - AB = \frac{1 + \dfrac{Z_o}{Z_L}\left(1 + \dfrac{Z_L}{Z_1 + Z_F}\right)}{1 + \dfrac{Z_o}{Z_L}\left(1 + \dfrac{Z_L}{Z_F}\right)} \left[1 - A_o \frac{\dfrac{Z_1}{Z_1 + Z_F}}{1 + \dfrac{Z_o}{Z_L}\left(1 + \dfrac{Z_L}{Z_F}\right)}\right] \tag{6.14}$$

Putting this into the overall gain equation of Eq. 6.12 yields

$$\frac{V_o}{V_s} = \frac{Z_i}{Z_i + Z_S} \frac{Z_F}{Z_1 + Z_F} \frac{A_o + (Z_o/Z_F)}{\left[1 + \dfrac{Z_o}{Z_L}\left(1 + \dfrac{Z_L}{Z_1 + Z_F}\right)\right]\left[1 - \dfrac{A_o[Z_1/(Z_1 + Z_F)]}{1 + \dfrac{Z_o}{Z_L}\left(1 + \dfrac{Z_L}{Z_1 + Z_F}\right)}\right]} \tag{6.15}$$

Open Loop Gain. The open-loop gain as defined by Eq. 6.9 can be determined (taking output impedance into account) from the equivalent circuit of Fig. 6.7. The result is

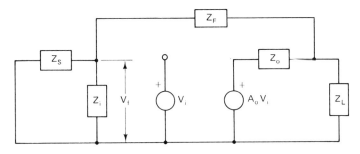

Figure 6.7 Circuit model for open-loop gain.

$$\frac{V_f}{V_i} = A_o \frac{Z_1/(Z_1 + Z_F)}{1 + (Z_o/Z_L)\{1 + [Z_L/(Z_1 + Z_F)]\}} \qquad (6.16)$$

which is seen to be identical to the comparable term in Eq. 6.15.

Output Impedance of Voltage (Shunt) Feedback Amplifier. The effective output impedance of an amplifier with voltage feedback, as seen looking back from the load, can be determined from the partial circuit of Fig. 6.8 with Z_L replaced by the voltage source V_2. The current, I_2, flowing into the amplifier output port is

$$I_2 = \frac{V_2 - A_o B V_2}{Z_o} \qquad (6.17)$$

where $B = V_i/V_2$ as previously defined. The effective output impedance, Z_o', is

$$Z_o' = \frac{V_2}{I_2} = \frac{Z_o}{1 - A_o B} \qquad (6.18)$$

Thus, for negative feedback (A_o or B negative), the output impedance is reduced by feedback in the same manner that the gain is reduced, except that the pertinent gain is the open-circuit gain A_o rather than terminal gain A. If the feedback is positive, the output impedance is increased. Z_o' is dependent only on B and not on the manner of

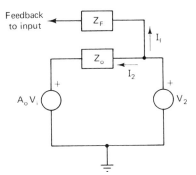

Figure 6.8 Circuit for determination of output impedance of feedback amplifier.

connection to the input. However, the total output impedance includes that of the parallel feedback path, which does depend on the input connection. For example, for the shunt–shunt configuration that has been discussed, the impedance $Z'_F = Z_F + Z_1$, which is in parallel with that given by Eq. 6.18.

There are many instances for very high gain amplifiers when the shunting impedance Z'_F is so large that it can be neglected, and Eq. 6.18 can be used as the total output impedance with negligible error. There are, however, important exceptions to this approximation.

Input Admittance of Feedback Amplifiers with Shunt-Connected Input. The effective input admittance of the feedback amplifier shunt connected at the input can be determined from the circuit of Fig. 6.9. The current in the feedback path is

$$I_f = \frac{V_1(1 - A)}{Z_F}$$

and

$$I_i = \frac{V_1}{Z_i}$$

Hence the total input admittance $Y_i = I_1/V_1$ is

$$Y_i = \frac{1}{Z_i} + \frac{1}{Z_F}(1 - A) \tag{6.19}$$

Thus, for negative feedback ($A = -|A|$), there is a component of input admittance that increases with gain; or, stated differently, there is a component of input impedance Z_{if} involving the feedback path alone given by

$$Z_{if} = \frac{Z_F}{1 - A} \tag{6.20}$$

that decreases with increasing gain for negative feedback and increases with increasing gain for positive feedback. This component is in parallel with the normal input impedance Z_i of the amplifier.

In the case of a high-gain amplifier, the feedback component of input impedance may be dominant, contrary to the situation with respect to output impedance.

It may be recalled that the consequences of capacitive feedback, $Z_F = 1/j\omega C_F$ for

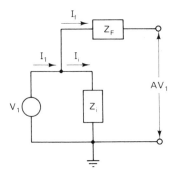

Figure 6.9 Circuit for input impedance determination.

Properties of Amplifiers with Feedback   Chap. 6

specific amplifier configurations, were analyzed in detail in Chapter 4 for single devices as amplifiers and in Chapter 5 for composite compound devices.

Shunt–Series Feedback

This form of feedback may be identified as voltage (shunt) feedback where the feedback voltage appears in series with the input signal, which is illustrated by the circuit of Fig. 6.10(a). Solving this circuit for K_1 and B either directly or by using superposition yields

$$K_1 = \frac{Z_i}{Z_S + Z_i + [Z_T Z_F/(Z_T + Z_F)]} \qquad (6.21)$$

$$B = \frac{Z_i}{Z_S + Z_i + Z_F + \{[Z_F(Z_S + Z_i)]/Z_T\}} \qquad (6.22)$$

and as before

$$\frac{V_2}{V_s} = \frac{K_1 A}{1 - AB}$$

Input Impedance. The effective input impedance, $Z_i' = V_1/I_1$, may be determined by solving the circuit of Fig. 6.10(b). The result is

$$Z_i' = \left(Z_i + \frac{Z_F Z_T}{Z_F + Z_T}\right)\left(1 - \frac{AZ_i}{Z_i + Z_F + \{[Z_F(Z_S + Z_i)]/Z_T\}}\right) \qquad (6.23)$$

The effective input impedance is thus increased for $A = -|A|$, which identifies negative feedback, which is opposite to the decrease in effective input impedance for the shunt-connected input. Conversely, for positive series feedback, the input impedance is decreased, which is also opposite to the situation for shunt feedback.

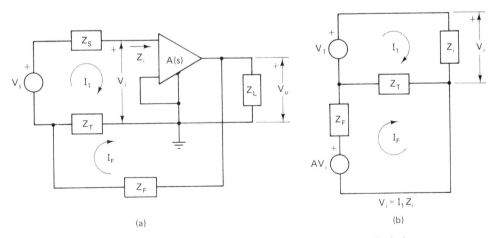

Figure 6.10 Shunt–series feedback connections. (a) Actual circuit. (b) Equivalent network.

For a low value of source impedance, that is, for

$$Z_S \ll Z_i + \frac{Z_T Z_F}{Z_T + Z_F}$$

it may be observed from Eqs. 6.21 and 6.22 that

$$Z_i' \simeq \frac{Z_i}{K_1} (1 - AB) \qquad (6.24)$$

This result would be exactly true if Z_S were incorporated as part of the input impedance, that is, for $Z_i = V_s/I_1$.

It may be observed that the model for series feedback described in this section requires that the signal voltage be isolated from the common input–output voltage reference, which in theory for ac signals can be achieved by transformer coupling, but in practice may be awkward to achieve.

Output Impedance. The output impedance of the shunt-connected output (Eq. 6.18) was derived independent of the manner of input connection. Therefore, the value of B given by Eq. 6.22 for the shunt–series amplifier may be used directly in Eq. 6.18 to determine the exact value for effective output impedance.

6.4 PROPERTIES OF CURRENT FEEDBACK: FEEDBACK DERIVED IN SERIES WITH LOAD

A general case of feedback where the feedback signal is derived from a sampling of current through the impedance rather than the voltage across it is shown in Fig. 6.11(a). This is generally referred to as *series feedback*. As before, the system is described by the general equations

$$V_i = K_1 V_s + B V_2 \qquad (6.25)$$

and

$$V_o = \frac{K_1 A V_s}{1 - AB} \qquad (6.26)$$

where $A \equiv V_o/V_i$ with everything in place.

The effective ouput impedance can be determined from the circuit of Fig. 6.11(b). It is convenient to define a feedback transfer impedance

$$Z_{tf} = -\frac{V_f}{I_2} \qquad (6.27)$$

somewhat analogous to the voltage feedback factor B.

Then, using Eq. 6.27 with the circuit of Fig. 6.11(b), the effective output impedance given by $Z_o' = V_2/I_2$ is

$$Z_o' = (Z_o + Z_F) \left(1 - \frac{A_o Z_{tf}}{Z_o + Z_F} \right) \qquad (6.28)$$

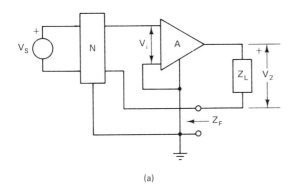

(a)

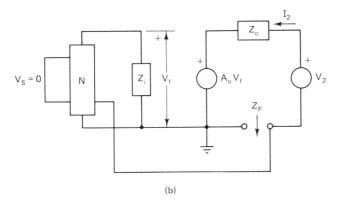

(b)

Figure 6.11 Amplifier with series (current) feedback. (a) General circuit model. (b) Circuit for output impedance determination.

or using the conventional definition of

$$B = \frac{V_f}{V_2} = -\frac{I_2 Z_{tf}}{-I_2 Z_L} = \frac{Z_{tf}}{Z_L} \tag{6.29}$$

$$Z_o' = (Z_o + Z_F)\left(1 - \frac{A_o B Z_L}{Z_o + Z_F}\right) \tag{6.30}$$

Thus the effective output impedance for the series-connected feedback is increased for negative feedback by approximately the same value as it is decreased by negative shunt feedback modified by the relative values of Z_L and $(Z_o + Z_F)$.

The nature of the input impedance and the gain A in terms of A_o are functions of the specific configuration of N and can best be illustrated by a specific example as shown in Fig. 6.12, where the generalized network N is replaced by the simple series source impedance. The switch SW indicates that either the source and load or the amplifier itself can be isolated from the common ground reference in order to derive the sampling current.

In this circuit

$$V_s = I_1(Z_S + Z_i + Z_T) + I_2 Z_T, \qquad I_2 = -\frac{V_2}{Z_L}, \qquad I_1 = \frac{V_i}{Z_i}$$

Then

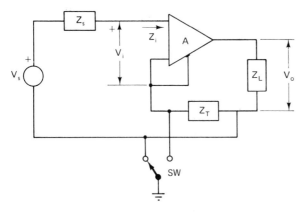

Figure 6.12 Alternative connections for series feedback.

$$K_1 = \frac{Z_i}{Z_S + Z_i + Z_T} \tag{6.31}$$

$$B = \frac{Z_T}{Z_L} \frac{Z_i}{Z_S + Z_i + Z_T} \tag{6.32}$$

$$\frac{V_o}{V_s} = \frac{Z_i}{Z_S + Z_i + Z_T} \frac{A}{1 - [(AZ_i)/(Z_S + Z_i + Z_T)][Z_T/Z_L]} \tag{6.33}$$

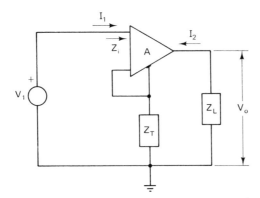

Figure 6.13 Input and output impedances for series feedback.

The effective input impedance may be obtained from Fig. 6.13 using

$$V_1 = I_1(Z_i + Z_T) + I_2 Z_T$$

$$I_2 = -\frac{V_2}{Z_L}$$

$$V_o = AV_i$$

$$V_i = I_1 Z_i$$

The result is

Properties of Amplifiers with Feedback   Chap. 6

$$Z_i' = (Z_i + Z_T)\left[1 - \frac{AZ_iZ_T}{Z_L(Z_i + Z_T)}\right] \tag{6.34}$$

or
$$Z_i' = \frac{Z_i}{K_1'}(1 - AB') \tag{6.35}$$

where K_1' and B' are defined by Eqs. 6.31 and 6.32 for $Z_S = 0$. This result agrees basically with Eq. 6.24 for the shunt–series amplifier, showing that for both cases the input impedance changes in the same manner and hence is essentially independent of how the feedback is derived from the output (series or shunt).

The value of A in terms of A_o can be determined from Fig. 6.14. The result is

$$A = \frac{A_o + [Z_T/(Z_S + Z_T)]}{1 + (Z_o/Z_L) + (1/Z_L)[Z_TZ_S/(Z_T + Z_S)]} \tag{6.36}$$

This value may be substituted into Eq. 6.33 for a complete overall gain equation leading to a relationship similar to that of Eq. 6.15 for the shunt–shunt amplifier.

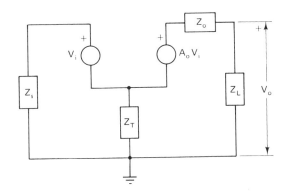

Figure 6.14 Determination of gain in terms of open-circuit gain and output impedance for series–series feedback connection.

6.5 SUMMARY OF FEEDBACK CLASSIFICATION AND PROPERTIES

Feedback has in general been classified as *positive feedback* or *negative feedback*. Positive feedback has generally been defined as feedback that *increases* the magnitude of the gain of an amplifier compared with its value without feedback, whereas negative feedback *decreases* the magnitude of the gain. As formulated by the general feedback equation, this depends on whether the magnitude of the term $1 - AB$ is less than or greater than unity, which in terms of a frequency-dependent amplifier depends on whether $AB(j\omega)$ has a positive or negative real component.

Feedback amplifiers have been further classified according to the method of connection of the feedback path to the output and input, as follows: *Voltage feedback* is feedback in which the feedback signal is derived as a sampling of the voltage across the load and is referred to also as *shunt feedback*, whereas *current feedback* is feedback in which the feedback signal is derived as a sampling of the *current* through the load impedance and is also referred to as *series feedback*.

Voltage (shunt) and current (series) feedback are further classified according to the method of connection at the input, as follows:

1. Shunt–shunt feedback is feedback introduced in shunt with both output and input.
2. Shunt–series feedback is feedback introduced in shunt with the output and in series with the input.
3. Series–shunt feedback is feedback introduced in series with the output and in shunt with the input.
4. Series–series feedback is feedback introduced in series with both the output and input.

These relationships are sometimes illogically stated in reverse order. It is preferable to use the first word in the two-word designations as the connection to the output because that is where the feedback signal is derived. Thus, if we refer to shunt or series feedback, it can be done without reference to how it is applied at the input.

Furthermore, input–output impedance properties for the various classifications of feedback may be summarized for negative and positive feedback as follows:

1. Feedback connected in shunt at the output decreases the effective output impedance of the amplifier if the feedback is negative and increases it if the feedback is positive.
2. Feedback connected in series with the output increases the effective output impedance for negative feedback and decreases it for positive feedback.
3. Feedback connected in series with the input increases the effective input impedance if the feedback is negative and decreases it if the feedback is positive.
4. Feedback connected in shunt with the input decreases the effective input impedance if the feedback is negative and increases it if the feedback is positive.

Examples from Single-Device Amplifiers

The single-device amplifiers analyzed in Chapter 4 offer specific examples of two distinct feedback paths, internal through the device involving the $h_r h_f$ term and external through the common impedance, Z_T. For example, the common-emitter amplifier shown in Fig. 4.4 with the specific defining equations for low-frequency operation, Eqs. 4.15 through 4.17, demonstrates two distinct feedback modes, positive series–series feedback involving μ_{re} and negative feedback involving R_E. For $R_E = 0$, $B = -\mu_{re}$, with the gain without feedback being

$$A = \frac{-\beta R_L}{R_S + r_{bes}}$$

If the term involving μ_{re} is negligible, the feedback is entirely negative feedback through the common resistance R_E, which provides feedback in series with the input signal. The feedback signal is a sampling of the collector current and represents series–series feedback if the output signal is taken from the collector. However, if the output is taken from the emitter as an emitter follower, the feedback is a sampling of the load voltage itself and is seen to be shunt–series feedback.

For the common-base amplifier shown in Fig. 4.6 with low-frequency properties defined by Eqs. 4.39 through 4.41, the feedback introduced through the μ_{rb} term internally and by the term involving R_B externally exhibit the properties of negative series–series feedback.

6.6 FEEDBACK IN DIFFERENTIAL FORM

The two-input differential amplifier may be used as a feedback amplifier where the input signal is applied to one input terminal and the feedback to the other, as indicated in Fig. 6.15, which assumes that $|A|$ is the same for both inverting and noninverting modes. In this model, for convenience, the input impedances for the signal and feedback inputs are incorporated as part of the impedance Z_{is} and Z_{if}. The appropriate feedback equations analogous to those for the single-input amplifier are

$$V_i = K_1 V_s + B V_o \qquad (6.37)$$

$$K_1 = \frac{Z_{is}}{Z_s + Z_{is}} \qquad (6.38)$$

$$B = \frac{-Z_{iF}}{Z_F + Z_{iF}} \qquad (6.39)$$

$$\frac{v_o}{v_s} = \frac{K_1 A}{1 - AB} \qquad (6.40)$$

The minus sign in Eq. 6.39 makes the defining equations exactly the same as those for the single-input amplifier.

If the amplifier is noninverting for the input signal and inverting for the feedback signal, A is positive and the feedback is negative because B is negative, which is opposite to the case for the single-input amplifier where an inverting amplifier results in negative feedback. For the two-input amplifier shown, an inverting amplifier to the signal results in positive feedback.

Output Impedance. The effective output impedance of the differential feedback amplifier may be determined in the same manner as was done for the single-input amplifier.

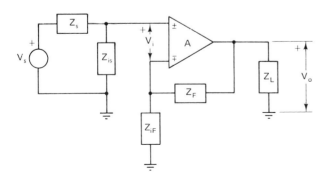

Figure 6.15 Differential amplifier with feedback.

With the signal source short-circuited and feedback applied to the remaining input terminal, the basic equations for output impedance are the same as those determined for the single amplifier, with the active input terminal being the one to which feedback is applied.

Input Impedance.   Normally, in a differential amplifier the impedance between terminals is so high that it may be ignored. Then, since feedback is applied to the opposite terminal from the signal, feedback has no effect on the input impedance of the signal-connected terminals.

However, if there is some terminal-to-terminal input impedance, as shown in Fig. 6.16, it affects not only the effective input impedance but the overall gain as well. The basic equations for K_1, B, A, and Z_i may be developed in the same manner as before taking this additional component into account.

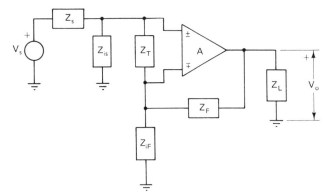

Figure 6.16   Differential feedback amplifier with impedence between input terminals.

Multiple Feedback Paths.   The differential amplifier may have more than one feedback path, as indicated in Fig. 6.17. The feedback equations may be generalized to include both feedback paths. Each feedback mode affects the properties of the amplifier in the same general way as it would if the other mode were not present.

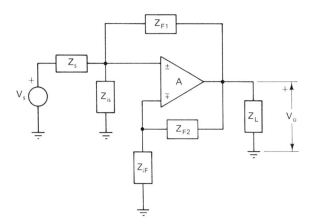

Figure 6.17   Amplifier with multiple feedback paths.

6.7 FREQUENCY CHARACTERISTICS OF NEGATIVE FEEDBACK AMPLIFIERS

The frequency characteristics of electronic devices and their associated networks result in overall amplifier responses that are also frequency dependent. The frequency characteristics of an amplifier with external feedback are dependent as well on the external impedance elements that are used to introduce the feedback, with the dominant term controlling the overall characteristics being the $(1 - AB)$ term common to the mathematical representation of most feedback systems. This term may be structured in such a way as to improve the bandwidth compared to the nonfeedback amplifier, which is included to produce a prescribed frequency characteristic such as that of the active filter, or to produce arbitrary responses with components independent of the input signal itself, such as the sinusoidal oscillator. Some of the specific properties depend on the nature of input and output impedances; however, some of the basic concepts can be introduced with structures independent of such internal detail.

Gain–Bandwidth Product

A general relationship of gain and bandwidth of an amplifier when feedback is introduced compared to such an amplifier without feedback may be illustrated using the basic relationships introduced in Sec. 6.2 and discussed again in Sec. 6.6.

If K_1 and B are both positive real numbers (e.g., composed of resistive elements and structured in such a way that they are independent of each other), and if A is inverting and internally frequency dependent with a single-pole response analogous to that of the single-pole low-pass RC circuit limited at high frequencies by shunt capacitance, then the response may be written in the alternative forms

$$A(j\omega) = \frac{-A_{0M}}{1 + j(\omega/\omega_2)} \tag{6.41}$$

$$A(j\omega) = \frac{-A_{0M}}{[1 + (\omega/\omega_2)^2]^{1/2}} \Big/ \underline{\tan \frac{-\omega/\omega_2}{1}} \tag{6.42}$$

$$A(s) = \frac{-A_{0M}\,\omega_2}{s + \omega_2} \tag{6.43}$$

where A_{0M} is the magnitude of the gain at zero frequency, $s = j\omega$, and ω_2 is the angular frequency at which the response is $1/\sqrt{2}$ of its magnitude at its low-frequency value. The magnitude and relative phase characteristics are plotted in Fig. 6.18 from Eq. 6.42.

The complex expression for overall voltage gain with feedback found by using Eq. 6.41 in Eq. 6.6 is

$$\frac{V_o}{V_s} = \frac{-K_1 A_{0M}}{(1 + A_{0M}B)\{1 + j(\omega/\omega_2)[1/(1 + A_{0M}B)]\}} \tag{6.44}$$

This equation shows that the low-frequency gain $-K A_{0M}$ without feedback is reduced by the factor $1/(1 + A_{0M}B)$, and the bandwidth given by

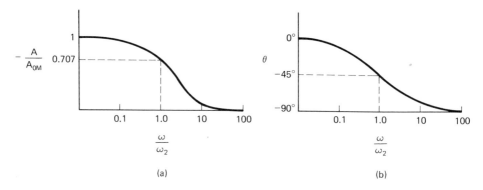

Figure 6.18 Single-pole high-frequency response. (a) Amplitude response. (b) Phase response.

$$\omega_{2f} = \omega_2(1 + A_{0M}B) \qquad (6.45)$$

is increased by the same factor, yielding a constant gain–bandwidth product.

If K_1 or B is not resistive or if they are mutually dependent, the results are subject to some modification, as illustrated in the following section using the shunt–shunt configuration as an example.

6.8 FREQUENCY CHARACTERISTICS OF NEGATIVE SHUNT–SHUNT FEEDBACK

To illustrate the effects of the frequency dependency of amplifier gain and feedback circuit elements, several variations will be considered:

CASE I: Inverting Amplifier with Single-Pole Response and Resistive Feedback.

In Fig. 6.19,

$$A(j\omega) = \frac{-A_{0M}}{1 + j(\omega/\omega_2)} \qquad (6.46)$$

where

$$K_1 = \frac{R_i}{R_i + R_S} \frac{R_F}{R_1 + R_F}$$

$$B = \frac{R_1}{R_1 + R_F}$$

with

$$R_1 = \frac{R_i R_s}{R_i + R_S}$$

Using these terms in Eq. 6.12, the overall gain can be expressed in the form

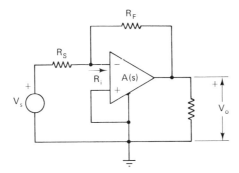

Figure 6.19 Single-pole amplifier with shunt–shunt feedback.

$$\frac{V_o}{V_s} = \frac{R_i}{R_i + R_s} \frac{R_F}{R_1 + R_F} \frac{-A_{OM}}{\left(1 + A_{OM}\dfrac{R_1}{R_1 + R_F}\right)\left(1 + j\dfrac{\omega}{\omega_2}\dfrac{1}{1 + A_{OM}\dfrac{R_1}{R_1 + R_F}}\right)}$$

(6.47)

The overall gain–bandwidth product is constant, as suggested in Sec. 6.7, except for the slight modification introduced by the interaction of K_1 and B, where the gain without feedback cannot be precisely defined because of the direct input–output path.

CASE II: Ideal Inverting Amplifier and Capacitance Loading at Input.

With all elements resistive except for the shunting capacitance at the input shown external to the amplifier, as shown in Fig. 6.20,

$$Z_i = \frac{R_i}{1 + j\omega R_i C_1}$$

$$Z_1 = \frac{R_1}{1 + j\omega R_1 C_1} \quad \text{where } R_i = \frac{R_S R_i}{R_S + R_i}$$

$$A = -A_{OM}$$

Again using these values in the general feedback equation, the overall gain is

$$\frac{V_o}{V_s} =$$

$$\frac{R_i}{R_i + R_s} \frac{R_F}{R_1 + R_F} \frac{-A_{OM}}{\left(1 + A_{OM}\dfrac{R_1}{R_1 + R_F}\right)\left[1 + j\omega\left(\dfrac{R_1 R_F}{R_1 + R_F}\right)C_1 \dfrac{1}{1 + A_{OM}\dfrac{R_1}{R_1 + R_F}}\right]}$$

(6.48)

This equation may be interpreted as follows: Without feedback, but taking R_F loading into account, the effective input low-pass time constant is

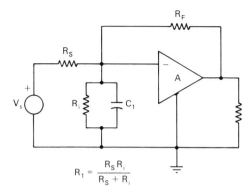

$$R_1 = \frac{R_S R_i}{R_S + R_i}$$

Figure 6.20 Shunt–shunt feedback with input capacitance.

$$\frac{1}{\omega_1} = \left(\frac{R_1 R_F}{R_1 + R_F}\right) C_1 \tag{6.49}$$

and the bandwidth with feedback can be expressed as

$$\omega_{1f} = \omega_1 \left(1 + A_{0M} \frac{R_1}{R_1 + R_F}\right) \tag{6.50}$$

Thus, the effective input time constant has the same effect as the internal low-pass time constant of the frequency-dependent amplifier. This is a reasonable expectation, because examination of the circuit from the open-loop gain concept puts the input time constant within the feedback loop and involves the AB term in the same manner, even though the frequency dependency is associated with B rather than A.

CASE III: Ideal Inverting Amplifier, Capacitance Shunting Feedback Resistance.

With an ideal amplifier and all elements resistive except for the feedback impedance, as shown in Fig. 6.21,

$$Z_S = R_S$$

$$Z_i = R_i$$

$$Z_F = \frac{R_F}{1 + j\omega R_F C_F} = \frac{R_F}{1 + j(\omega/\omega_f)}$$

where $\omega_f = 1/R_F C_F$. The overall response is

$$\frac{V_o}{V_s} = \frac{R_i}{R_i + R_s} \frac{R_F}{R_1 + R_F} \frac{-A_{0M}}{1 + \{A_{0M}[R_1/(R_1 + R_F)][1 + j(\omega/\omega_f)]\}} \tag{6.51}$$

In this case, the bandwidth limitation would seem to be due to the $R_F C_F$ time constant alone and not explicitly to the reduction in gain. However, the gain reduction factor at low frequencies, which is

$$B_0 = \frac{R_1}{R_1 + R_F}$$

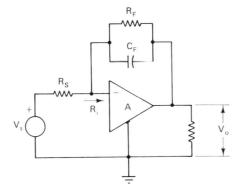

Figure 6.21 Shunt–shunt feedback with RC feedback impedance.

controls the value of R_F to be used. From the preceding equation

$$R_F = \frac{R_1(1 - B_0)}{B_0}$$

and ω_f can be written as

$$\omega_f = \left(\frac{1}{R_1 C_F}\right) \frac{B_0}{1 - B_0}$$

showing that ω_f increases with B_0 and for a prescribed B_0; it is seen that the $R_1 C_F$ time constant controls the frequency response.

For pure capacitance feedback, as shown in Fig. 6.22,

$$Z_F = \frac{1}{j\omega C_F}$$

and the overall gain can be written as

$$\frac{V_o}{V_s} = \frac{R_i}{R_i + R_s} \frac{-A_{0M}}{1 + j\omega R_1 C_F(1 + A_{0M})} \tag{6.52}$$

This equation shows that the overall bandwidth is determined by the effective time constant

$$R_1 C_F(1 + A_{0M}) = \frac{1}{\omega_f}$$

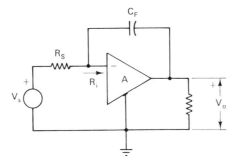

Figure 6.22 Shunt–shunt feedback with capacitor as feedback impedance.

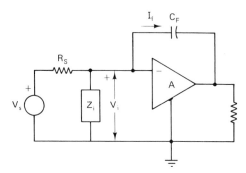

Figure 6.23 Circuit for determination of input admittance with capacitance feedback.

which is dependent on amplifier gain. The bandwidth is reduced for large gains even for very small values of limiting capacitance.

It is useful to consider the effect of C_F on the input impedance from consideration of Fig. 6.23, where Z_i, the input impedance, is for convenience placed external to the amplifier and in parallel with any that is introduced externally. Due to C_F, there is an additional component of input admittance given by $Y_{if} = I_f/V_i$; then since $V_o = A_{0M}V_i$, this component can be written as

$$Y_{if} = j\omega C_F(1 + A_{0M}) \tag{6.53}$$

Thus the effect of the feedback capacitance C_F is to create an effective input capacitance component of $C_F(1 + A_{0M})$. This is the general case of the Miller capacitance discussed in Chapter 4 for single-device amplifiers.

In most amplifiers employing shunt–shunt feedback the frequency dependencies discussed in the preceding examples exist simultaneously, and each individually has the effect shown, but not precisely to the same degree because they do not act independently. The more complex frequency characteristics that result will be discussed in a later section.

6.9 FREQUENCY CHARACTERISTICS OF THE DIFFERENTIAL FEEDBACK AMPLIFIER

The differential feedback amplifier discussed in Sec. 6.6, in general, is a negative feedback amplifier if the signal is applied to the noninverting input terminal and the feedback to the inverting input terminal. The frequency dependencies resulting from amplifier frequency characteristics and capacitances in the network elements are similar in some respects and different in others. For example, like Case I of the shunt–shunt amplifier with the amplifier itself being the only source of frequency limitation, as indicated in Fig. 6.24 the defining equations are:

$$A(j\omega) = \frac{A_{0M}}{1 + j(\omega/\omega_2)}$$

$$K_1 = \frac{R_{is}}{R_{is} + R_S}$$

$$B = \frac{-R_{iF}}{R_F + R_{iF}}$$

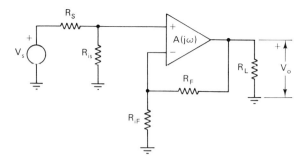

Figure 6.24 Frequency-dependent amplifier with differential feedback.

The response is like that of Eq. 6.47 except that it is noninverting at zero frequency, and there is no K_1 dependency on B; that is, the term $R_F/(R_1 + R_F)$ does not exist. The overall gain equation is

$$\frac{V_2}{V_s} = \frac{R_{is}}{R_s + R_{is}} \frac{A_{OM}}{1 + A_{OM} \dfrac{R_{if}}{R_F + R_{if}}} \frac{1}{1 + j\dfrac{\omega}{\omega_2} \dfrac{1}{1 + A_{OM}[R_{if}/(R_F + R_{if})]}} \quad (6.54)$$

The characteristic comparable to Case II of the shunt–shunt amplifier with the existence of shunt capacitance is shown in Fig. 6.25, where it is necessary to consider the shunting capacitance at both inputs. In this case

$$K_1 = \frac{R_{is}}{R_s + R_{is}} \frac{1}{1 + j(\omega/\omega_{is})}$$

where

$$\omega_{is} = \frac{1}{C_{is}[(R_s R_{is})/(R_s + R_{is})]}$$

and

$$B = -\frac{R_{if}}{R_F + R_{if}} \frac{1}{1 + j(\omega/\omega_{if})}$$

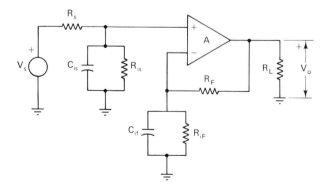

Figure 6.25 Differential amplifier with RC input and feedback networks.

where

$$\omega_{if} = \frac{1}{C_{if}[R_F\,R_{if}/(R_F\,+\,R_{if})]}$$

The overall gain is

$$\frac{V_o}{V_s} = \frac{R_{is}}{R_s\,+\,R_{is}}\,\frac{1\,+\,j(\omega/\omega_{if})}{1\,+\,j\dfrac{\omega}{\omega_{is}}}\,\frac{A_{0M}}{\left(1\,+\,\dfrac{A_{0M}\,R_{if}}{R_F\,+\,R_{if}}\right)\left(1\,+\,j\dfrac{\omega}{\omega_{if}}\dfrac{1}{1\,+\,\dfrac{A_{0M}\,R_{if}}{R_F\,+\,R_{if}}}\right)} \tag{6.55}$$

This is comparable to Eq. 6.48 for the shunt–shunt amplifier, showing the same gain–bandwidth relationship except for the additional term due to the two time constants. For the special case of $\omega_{is} = \omega_{if}$, the two equations are identical in form and differ only in the matter of polarity inversion and the absence of the $R_F/(R_1 + R_F)$ term.

Similarly, if capacitance C_F is introduced in parallel with the feedback resistance R_F comparable to Case III of the shunt–shunt amplifier, the results are comparable, again except for the interactive term $R_F/(R_1 + R_F)$.

In general, then, for the differentially introduced feedback, the statement of a constant gain–bandwidth is more literally true than it is for the shunt–shunt feedback amplifier because of the isolation of the feedback path from the signal path.

6.10 FEEDBACK AMPLIFIERS WITH MULTIPLE POLE RESPONSES

For the amplifiers that have been discussed, the various frequency limitations may exist simultaneously. The overall forms of response that may result and their implications will be analyzed, starting with two time constants (i.e., a two-pole response function) and proceeding to more general cases.

Two-Time-Constant Amplifier (Two-Pole Response, No Zeros)

For the amplifier with two limiting time constants and none in the feedback loop; i.e.

$$A(j\omega) = \frac{A_{0M}}{[1\,+\,j(\omega/\omega_1)]\,[1\,+\,j(\omega/\omega_2)]}$$

$$B = -B_M$$

$$K_1 = K_{1M}$$

the overall gain may be expressed as

$$\frac{V_o}{V_s} = \frac{K_{1M}A_{0M}}{(1\,+\,A_{0M}B_M)\,-\,k_2(\omega/\omega_2)^2\,+\,j(\omega/\omega_2)(1\,+\,k_2)} \tag{6.56}$$

where $\omega_2 = k_2\omega_1$. This response can be written in terms of its magnitude and phase as

$$\left(\left|\frac{V_o}{V_s}\right|\right)^2 = \frac{(K_{1M}A_{0M})^2}{(1 + A_{0M}B_M)^2 + [(1 + k_2)^2 - 2k_2(1 + A_{0M}B_M)](\omega/\omega_2)^2 + k_2^2(\omega/\omega_2)^4} \quad (6.57)$$

$$\theta = -\tan^{-1}\frac{(\omega/\omega_2)(1 + k_2)}{1 + A_{0M}B_M - k_2(\omega/\omega_2)^2} \quad (6.58)$$

The phase shift as a function of frequency progresses negatively, reaching $-90°$ at $k_2(\omega/\omega_2)^2 = (1 + A_{0M}B_M)$ and approaches $180°$ as $\omega \to \infty$, but never reaching it. Such a phase characteristic can never permit instability.

The precise form of the magnitude of the overall response can be seen from Eq. 6.57 to depend on the relationship among ω_1, ω_2, and $A_{0M}B_M$.

For example, for

$$(1 + k_2)^2 = 2k_2(1 + A_{0M}B_M) \quad (6.59)$$

the intermediate term involving $(\omega/\omega_2)^2$ vanishes, and the overall response is the maximally flat form of

$$\left|\frac{V_o}{V_s}\right| = \frac{K_{1M}A_{0M}}{1 + A_{0M}B_M}\frac{1}{\sqrt{1 + \dfrac{k_2^2}{(1 + A_{0M}B)^2}\left(\dfrac{\omega}{\omega_2}\right)^4}} \quad (6.60)$$

For the special case of $\omega_1 = \omega_2$, that is, $k_2 = 1$, $A_{0M}B_M = 1$ and Eq. 6.60 becomes

$$\left|\frac{V_o}{V_s}\right| = \frac{K_1A_{0M}}{2}\frac{1}{\sqrt{1 + \frac{1}{4}(\omega/\omega_2)^4}} \quad (6.61)$$

Alternatively, for the general case, Eq. 6.56 can be written in terms of $s = j\omega$ as

$$\frac{V_o}{V_s} = \frac{K_1A_{0M}\omega_2^2/k_2}{s^2 + \{[\omega_2(1 + k_2)]/k_2\}s + [(1 + A_{0M}B_M)/k_2]\omega_2^2} \quad (6.62)$$

which is of the form

$$\frac{K_1A_{0M}\omega_2^2/k_2}{(s - s_1)(s - s_2)}$$

where s_1 and s_2, which are the roots of the quadratic, are known as the poles of the overall response function. The poles are

$$s_1, s_2 = -\frac{\omega_2(1 + k_2)}{2k_2} \pm \frac{\omega_2}{2k_2}\sqrt{(1 + k_2)^2 - 4(1 + A_{0M}B_M)k_2} \quad (6.63)$$

Regardless of the values of k_2 and $A_{0M}B_M$, there are two poles always in the left half of the complex frequency plane. For example, for

$$(1 + k_2)^2 > 4(1 + A_{0M}B_M)k_2$$

the poles are on the negative real axis, and for

$$(1 + k_2)^2 = 4(1 + A_{0M}B_M)k_2$$

the two poles are coincident at

$$s_1, s_2 = -\frac{\omega_2(1 + k_2)}{2k_2}$$

But for $4(1 + A_{0M}B) > (1 + k_2)^2$, the poles are complex. Specifically, for the maximally flat response defined by Eq. 6.60, the poles are

$$s_1, s_2 = -\frac{\omega_2(1 + k_2)}{2k_2} \pm j\frac{\omega_2(1 + k_2)}{2k_2} \qquad (6.64)$$

These poles are plotted at the points indicated in Fig. 6.26.

Furthermore, for all values for complex poles, the two poles always lie equidistant from the negative real axis along the dashed line, and they all have negative real parts but move closer to the j axis as ω_2 becomes smaller, but never move into the right half-plane. This is analogous to the observation from Eq. 6.58 that the phase shift never reaches 180°.

The *transient response,* usually defined as the response to a step function as indicated in Fig. 6.27(a), of

$$v_s(t) = VU(t)$$

having the Laplace transform of

$$V(s) = V\mathcal{L}U(t) = \frac{V}{s}$$

Then the output can be written as

$$\frac{V_o}{V_s}(s) = \frac{K_{1M}A_{0M}\omega_2^2/k_2}{s[(s + \alpha)^2 + \beta^2]} \qquad (6.65)$$

where from Eq. 6.62

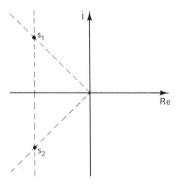

Figure 6.26 Pole locations for two-pole maximally flat response.

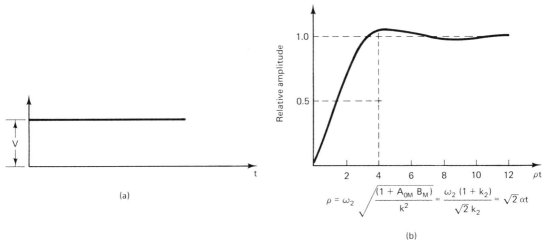

(a)

$$\rho = \omega_2 \sqrt{\frac{(1 + A_{0M}\,B_M)}{k^2}} = \frac{\omega_2\,(1 + k_2)}{\sqrt{2}\,k_2} = \sqrt{2}\,\alpha t$$

(b)

Figure 6.27   Transient response for two-pole maximally flat frequency response.

$$\alpha = \frac{\omega_2(1 + k_2)}{2k_2}$$

$$\beta = \frac{\omega_2}{2k}\sqrt{4(1 + A_{0M}B)k_2 - (1 + k_2)^2}$$

for complex poles or

$$\frac{V_o}{V_s}(s) = \frac{K_{1M}A_{0M}\omega_2^2/k_2}{s[s^2 + 2\alpha s + \beta_o^2]} \qquad (6.66)$$

where

$$\beta_o^2 = \alpha^2 + \beta^2) = \frac{1 + A_{0M}B_M}{k_2}\omega_2^2$$

The inverse transform of Eq. 6.65 is

$$\frac{V_o}{V_s}(t) = K_{1M}A_{0M}\omega_2^2/k_2\left[1 + \frac{\beta_o}{\beta}e^{-\alpha t}\sin(\beta t - \psi)\right] \qquad (6.67)$$

where $\psi = \tan^{-1}(\beta/-\alpha) = 180° - \sin^{-1}(\beta/\beta_o)$.

For the maximally flat case defined by Eqs. 6.59 and 6.64,

$$\alpha = \beta = \frac{\omega_2(1 + k_2)}{2k_2}$$

and Eq. 6.67 becomes

$$\frac{V_o}{V_s}(t) = \frac{K_{1M}A_{0M}}{1 + A_{0M}B_M}\left[1 + \sqrt{2}\,e^{-\alpha t}\sin\left(\alpha t - \frac{3\pi}{4}\right)\right] \qquad (6.68)$$

This is plotted in Fig. 6.27(b), showing that there is some overshoot (rise above the final steady value) even for maximal flatness. For an increasing spread of $\pm\beta$, the overshoot will increase, but will always decay to its final steady value.

Three-Pole Feedback Factor

If there are three frequency limiting time constants for the amplifiers that have been discussed, which may be associated with A and/or B such that

$$AB = \frac{-A_{0M}B_M}{[1 + j(\omega/\omega_1)][1 + j(\omega/\omega_2)][1 + j(\omega/\omega_3)]} \tag{6.69}$$

there will be a factor appearing in the overall response equation of the form

$$F(j\omega) = \frac{1}{(1 + A_{0M}B_M) - (\omega/\omega_2)^2 (M + N) + j(\omega/\omega_2)[(M + 1) - N(\omega/\omega_2)^2]} \tag{6.70}$$

where

$$M = \frac{\omega_2}{\omega_1} + \frac{\omega_2}{\omega_3} \quad \text{and} \quad N = \frac{\omega_2^2}{\omega_1\omega_3}$$

The magnitude function can be written as

$$|F(j\omega)|^2 =$$

$$\frac{1}{(1 + A_{0M}B_M)^2 + [(1 + M)^2 - 2(1 + A_{0M}B_M)(M + N)]\left(\dfrac{\omega}{\omega_2}\right)^2 + [(M + N)^2 - 2N(M + 1)]\left(\dfrac{\omega}{\omega_2}\right)^4 + N^2\left(\dfrac{\omega}{\omega_2}\right)^6} \tag{6.71}$$

It is apparent from examination of the imaginary part of Eq. 6.70 that at some frequency the phase shift reaches 180° and exceeds it at frequencies beyond. It is also apparent from examination of Eq. 6.71 that the maximally flat response cannot be obtained inasmuch as the coefficient of $(\omega/\omega_2)^4$ cannot vanish for any realizable relationship among ω_1, ω_2, and ω_3.

The nature of the response can best be examined by rewriting Eq. 6.70 in terms of a normalized complex frequency, $s_n = j(\omega/\omega_4)$ and including the step function of excitation, $1/s_n$. The result is

$$\frac{F(s_n)}{s_n} = \frac{1}{Ns[s^3 + [(M + N)/N]s^2 + [(M + 1)/N]s + (1 + A_{0M}B_M)]} \tag{6.72}$$

Most of the important points concerning possible responses can be illustrated by considering the coincident time-constant case of $\omega_1 = \omega_2 = \omega_3$. In this case, $M = 2$, $N = 1$, and Eq. 6.72 reduces to

$$\frac{F(s_n)}{s_n} = \frac{1}{s[s^3 + 3s^2 + 3s + (1 + A_{0M}B_M)]} \tag{6.73}$$

Three examples will be given.

(a) Choice of $(1 + A_{0M}B_M)$ to place one pair of complex poles on the j axis; for example, $(1 + A_{0M}B_M) = 9$. For this case, the poles are at $s_1 = -3$, $s_2 = -j\sqrt{3}$, and $s_3 = +j\sqrt{3}$, as indicated in Fig. 6.28(a). The inverse transform scaled to include ω_2 is

$$\frac{v_o(t)}{V} = 1 - \frac{1}{4}e^{-3\omega_2 t} + \frac{\sqrt{3}}{2}\sin(\sqrt{3}\omega_2 t - 120°) \tag{6.74}$$

and is plotted in Fig. 6.28(b). There is a residual sinusoidal response remaining after the initial transient, which is not a function of the excitation. This response would thus be classified as *unstable*.

(b) Choice of $(1 + A_{0M}B_M)$ to place poles in left half of complex s plane; for example, $(1 + A_{0M}B_M) = 6$. For this case, $s_1 = -2.72$, $s_2 = -0.145 + j1.48$, and $s_3 = 0.145 - j1.48$, as shown in Fig. 6.29(a), and the real-time response is

$$\frac{v_o(t)}{V} = \left[1 - \frac{1}{4}e^{-2.72\omega_2 t} + \frac{12}{13}e^{-0.145\omega_2 t}\sin(1.48\omega_2 t - 126°)\right] \tag{6.75}$$

This response, plotted in Fig. 6.29(b), shows an initially large but decaying ripple corresponding to the exponentially decaying sinusoidal term in Eq. 6.75. After the decay of this initial perturbation, the final amplitude as determined by the low-frequency gain is reached. The amplifier is classified as stable because there are no sustained or growing oscillations that are not related to the excitation. However, because of the excessive overshoot, such a response would not be satisfactory for most amplifier applications.

(c) Choice of $(1 + A_{0M}B_M)$ to place complex poles in right half of the complex s plane; for example, $(1 + A_{0M}B_M) = 10$. For this case, $s_1 = -3.154$, $s_2 = +0.077 - j1.866$, and $s_3 = +0.077 + j1.866$, as indicated in Fig. 6.30(a). The resultant real-time response is

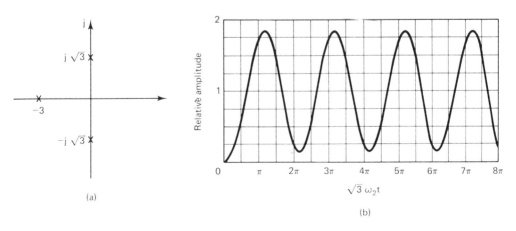

Figure 6.28 Three-pole response for two poles on j axis.

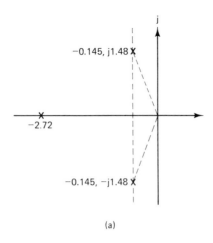

(a)

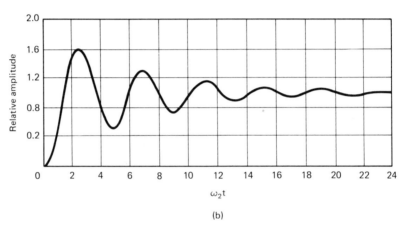

(b)

Figure 6.29 Three-pole response with all poles in left half-plane.

$$\frac{v_o(t)}{V} = \left[1 - \frac{1}{4} e^{-3.154\omega_2 t} + \frac{11}{13} e^{0.077\omega_2 t} \sin(1.866\omega_2 t - 117.6°) \right] \quad (6.76)$$

This response, plotted in Fig. 6.30(b), shows an exponentially growing oscillation independent of the excitation, which classifies it as unstable.

In a practical amplifier, this increasing amplitude could progress only until some limiting value corresponding to the limiting amplitude range of the amplifier itself is reached. Such an unstable response is useless for a linear amplifier. However, this form of response, together with certain amplifier nonlinearity conditions, permits it to form the basis for sinusoidal oscillator design.

Information derived from the pole locations of examples (a), (b), and (c) can be correlated with that which can be obtained from a complex frequency-phase plot of the return ratio, $T_r = -(AB)(j\omega)$. From Eq. 6.69 for $\omega_1 = \omega_2 = \omega_3$, T_r in normalized form is

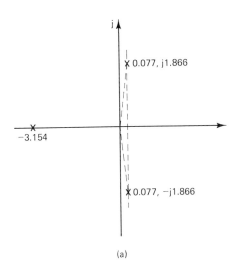

(a)

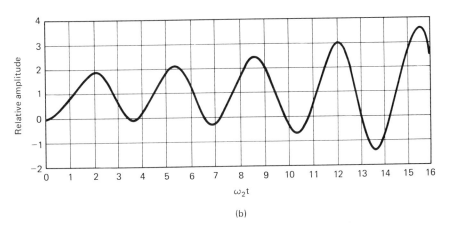

(b)

Figure 6.30 Three-pole response with two poles in right half-plane.

$$\frac{T_r}{A_{0M}B_M} = \frac{1}{1 - 3(\omega/\omega_2)^3 + j(\omega/\omega_2)[3 - (\omega/\omega_2)^2]} \qquad (6.77)$$

The solid line in Fig. 6.31, starting with $\omega = 0$ and progressing clockwise to $\omega = \infty$, shows that 180° phase shift is reached at $\omega = \sqrt{3}\omega_2$ and exceeds it at frequencies beyond. If the plot is started at $\omega = -\infty$ and progresses to $\omega = 0$, the dashed curve is traced.

If $A_{0M}B_M = 8$ as in example (a), the phase shift is $-180°$ when $T_r = -8(0.125) = -1$. If $A_{0M}B_M = 5$ as in example (b), it is $-180°$ when $T_r = -5(0.125) = 0.625$. If $A_{0M}B_M = -9$ as in example (c), it is $-180°$ when $T_r = -9(0.125) = -1.125$. The results can be stated in terms of a clockwise plot of the T_r function in the complex plane progressing from $\omega = -\infty$ to $\omega = +\infty$ and passing

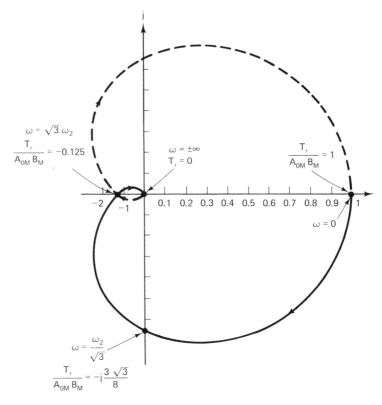

Figure 6.31 Nyquist plot for three-pole response.

through the -1, $j0$ point for the continuously oscillating response and completely enclosing it (in a clockwise direction) for the exponentially growing response, but passing to the right of it (between 0 and -1) for the damped oscillatory or stable response.

Comments on Three-Pole Response Forms. The discussion following Eq. 6.69, which can alternatively be written as

$$AB = \frac{-A_{0M}B_M\omega_1\omega_2\omega_3}{(s + \omega_1)(s + \omega_2)(s + \omega_3)} \qquad (6.78)$$

with   (6.79)

$$\frac{1}{1 - AB} = \frac{(s + \omega_1)(s + \omega_2)(s + \omega_3)}{(s + \omega_1)(s + \omega_2)(s + \omega_3) + A_{0M}B_M\,\omega_1\omega_2\omega_3}$$

dealt only with the portion

$$F(\delta) = \frac{1}{(s + \omega_1)(s + \omega_2)(s + \omega_3) + A_{0M}B_M\,\omega_1\omega_2\omega_3} \qquad (6.80)$$

The overall response may contain the zeros of Eq. 6.79, which modify the overall transient response but are not involved in the conditions relating to stability. Hence, in evaluating conditions for stability, it is only necessary to deal with the poles of the overall transfer function, which are the *zeros* of the $1/(1 + T_R)$ function, and the statements for stability conditions with respect to $F(s)$ apply to the zeros of the function $1 + T_r(s)$, which are identical.

Extension to n-Pole Response Forms

For n time constants in the AB loop, an equation analogous to Eq. 6.78 can be written as

$$AB = \frac{-A_{0M}B_M\omega_1\omega_2\omega_3 \cdots \omega_n}{(s + \omega_1)(s + \omega_2)(s + \omega_3) \cdots (s + \omega_n)} \tag{6.81}$$

and $F(s)$ analogous to Eq. 6.80 is

$$F(s) = \frac{1}{(s + \omega_1)(s + \omega_2)(s + \omega_3) \cdots (s + \omega_n) + A_{0M}B_M\omega_1\omega_2\omega_3 \cdots \omega_n} \tag{6.82}$$

which has the general form

$$F(s) = \frac{1}{s^n + Q_1s^{n-1} + Q_2s^{n-2} + \cdots + Q_n} \tag{6.83}$$

or

$$F(s) = \frac{1}{(s - s_1)(s - s_2)(s - s_3) \cdots (s - s_n)} \tag{6.84}$$

Criteria for the stability of such n-pole responses can be extrapolated from those determined for the three-pole response.

Effect of Right Half-Plane Zeros in the AB Function

The examples for the two- and three-pole responses did not include the effect of right half-plane zeros (such as might exist in a single-stage inverting amplifier as discussed in Chapter 4). Such a response exhibits a more rapid change of phase than that due to the poles alone. The possible consequences of such right half-plane zeros can be illustrated by the following examples using a two-pole, one-zero gain function.

$$A = \frac{-(1 - j\omega/\omega_3) A_{OM}}{[1 + j(\omega/\omega_1)] [1 + j(\omega/\omega_2)]} \tag{6.85}$$

or

$$A = \frac{(\omega_1\omega_2/\omega_3)(s - \omega_3)A_{0M}}{(s + \omega_1)(s + \omega_2)} \tag{6.86}$$

Then if B is positive and nonfrequency dependent, the poles of the response function or the zeros of the $1 - AB$ function are the solutions of

$$s^2 + [(\omega_1 + \omega_2) - \frac{\omega_1\omega_2}{\omega_3} A_{0M}B]s + A_{0M}B\omega_1\omega_2 \qquad (6.87)$$

Without pursuing the matter further, it is quite obvious that there is a possibility for instability *even though* the AB function has only two poles. For example, the s coefficient in Eq. 6.87 is zero when

$$A_{0M}B = \frac{\omega_1 + \omega_2}{\omega_1\omega_2} \omega_3 \qquad (6.88)$$

which is the condition for a sinusoidal component in the output.

Usually, unless deliberately made otherwise,

$$\omega_3 >> \frac{\omega_1\omega_1}{\omega_1 + \omega_2}$$

which means that the AB value can be quite high before the onset of instability; nevertheless, the potential for instability does exist.

Thus, for the all-pole response, instability cannot occur with less than three poles in the AB function; however, it can occur for a two-pole response if a right half-plane zero exists because it can result in imaginary axis or right half-plane poles in the $1 - AB$ function.

6.11 FORMAL EXPRESSIONS OF STABILITY CRITERIA FOR NEGATIVE FEEDBACK AMPLIFIERS

The discussion in Sec. 6.10 correlating various approaches for evaluating the stability of the three-pole (or the two-pole) response with one-right half-plane zero response fit general conditions for stability of feedback amplifiers with n-pole responses, which can be stated as follows if they are restricted to linear systems with parameters that do not vary with time:

(1) A necessary and sufficient condition that a system having linear, time-invariant parameters be stable is that all poles of the overall transfer function of $1/1 - AB$ have negative real parts.

The proof of this criterion is very simple. Any transfer function having the Laplace transform

$$F(s) = \frac{1}{(s - s_1)(s - s_2) \cdots (s - s_n)} \qquad (6.89)$$

has an inverse containing a term of the form

$$f(t) = K_1 e^{s_1 t} + K_2 e^{s_2 t} + \cdots + K_n e^{s_n(t)} \qquad (6.90)$$

Thus, any pole $s_k = \alpha_k + j\omega_k$ that has a positive real part (i.e., $\alpha_k = $ positive) results in a positively rising exponential in the term containing s_k, which ultimately controls the response for large values of t. Any pole that has the part $\alpha_k = 0$ results in a continuously oscillatory component. Both of these conditions are classified as unstable.

(2) A necessary and sufficient condition that a linear, time-invariant feedback system be stable is that all zeros of the return difference function $(1 - AB)$ or $(1 + T_r)$ have negative real parts. This condition follows from condition (1) since the zeros of the return difference function become poles of the overall transfer function, which involves the feedback loop.

(3) A necessary and sufficient condition that a linear, time-invariant feedback system be stable is that a plot of the complex return ratio T_r not pass through or completely enclose the point $(-1, j0)$ in a clockwise direction, as ω is continuously varied from $-\infty$ through 0 to $+\infty$. Such a plot is referred to as a *Nyquist diagram* and the statement of stability as *Nyquist's criterion*. The examples of the previous section illustrate, but do not constitute, a general proof.

6.12 REDUCTION OF AMPLIFIER DISTORTION WITH NEGATIVE FEEDBACK

The bandwidth limitation of an amplifier can be thought of as a form of distortion, and to the extent that the bandwidth can be extended by the use of negative feedback, it can be considered as a reduction of such distortion, but at the expense of decreasing the gain in the same proportion. It would therefore seem intuitively that other forms of distortion (such as amplifier nonlinearity) might be reduced in the same manner. This perception is explored as follows:

A very simple feedback amplifier, without being complicated by input or output impedance considerations or by input summing networks, is shown in Fig. 6.32. Within this framework, let us assume that somewhere within the amplifier with gain $A = A_1 A_2$ is a distortion voltage, V_d, with A_2 being the gain following the point at which V_d is introduced. The output voltage is

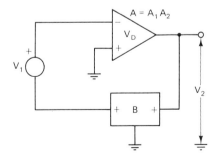

Figure 6.32 Feedback model for distortion representation.

$$V_2 = -\frac{A_1A_2V_1 + A_2V_d}{1 + A_1A_2B} \tag{6.91}$$

The degree of reduction of distortion by the $(1 + A_1A_2B)$ factor is a function of where in the amplifier the distortion voltage V_d occurs, with the output signal arising from the distortion voltage being

$$V_{2d} = \frac{-A_2V_d}{1 + A_1A_2B} \tag{6.92}$$

If the distortion voltage is at the output of the final amplifier stage itself, $A_2 = 1$, and the full V_d is reduced by the $1 + A_1A_2B$ factor. If it occurs within the amplifier and is of the same magnitude, it is multiplied by A_2, so it is not reduced by the same factor unless A_1 is increased, which in some cases can be accomplished by adding distortion-free stages at the input.

Random Noise as Distortion

Random or quasi-random noise generated within an amplifier can be considered as distortion. However, the most significant components of noise are those involved in the low-level input stages before the input voltage is amplified significantly. In this case, feedback would not in general improve the output signal-to-noise ratio because the signal and noise would be reduced in the same proportion. As a matter of fact, if feedback is employed in an amplifier that is used to amplify extremely small signals, it is possible that resistive elements incorporated in the feedback network will add significant sources of noise not otherwise present.

For example, a current-drive video amplifier such as might be used in a television camera with various noise sources present is shown in Fig. 6.33. To reduce its noise contribution, R_i is increased to the point that considerable bandwidth degradation might take place at the input due to C_i. This is compensated for in a later stage of the amplifier after the signal has been amplified substantially, which might be done without feedback $(R_F \rightarrow \infty)$. However, if such compensation is achieved using negative feedback, the noise introduced by R_F itself must be included.

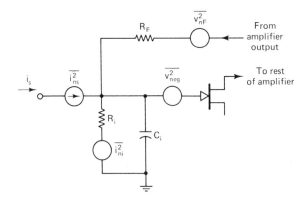

Figure 6.33 Representation of noise sources in feedback amplifier.

6.13 FEEDBACK AMPLIFIERS AS SINUSOIDAL OSCILLATORS

A clue to the generation of sinusoidal oscillations using a *negative feedback* amplifier with three time constants in the loop was given in the previous section as indicated in Fig. 6.28, where two of the three poles were placed on the j axis of the s plane. In that example, the poles were assumed to be within the amplifier structure.

A similar result can be achieved with an ideal amplifier and a three-section RC filter in the feedback loop, as shown in Fig. 6.34.

Solving this circuit for the return difference function $(1 - AB)$, its numerator contains a polynomial whose zeros can be found from

$$s^3 + \frac{R_2C_2 + R_1C_1 + R_1C_2}{R_1C_1R_2C_2} s^2 + \frac{R_3C_2 + (R_2C_3 + R_1C_1 + R_1C_2)[1 + (C_2/C_3)]}{R_1C_1R_2C_2R_3C_2} s$$

$$+ \left[\frac{C_3}{C_2}\left(1 + \frac{C_2}{C_3}\right) - A\right] \frac{C_2/C_3}{R_1C_1R_2C_2R_3C_3} = 0 \qquad (6.93)$$

This should be of the form

$$(s + \gamma)(s^2 + \beta^2) = 0 \qquad (6.94)$$

for two poles at $\pm j\beta$ on the j axis, with one at $s = -\gamma$ on the negative real axis.

Many relationships among the RC time-constants and A allow such a placement of zeros. This will be illustrated by the simplest but not necessarily the optimum case, using all C's equal, all R's equal, and hence all time constants equal.

For this special case, Eq 6.93 reduces to

$$s^3 + 3\omega_2 s^2 + 7\omega_2^2 s + (2 - A)\omega_2^3 \qquad (6.95)$$

which from Eq. 6.90 should be of the form

$$s^3 + \gamma s^2 + \beta^2 s + \gamma\beta^2$$

where $\gamma = 3\omega_2$, $\beta^2 = 7\omega_2^2$, and $\gamma\beta^2 = 21\omega_2^3$. Hence

$$21\omega_2^3 = 2 - A\omega_2^3$$

and $A = -19$.

Other choices of relative time constants would lead to other values of A for conditions of sinusoidal oscillations.

In similar fashion, it can be shown that conditions for oscillation can be established

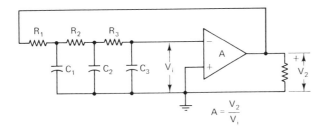

Figure 6.34 Phase-shift oscillator using low-pass RC feedback network.

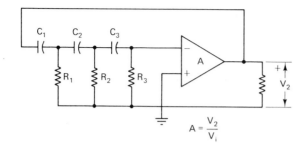

$$A = \frac{V_2}{V_i}$$

Figure 6.35 Phase-shift oscillator using high-pass *RC* feedback network.

using a three-section high-pass *RC* filter, as shown in Fig. 6.35. Such oscillators, based on the use of a negative feedback amplifier with a phase-shifting network to restore the required positive feedback condition, are referred to as *phase-shift* oscillators.

Sinusoidal Oscillators Using Noninverting Amplifiers

If, as a starting point, a noninverting amplifier is used in a positive feedback mode, the initial 180° phase shift, which is equivalent to an amplifier polarity reversal, is not required, and conditions for oscillation can be established using somewhat simpler feedback networks.

One well-known such configuration is that shown in Fig. 6.36. In this case, the zeros of the $1 - AB$ function can be obtained from

$$s^2 + \left[\frac{1}{R_1 C_1} + \frac{1}{R_2 C_2} + \frac{1}{R_2 C_1} (1 - A) \right] s + \frac{1}{R_1 C_1 R_2 C_2} = 0 \qquad (6.96)$$

and the condition for sinusoidal oscillations can be obtained by setting the *s* coefficient equal to zero, with the result that the condition for oscilation is

$$A = 1 + \frac{R_2}{R_1} + \frac{C_1}{C_2} \qquad (6.97)$$

with the frequency of oscillation being

$$\omega_2 = \frac{1}{\sqrt{R_2 C_1 R_2 C_2}} \qquad (6.98)$$

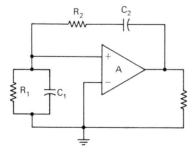

Figure 6.36 *RC* oscillator using noninverting amplifier.

Any oscillator using this particular feedback configuration is referred to as a Wien bridge oscillator because the particular form of the RC's are the two arms of such a bridge circuit.

Practical Considerations for Oscillator Design

The equations that have been developed in this section for sinusoidal oscillations can only be regarded as approximate since they would depend on a linear amplifier with absolute gain stability. A slight decrease in gain would result in a decay of oscillations similar to that shown in Fig. 6.29, while an increase in gain would result in increasing amplitude of oscillation, as illustrated in Fig. 6.30, until limited by the output voltage limits of the amplifier itself; this, if the amplifier is linear up to the point of such limiting, would result in a flattening of the positive and negative peaks of the wave form or prevention of continuation of oscillations.

As a matter of practical oscillator design, the amplifier gain should be made sufficiently greater than the design center value so that nominal variation in such gain could not bring it below the design center value, and it should incorporate a controlled nonlinearity (decreasing gain) in the vicinity of the peaks such that peak flattening is minimized.

PROBLEMS

6.1 From an examination of Eq. 5.2:
   (a) Identify the components A and B in the representation $A/(1 - AB)$.
   (b) From an examination of the equation and processes within the circuit itself, classify the types of feedback the terms involving μ_{re} and R_E represent (i.e., positive or negative, voltage or current, and more specifically shunt–shunt, series–shunt, etc.).

6.2 Starting with the h-parameter model of the transistor with μ_{re} neglected (i.e., using r_{bes}, β, and r_{ceo}, change it to the model of Fig. 6.2 and identify R_i, and R_o, and A_o.

6.3 In the circuit shown, the amplifier with gain A is a single transistor represented by its low-frequency common-emitter h parameters, but with R_L and r_{ceo} lumped in parallel as R'_L.

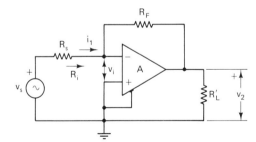

Prob. 6.3

   (a) Draw the circuit and obtain expressions for K_1, A, and B.
   (b) Write an equation for $R_i = v_i/i_1$, separating the components involving μ_{re} and R_F.
   (c) Write an equation for the return difference, separately identifying components containing μ_{re} and R_F, and classify each as representing positive or negative feedback.

6.4 The circuit shown is obviously drawn in terms of the common-emitter low-frequency h parameters of a transistor.

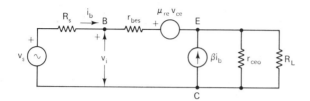

Prob. 6.4

(a) Using $R'_L = r_{ceo}R_L/(r_{ceo} + R_L)$, determine K_1, A, and B.

(b) Using the results of part (a), obtain v_2/v_s, putting your results in such form that the $1/(1 - AB)$ factor is identified.

(c) Change the form of your result from part (b) or Eq. 4.32 with $R_c = 0$ to see if they are the same.

6.5 Consider the circuit shown, assuming that it is properly biased, and use the common-emitter h parameters with μ_{re} and R_E neglected.

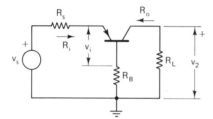

Prob. 6.5

(a) Determine K_1, A, and B for the feedback equation.

(b) Determine R_i and R_o.

(c) Write an equation for overall gain, identifying the term $(1 - AB)$.

(d) Classify the feedback through R_B in all possible ways.

6.6 The feedback amplifier of Fig. 6.5 incorporates an amplifier whose gain $A = -1000$ and whose output impedance is sufficiently low such that it is practically independent of loading conditions and whose input resistance R_i is so high as to be negligible.

(a) Write an equation for R_F in terms of a design value for v_o/v_s and determine its value for $R_s = 2000\ \Omega$ and for $v_o/v_s = 100$.

(b) Instead of the ideal amplifier of part (a), an amplifier is used whose output resistance $R_o = 2000\ \Omega$; determine the required open-circuit gain, A_o, for the overall gain specification of part (a) and for $R_L = 1000\Omega$.

(c) For part (b), what is the effective output impedance R'_o looking back into the amplifier terminals? Compare its value with that looking back into the feedback path itself.

(d) What is the effective input impedance looking from R_s into the amplifier, including the feedback path?

6.7 It is suggested that the circuit shown be solved using the model of Fig. 6.12 with the approximation that R_L is sufficiently low that μ_{re} and r_{ceo} can be neglected.

(a) Write equations for K_1, A, and B.

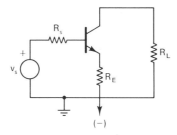

Prob. 6.7

(b) Write the equation for overall gain using the components of part (a) and compare it with Eq. 4.17.

6.8 Examine the general characteristics of the circuit shown.

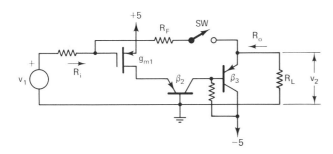

Prob. 6.8

(a) With SW open, write an equation for v_2/v_1 using device models involving only g_{m1}, β_2, and β_3. Explain the basis for neglecting μ_{re} and r_{ceo}.

(b) With SW closed, express K_1, A, and B in the conventional feedback equations.

(c) Write an equation for v_2/v_1 with $R_F \gg R_L$.

(d) Without making the assumption of part (c), derive equations for R_i and R_o.

(e) Determine the relative values for R_F to make $R_o = R_L = R_1 = R_i$ and the voltage gain under such conditions.

6.9 The amplifier shown is to be designed as a unity gain voltage follower (i.e., $v_2/v_s = 1$). Assume that negligible current flows into the amplifier input terminals.

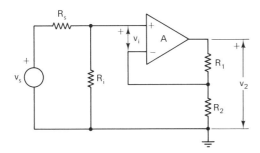

Prob. 6.9

(a) For an amplifier having gain $A = v_2/v_i = 10$, find the ratio $R_2/(R_1 + R_2)$ for $R_s = 1000\ \Omega$ and $R_i = 1000\ \Omega$.

(b) Substitute an amplifier having a very large A_i (e.g., $A \to \infty$). Find v_2/v_s for the $R_1(R_1 + R_2)$ ratio determined in part (a).

(c) For the case of $A \to \infty$, what value of $R_2(R_1 + R_2)$ is needed to make $v_2/v_s = 1$?

6.10 The amplifier with input and output stages shown and with intermediate stages μn specified has an open-circuit gain $v_e/v_i = 140$; that is, the load R_L and the feedback resistance R_F are not connected.

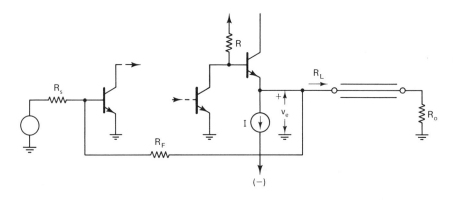

Prob. 6.10

(a) Neglecting μ_{re} and r_{ceo} for the transistors involved, what are R_i and R_o for the amplifier with feedback resistor and R_L in place?

(b) With R_L being a transmission line properly terminated in its characteristic impedance, $Z_o = 75\ \Omega$ and with $R_s = 2000\ \Omega$ and $R = 9200\ \Omega$, determine the required value of R_F to properly match the transmission line.

6.11 The circuit shown contains an amplifier with $Z_i \to \infty$, $Z_o \to 0$, and $A = v_2/v_i \gg 1$.

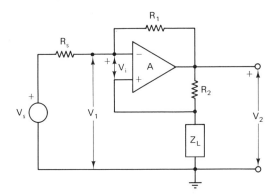

Prob. 6.11

(a) Write equations for K_1 and B.

(b) Determine v_2/v_s.

(c) Obtain an expression for $Z_i = V_1/I_1$.

(d) Under what conditions will $Z_i = -(R_1/R_2)\,Z_L$?

6.12 A signal source V_1, with series resistance R_1, is loaded by capacitance C_1, limiting the bandwidth at the input of the amplifier to $f_2 = 1/(2\pi R_1 C_1)$. The differential amplifier shown is suggested as a means of increasing the bandwidth at the output of the amplifier.

(a) Specify from inspection whether A is to be inverting or noninverting for the input signal.

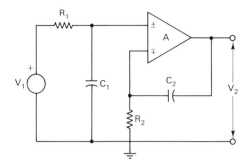

Prob. 6.12

(b) For $R_1C_1 = R_2C_2$, find $V_2(j\omega)/V_1(j\omega)$, and specify a numerical value for A to make the bandwidth at the output of the amplifier greater than f_1 by a factor of 10. What is V_2/V_1 in this case?

(c) What will be the maximum value A can have before the circuit becomes unstable? And where will the pole of the response function be?

6.13 The multistage feedback amplifier shown has an internal structure such that the effective RC time constant at the input is clearly the dominant factor in limiting the high-frequency response. All transistors are biased at the same current levels at which $r_{bes} \cong 3000 \ \Omega$ and $\beta \cong 100$.

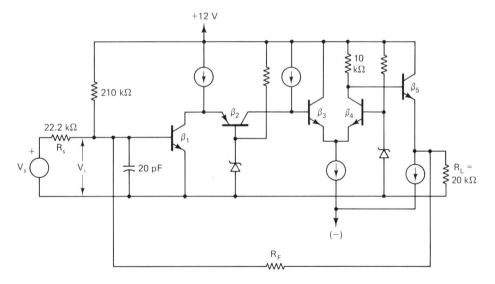

Prob. 6.13

(a) Using the simplest possible approximation, determine an approximate value for $A = V_2/V_i$.

(b) What is the bandwidth without R_F present?

(c) For $R_F = 100 \ \text{k}\Omega$, what is the overall bandwidth?

(d) What is the overall gain V_2/V_s with R_F present?

6.14 The circuit of Prob. 6.13 is revised somewhat as follows: The input transistor, Q-1, is operated at a much lower current level such that its diffusion capacitance is much less, while at the same time it is derived from a much lower impedance source (i.e., $R_s \cong 2\text{k}\Omega$). Then a capacitance $C_5 = 40$ pF is connected from the base of Q-5 to ground such that the effective time-constant at this point is dominant.

(a) Considering the frequency response to be that at the point where C_5 is connected, what approximately is the bandwidth of the amplifier without R_F present?

(b) With $R_f = 100 \text{ k}\Omega$, what is the bandwidth?

(c) What is the overall gain V_2/V_s with R_F present?

6.15 The noninverting amplifier shown is characterized by its input impedance Z_i (R_i in parallel with C_i) and its open-circuit voltage A_oV_i, with series output resistance R_o. The amplifier was connected to its voltage supplies as shown without any input signal applied (open-circuit input), and it was found that a time-varying voltage appeared at the output. The resulting voltage instability was finally traced to a small stray capacitance, C_F, between input and output terminals as shown.

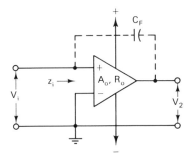

Prob. 6.15

(a) Could the same thing happen if the amplifier had been used as an inverting amplifier? Explain.

(b) Examine the return difference function $(1 - AB)$ for the amplifier, and determine a relationship between A_o and the circuit time constants to indicate a possibility of sinusoidal oscillations.

6.16 Apply the basic gain equation developed for the shunt–shunt amplifier to the circuit shown:

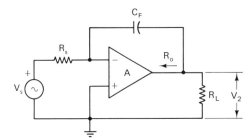

Prob. 6.16

(a) Find $v_2(t)$ for $V_s(s) = V_s$.

(b) What should be the nature of A to make the circuit a perfect negative-going ramp generator?

(c) If the output resistance of the amplifier must be considered, it can be represented by its open-circuit voltage generator A_oV_i in series with its output resistance, R_o. Using this representation, determine $A(s) = V_2(s)/V_i(s)$; substitute it in your original equation, find $v_2(t)$, and discuss the effect of R_o on the result.

6.17 The two FET in the accompanying circuit are described by their g_m's alone (i.e., $r_d \to \infty$).

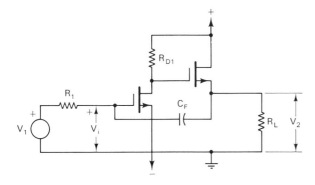

Prob. 6.17

(a) Draw the simplest possible equivalent circuit containing only one internal current generator.

(b) Determine K_1, A, and B.

(c) For $v_1(t) = V_1 U(t)$ (i.e., a step function), write an equation for $v_2(t)$ and sketch the waveform.

6.18 The idealized nonfrequency-dependent amplifier is shown in conjunction with an FET output transistor.

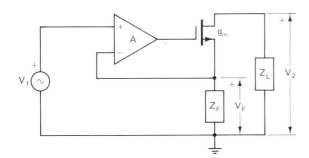

Prob. 6.18

(a) Determine $V_f(V_1)$ and show its limiting value for $A \to \infty$.

(b) Determine V_2/V_1 and establish the condition for which $V_2/V_1 = f(Z_L, Z_F)$ alone.

6.19 For the feedback amplifier show:

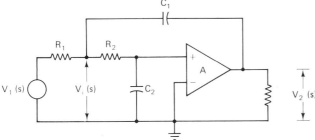

Prob. 6.19

(a) With $V_i(s)$ taken at the terminals shown, determine $A'(s) = V_2(s)/V_i(s)$.

(b) Determine K_1 and B using $V_i(s)$ as the input reference.

(c) Obtain an expression for $V_2(s)/V_1(s)$; be sure to reduce the denominator to a polynomial in s.

(d) Find a relationship between A and the passive circuit components that would lead to the possibility of sinusoidal oscillations.

6.20 For the feedback circuit shown, incorporating the ideal amplifier with gain A:

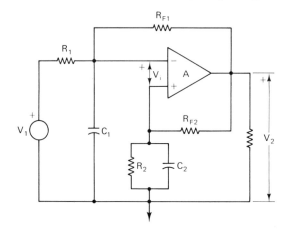

Prob. 6.20

(a) Write equations for K_1 and B.

(b) Determine conditions to make $B = 0$.

(c) For the condition of $B = 0$, write the equation for V_2/V_1 and show how it varies with frequency.

REFERENCES

BODE, H.W. *Network Analysis and Feedback Amplifier Design,* Van Nostrand Reinhold, New York, 1945.

EVELEIGH, V. W. *Introduction to Control Systems Design,* McGraw-Hill, New York, 1972.

GHAUSI, M. S. *Electronic Devices and Circuits,* Holt, Rinehart and Winston, New York, 1985.

GLASFORD, G. M. *Linear Analysis of Electronic Circuits,* Addison-Wesley, Reading, Mass., 1965.

GRAY, P. R. AND ROBERT G. MEYER. *Analog Integrated Circuits,* Wiley, New York, 1984.

MILLMAN, J. *Microelectronics,* McGraw-Hill, New York, 1979.

Chapter 7

Operational Amplifiers: Specifications, Analysis, and Applications

INTRODUCTION

Historically, an operational amplifier generally has been understood to be a differential-input, single-output, extremely high gain multistage direct-coupled amplifier. When negative feedback is applied to such an amplifier, reasonably large gains can still be achieved when the amplifier gain itself is so large that the overall transfer function is essentially independent of amplifier gain and any variations in it. Also, it has been generally understood that such an amplifier has a sufficiently low output impedance that the gain is essentially independent of it with reasonable values of load impedance and a sufficiently high input impedance as to present negligible loading to the signal source that drives it. The term *operational* is derived from the fact that, with appropriate frequency-dependent feedback and input networks, certain linear mathematical operations such as summing, differentiation, and integration can be modeled; with several amplifiers performing such functions connected together in an appropriate feedback configuration, high-order differential equations can be modeled, forming what is generally known as an analog computer.

Operational amplifiers are, in general, composed of composite-compound device configurations discussed in Chapter 5, with their analysis involving the general feedback equations developed in Chapter 6. To achieve internal voltage gains of the magnitude required, operational amplifiers have been restricted for the most part to relatively narrow bandwidths when operated in their open-loop (nonfeedback) mode, with any extended bandwidth requirement being achieved by feedback networks, reducing the gain and increasing the bandwidth in accordance with the overall gain–bandwidth product limitations discussed in Chapter 6.

Over time, the general term operational amplifier, or op-amp, has been extended to include (1) broad-band amplifiers (video amplifiers) at reduced but still relatively high gain whose required bandwidth, however, involves increased power consumption and special design configurations; (2) amplifiers with a current (high output impedance) output, rather than a voltage (low output impedance) output, such amplifiers being referred to as *transconductance amplifiers;* and (3) a special class of amplifiers whose gain can be varied over a wide range by varying the bias current and hence the effective transconductance of the input differential stage, with such amplifiers being referred to as *operational transconductance amplifiers.* Such amplifiers are extremely useful in extending the application of operational amplifiers to modeling nonlinear mathematical functions. Another class of operational amplifiers are those designed to function at extremely low voltage and current levels. These are referred to as *micropower operational amplifiers.*

The purpose of this chapter is threefold: (1) to summarize the semistandard definitions and performance specifications for operational amplifiers to permit circuit designs to use such op-amps without reference to their internal anatomy, (2) to analyze the internal structure (anatomy) of op-amps to determine factors that affect the external characteristics, and (3) to discuss the range of applications for operational amplifiers with various performance characteristics, including linear, nonlinear, and switching applications.

7.1 GENERAL DC AND LOW-FREQUENCY PROPERTIES AND SPECIFICATIONS

Ideally, an operational amplifier has extremely high voltage gain, extremely high input impedance, extremely low output impedance, a negligible dc input current for zero input signal voltage, and a gain that is independent of frequency over the frequency range of input signals to be applied.

Ideally, referring to Fig. 7.1(a), in the conventional negative shunt–shunt feedback mode using the conventional feedback equations of Chapter 6

$$R_1 = R_S$$

$$K_1 = \frac{R_S R_F}{R_S + R_F}$$

$$B = \frac{R_S}{R_S + R_F}$$

and hence

$$\frac{v_o}{v_s} = -\frac{R_F}{R_S + R_F} - \frac{A}{1 + A[R_S/(R_S + R_F)]} \tag{7.1}$$

This implies that $R_i >>> R_S$ and

$$R_o << \frac{R_L (R_S + R_F)}{R_L + R_S + R_F}$$

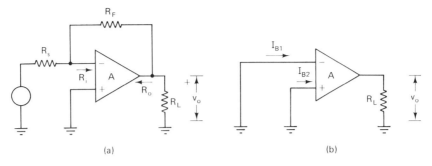

Figure 7.1 Operational amplifier with feedback. (a) Shunt–shunt feedback connection. (b) Amplifier showing input bias currents.

Then for

$$A\frac{R_S}{R_S + R_F} >> 1$$

$$\frac{v_o}{v_s} \cong -\frac{R_F}{R_S} \tag{7.2}$$

If these relationships are to hold at zero frequency, the dc input currents I_{B1} and I_{B2} shown in Fig. 7.1(b) should be negligible. However, this is not always the case.

The dc input current that does flow with no voltage applied at the input is referred to as the *input bias current*. In general, the two input currents, which may not be identical, are averaged to yield a single specification with the input bias current being defined as the average of the two:

$$I_{\text{Bias}} = \frac{I_{B1} + I_{B2}}{2} \tag{7.3}$$

A finite value for input bias current affects the dc output when the amplifier is used in a feedback configuration, as shown in Fig. 7.2(a), showing a worst case scenario of $I_{B2} = 0$ (leading to the greatest output unbalance) and I_{B1} independent of v_i. Solving for v_i,

$$v_i = \frac{-I_B}{(1/R_S) + [(1 + A_{0M})/R_F]} \tag{7.4}$$

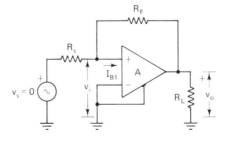

Figure 7.2 Circuit for determination of input offset current.

and
$$v_o = \frac{I_{B1}R_FA_{OM}}{1 + (R_F/R_S) + A_{OM}} \tag{7.5}$$

where A_{OM} is the magnitude of the gain at dc or zero frequency.

Errors due to the inequality of input bias currents can be evaluated from a specification of *input offset current* defined as

$$I_{OS} = |I_{B1} - I_{B2}| \tag{7.6}$$

The input bias current and input offset current are usually defined as the maximum possible value for the type of amplifier for which they are specified and will usually be less than the maximum for a particular amplifier of the class.

Another specification that describes amplifier unbalance is the *input offset voltage,* which is the differential voltage that is required to be applied across the input terminals to reduce the output voltage to zero, as indicated in Fig. 7.3.

Input offset current and input offset voltage are different but related ways of describing amplifier unbalance. The specification that is most pertinent depends on source and feedback loading conditions. For example, for an operational amplifier using emitter-coupled bipolar junction transistors as an input stage, as indicated in Fig. 7.4, with two voltage inputs, v_1 and v_2, the input bias currents, I_{B1} and I_{B2}, may be small but significant.

If $v_1 = v_2 = 0$, $I_{B1} = I_{B2}$ and $h_{FE1} = h_{FE2}$. I_{OS} and V_{OS} are both zero.

But if $h_{FE1} \neq h_{FE2}$ and/or $I_{B1} = f_1(v_{BE1})$ and $I_{B2} = f_2(V_{BE2})$ with $f_1(V_{BE1}) \neq f_2(V_{BE2})$, with the result that $I_{B1} \neq I_{B2}$, then V_{OS} and I_{OS} are both finite and specifically related to describe the unbalance correctly for source and load conditions.

On the other hand, with a very high input impedance FET amplifier, as shown in Fig. 7.5, I_{B1} and I_{B2} and hence I_{OS} are likely to be insignificant, and the unbalance lies within structural difference of the two FETs, resulting in different V_{OS}, I_D, V_{GS} relationships, with I_B not a factor. In this case, V_{OS} is the most significant specification for unbalance conditions.

Regardless of how the unbalances are best described (i.e., in terms of V_{OS} or I_{OS}), the output can be reduced to zero (or its specified value) by an additional compensating

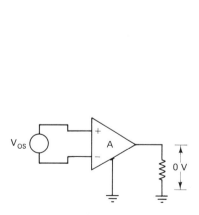

Figure 7.3 Input offset representation for operational amplifier.

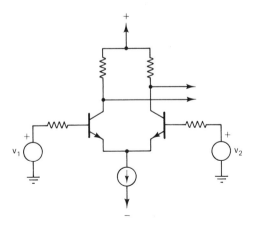

Figure 7.4 Simple differential input stage for defining input offset voltages and currents.

level shift somewhere within the op-amp itself. This can be accomplished by a separate terminal to which a compensating voltage can be applied, as indicated in Fig. 7.6.

However, even though V_{OS} or I_{OS} can be corrected for at a specified operating temperature, unbalances within the structure that are temperature dependent are not corrected for, and the output will vary from its normal value or be said to *drift* as the internal temperature varies.

The change in input voltage offset as a function of temperature is the *input offset voltage drift*, given by

$$\text{Voltage drift} = \frac{\Delta V_{OS}}{\Delta T} \tag{7.7}$$

Similarly, the input offset current drift is given by

$$\text{Current drift} = \frac{\Delta I_{OS}}{\Delta T} \tag{7.8}$$

Depending on how the bias adjustment is made to correct for V_{OS} or I_{OS}, the drift may be lesser or greater for the corrected amplifier than for the uncorrected amplifier.

An important differential amplifier specification that characterizes unbalances both with respect to dc offset voltage and small-signal voltage gain is called the *common-mode rejection ratio (CMRR)*, which measures the discrimination against unbalances in the two sides of the amplifier defined as

$$\text{CMRR} = \frac{(v_o/v_i)_d}{(v_o/v_i)_c} = \frac{A_{dm}}{A_{cm}} \tag{7.9}$$

where $(v_o/v_i)_d = A_{dm}$ is the small-signal output voltage when a differential voltage v_i is applied at the input, that is, $+(v_i/2)$ at the noninverting terminal and $-(v_i/2)$ at the inverting terminal.

These balanced inputs are indicated in Fig. 7.7(a). The common-mode voltage is the voltage appearing at the output when a voltage v_i of the same polarity is applied to

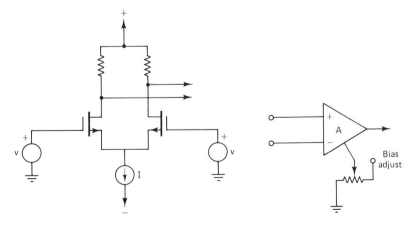

Figure 7.5 Input offset voltage for FET differential input stage.

Figure 7.6 Bias adjustment for compensation of voltage offset.

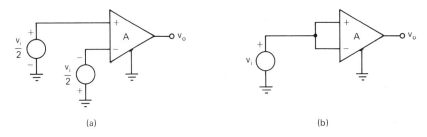

(a)   (b)

Figure 7.7 Op-amp circuits for determination of common-mode rejection ratio. (a) Differential-mode connection. (b) Common-mode connection.

both terminals, as indicated in Fig. 7.7(b). These measurements are usually made at some frequency $\neq 0$ to remove the independent effects of V_{OS}.

Alternatively, the CMRR can be expressed in terms of the change in differential/common voltage ratio required to maintain the output at zero:

$$\text{CMRR} = \left.\frac{\Delta v_{ic}}{\Delta v_{id}}\right|_{v_o \ = \ const.} \tag{7.10}$$

If the measurement is made at zero frequency (dc), it can be related to the input offset voltage V_{OS} as follows: Since V_{OS} is a differential input voltage required to reduce the output to zero, a common-mode signal may be applied, which will reduce the output to zero, and the CMRR may be defined as

$$\text{CMRR} = \left.\frac{\Delta V_{ic}}{V_{OS}}\right]_{v_o \ = \ 0} \tag{7.11}$$

where ΔV_{ic} is the common-mode signal required to neutralize the effect on the output of V_{OS}. In other words, V_{OS} itself may be used as the differential-mode input signal.

The *open-loop* gain of an operational amplifier is usually defined as the small-signal voltage gain at low frequencies, with a typical load resistance specified that is large compared to the output resistance of the amplifier itself, but with no external feedback. This agrees with the general feedback open-loop gain or return ratio $AB = -T_r$ only for the case in which the feedback presents no significant load to the amplifier and with $B = 1$.

The amplifier gain at low frequencies is often expressed in terms of decibels as

$$A_{\text{dB}} = 20 \log_{10} A_{OM} \tag{7.12}$$

and the common-mode rejection ratio in decibels is

$$\text{CMRR}_{\text{dB}} = 20 \left(\log_{10} A_{dm} - \log_{10} A_{cm}\right) \tag{7.13}$$

or

$$\text{CMRR}_{\text{dB}} = 20 \log_{10} \text{CMRR}_v \tag{7.14}$$

7.2 FREQUENCY- AND TIME-DOMAIN RESPONSE OF OPERATIONAL AMPLIFIERS

Operational amplifiers are frequency limited, sometimes in mathematically complex ways due to circuit capacitance and other device limitations, such as h_{fe} frequency dependence within the structure. For reasons of convenience in use, as well as simplicity of analysis, it is desirable insofar as possible to reduce the limitation to that of a single-pole response. We may regard such an amplifier as a semiideal amplifier whose response can be written as

$$A = \frac{\pm A_{OM}}{1 + j(\omega/\omega_2)} = \frac{V_o}{V_i} \qquad (7.15)$$

where ω_2 is the 0.707 response with normal load but with no feedback, as shown in Fig. 7.8.

The open-loop bandwidth is defined in terms of the open-loop gain for $A_M = 1$ given by

$$1 = \frac{A_{0M}^2}{1 + (\omega/\omega_2)^2} \quad \text{or} \quad \omega = \omega_2 \sqrt{A_{0M}^2 - 1} \qquad (7.16)$$

$$(BW)_{OL} = f_2 \sqrt{A_{0M}^2 - 1} \qquad (7.17)$$

For an operational amplifier, usually $A_{0M} \gg 1$ and

$$(BW)_{OL} = A_{0M} f_2 \qquad (7.18)$$

Thus, for the single-pole amplifier response, the open-loop bandwidth is the product of the open-loop gain and the conventional 0.707 response value.

Unity-gain (voltage-follower) mode bandwidth is defined as the 0.707 bandwidth for the voltage-follower mode, as indicated in Fig. 7.9 for the noninverting amplifier. If the amplifier has the single-pole response with $B = -1$ and $K = 1$, the overall gain is

$$\frac{V_o}{V_i} = \frac{A_{0M}}{1 + A_{0M}} \frac{1}{1 + j\{\omega/[\omega_2(1 + A_{0M})]\}} \qquad (7.19)$$

and the 0.707 bandwidth is

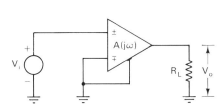

Figure 7.8 Frequency-dependent designation of op-amp.

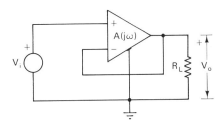

Figure 7.9 Frequency-dependent op-amp in voltage-follower mode.

$$\left. BW \right]_{0.707} = f_2(1 + A_{0M})$$

And for $A_{0M} \gg 1$,

$$\frac{A_{0M}}{1 + A_{0M}} \rightarrow 1$$

and

$$\left. BW \right]_{0.707} = A_{0M}f_2$$

Thus the 0.707 bandwidth of the unity gain mode amplifier, also referred to as the closed-loop bandwidth, is identical to the open-loop bandwidth (defined for unity gain) with no feedback. These relationships are not exact for an amplifier with an open-loop frequency response that departs from the single-pole form such as represented by two or more internal poles.

As a matter of fact, an amplifier with even a two-pole response that would not be inherently unstable with feedback might become unstable if another pole is introduced as the feedback loop is closed, even for overall gains substantially greater than unity.

Some specific aspects of the stabilization problems can be illustrated by the following hypothetical response function: An open-loop gain function is prescribed as having the form

$$A = \frac{A_{0M}}{[1 + j(\omega/\omega_2)][1 + j(\omega/a\omega_2)][1 + j(\omega/b\omega_2)]} \tag{7.20}$$

Furthermore, it is assumed that a and b both $\gg 1$. The pole defined by ω_2 is referred to as the *dominant pole*.

In normalized form, with $\omega/\omega_2 = \rho$,

$$A = \frac{A_{0M}}{(1 + j\rho)[1 + j(\rho/a)][1 + j(\rho/b)]} \tag{7.21}$$

which can be written as

$$A = \frac{A_{0M}}{1 - (\rho^2/ab)(1 + a + b) + j\rho[1 + (1/a) + (1/b) - (\rho^2/ab)]} \tag{7.22}$$

Examination of the $j\rho$ term shows that the phase shift reaches 180° when

$$\rho^2 = ab\left(1 + \frac{1}{a} + \frac{1}{b}\right) \tag{7.23}$$

The voltage gain at the frequency of 180° phase shift

$$A_{180°} = \frac{A_{0M}}{1 - (\rho^2/ab)(1 + a + b)}$$

which by combining the two preceding equations can be written as

$$A_{180°} = \frac{A_{0M}}{1 - [1 + (1/a) + (1/b)](1 + a + b)} \tag{7.24}$$

Suppose, in the circuit of Fig. 7.10(a), that the loop is closed as shown. Then

$$\frac{V_o}{V_i} = \frac{A}{1 - AB}$$

with $B = -1$ or

$$\frac{V_o}{V_i} = \frac{1}{1 + A}$$

which will indicate instability if $A \geq -1$, according to the Nyquist criterion.

If the open-loop gain and relative pole locations are such that $A_{180°} \geq -1$, the amplifier cannot operate in the unity-gain voltage-follower mode and remain stable.

If feedback is to be used, it must be limited to a value of $B < -1$, as shown in Fig. 7.10(b), such that $A_{180°}B < 1$, or the pole locations would have to be further separated to make $A_{180°} < -1$ at the frequency of 180° phase shift.

Specific Example

Consider an amplifier having the configuration of Fig. 7.10 with $A_{OM} = 2 \times 10^5$ and frequency-limiting pole locations of

$$f_2 = 20 \text{ kHz}$$
$$af_2 = 2 \text{ MHz} \quad \text{(i.e., } a = 100\text{)}$$
$$bf_2 = 10 \text{ MHz} \quad \text{(i.e., } b = 500\text{)}$$

Then for 180° phase shift

$$\rho^2 = 5 \times 10^4 (1 + 0.01 + 0.002)$$
$$= 5.06 \times 10^4$$

or $\rho = 224$ and $f = 20 \times 10^{-3} \times 224$.

$$f_{180°} = 4.5 \text{ MHz}$$
$$A_{180°} = \frac{2 \times 10^5}{1 - (1 + 0.01 + 0.002)(1 + 100 + 500)}$$
$$= -330$$

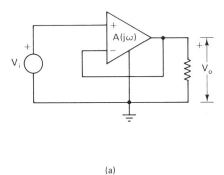 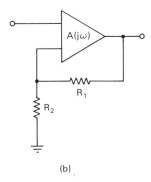

(a)   (b)

Figure 7.10 Negative feedback frequency-dependent op-amp for stability consideration. (a) Voltage-follower mode. (b) General feedback connection.

Thus the loop cannot be closed to allow the amplifier to function in the unity-gain mode because of instability. Rather, to prevent instability, AB must be $AB < 1$. Hence, for $AB_{180°} = 1$, $B = -(1/330) = -0.003$ which can be established by the appropriate resistance ratios in Fig. 7.10(b).

If it is desired to close the loop for $B = -1$, then assuming that the limitation is due to the higher-order poles, which cannot be changed, the only way to provide added pole separation is to degrade the response of the dominant pole. This calls for $A_{180°} = -1$ in the equation

$$A_{180°} = -1 = \frac{A_{OM}}{1 - [1 + (1/a) + (1/5a)][1 + a + 5a]}$$

If the same relationship between a and b is retained (i.e., $b = 5a$), then

$$a = -\frac{36 - 5A_{OM}}{60} \pm \frac{1}{60} \sqrt{(36 - 5A_{OM})^2 - (24 \times 30)}$$

For $A_{OM} = 2 \times 10^5$, $a \simeq 3.33 \times 10^4$. Then the new value for f_2 is given by

$$f_2 = \frac{2 \times 10^6}{a} = \frac{2 \times 10^6}{3.33 \times 10^4}$$

$$= 60 \text{ Hz}$$

This value of $f_2 = 60$ compared with the old value of $f_2 = 20$ kHz allows the amplifier barely to be stabilized (zero phase margin) for the closed-loop case.

Many aspects of the problem of stabilization can be visualized by plotting separately the several components of the gain function expressed in terms of the magnitude of the gain in decibels.

For the gain of the three-pole response just discussed, the magnitude is

$$A_M = \frac{A_{OM}}{(1 + \rho^2)^{1/2}[1 + (\rho/a)^2]^{1/2}[1 + (\rho/b)^2]^{1/2}} \tag{7.25}$$

which can be expressed in decibels as

$$A_{Mdb} = 20 \log_{10} A_{OM} - 10 \log (1 + \rho^2)$$
$$- 10 \log\left[1 + \left(\frac{\rho}{a}\right)^2\right] - 10 \log\left[1 + \left(\frac{\rho}{b}\right)^2\right] \tag{7.26}$$

The gain relative to that of zero frequency (subtracting out the zero frequency gain) is

$$A_{REL} = -10\left\{\log\left(1 + \rho^2\right) + \log\left[1 + \left(\frac{\rho}{a}\right)^2\right] + \log\left[1 + \left(\frac{\rho}{b}\right)^2\right]\right\} \tag{7.27}$$

The three individual components are plotted as shown in Fig. 7.11 for the specific example just described. Each pole contributes a decrease in gain of 6 dB/octave, which is the overall response as long as the dominant pole is the only contributor, reaching 12 dB/octave when two poles contribute, and 18 dB/octave when all three poles contribute. It is seen that 180° phase shift is reached when the slope becomes -12 dB/octave, which occurs at a relative gain of -56 dB, which agrees with the previous calculation of $A = 330$.

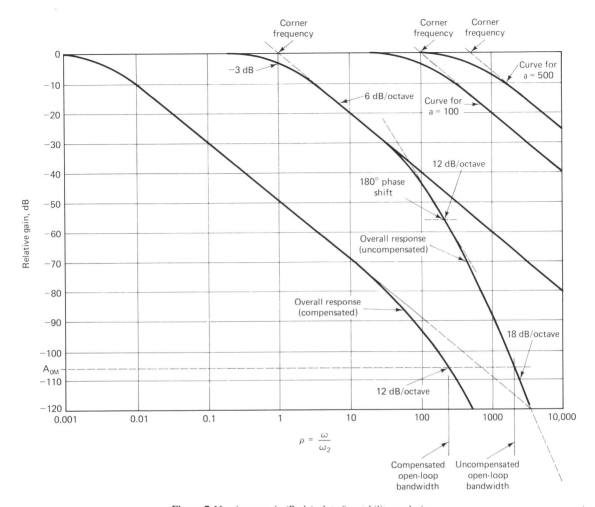

Figure 7.11 Asymptotic (Bode) plots for stability analysis.

The amplifier response with degraded f_2 is plotted as the compensated curve and is seen to reach 180° phase shift or 12 dB/octave at the unity gain value. Again the amplifier is stabilized (but with zero phase margin) for the unity-gain mode.

Plots of the sort shown are made using the asymptotic values, that is, the slope of ndB/octave projected as shown in the dashed curves. Thus the -3-dB points of the single-pole responses become the *corner frequencies* of the asymptotic plots. Plots of this sort are called *Bode plots* and can be used to estimate stability situations.

Whenever an overall Bode plot reaches 12 dB/octave, it shows the absolute limiting value of feedback that can be used without resulting instability (zero phase margin).

The general relationships between bandwidths and rise times that have been discussed in Chapters 4 through 6 are, of course, valid for the operational amplifier for very small signal inputs. In general, the response to a voltage step, whose transform V_i/s, is determined from the transform response of the amplifier using $s = j\omega$ and solving for the resultant time-domain response, which is the inverse Laplace transform of the product of the transform of the input step and that of the response.

There is, however, a limit to this relationship for large input signals where the dynamic range of one or more stages of the amplifier might be approached or exceeded because of nonlinearities that exist in such stages. For a rapid input change, such as a nonlinear stage (e.g., one having a decreasing g_m with current) might momentarily cut off. Then such shunt capacitances that may be involved will change voltage at a rate determined by associated RC time constants until the transistor in question again conducts. An accurate analysis incorporating such nonlinear properties becomes exceedingly cumbersome. However, the phenomenon does lead to another amplifier specification known as the *slew rate*, which is defined as

$$ \mathrm{SR} = \left. \frac{\Delta v}{\Delta t} \right]_{v_i\ max} \tag{7.28} $$

which is the maximum rate of changing output voltage as the input v_i reaches the maximum possible value. The slew rate will in general be found to be *related to* but somewhat slower than the rate of rise as the response to a step function as determined on a linear basis.

7.3 EVOLUTION OF OPERATIONAL-AMPLIFIER STRUCTURES

A very simple structure that illustrates the basic elements contained in a conventional operational amplifier is shown in Fig. 7.12. It pays some attention to frequency-limiting conditions and is made up of the basic single and composite building blocks discussed in Chapter 4 and Chapter 5.

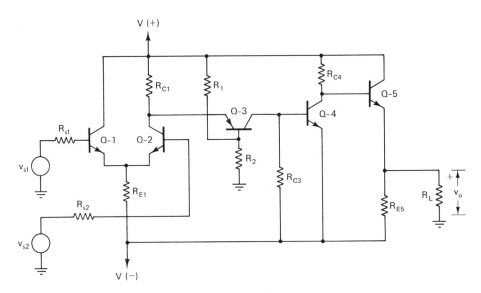

Figure 7.12 Primitive op-amp structure.

The transistors Q-1 and Q-2 comprise a differential-input, single-output emitter-coupled amplifier that drives the common-base stage, Q-3. From the viewpoint of v_{s1}, it is an emitter-coupled amplifier, and without a load resistance in the collector circuit of Q-1, there is no feedback capacitance loading (Miller effect) of the source v_{s1}. From the viewpoint of v_{s2}, Q-2 is a conventional common-emitter amplifier with source degeneration driving a common-base stage as a composite cascode stage, which also minimizes the capacitance loading of v_{s2} because of the low input impedance of the common-base stage, Q-3, provided that the effective resistance in its base circuit, R_1 and R_2, in parallel, used for biasing Q-3 is also low. The current-driven common-emitter amplifier, Q-4, is designed to provide the major portion of the overall gain, which because of the large feedback capacitance accounts for the major frequency limitation, which is approximately that of a single-pole response when the effective input capacitance as a load on Q-3 is considered, as illustrated by the example in Sec. 4.12. Other poles contributed by frequency limitations in other portions of the circuit, even though they may be much higher in frequency, have some effect on the overall response, which could lead to instability in a negative feedback configuration, as discussed in Chapter 6 and in Sec. 7.2. A low output impedance is provided by the emitter follower, Q-5, with the output established at ground potential by the appropriate choice of R_{E5}.

The circuit of Fig. 7.12, which may be viewed primarily as a conceptual skeleton prototype of an operational amplifier, has several limitations. First, the finite value of R_{E1} causes a gain unbalance with respect to v_{s1} and v_{s2}. Also, R_{C1} produces a waste of available gain because it diverts signal current from the input of Q-3. Furthermore, the load at the base of Q-3 (R_1 and R_2 in parallel) limits the frequency response of Q-2, as discussed in Chapter 5 for the cascode stage. The resistor, R_{C3}, necessary both to prevent excessive dc base current in Q-4 and also to maintain the output impedance of Q-4 low at the same time, diverts signal current from the input of Q-5. The single emitter follower also is not an ideal output stage. It does not minimize dc drift at the output and also causes an unbalanced slew rate for positive-going and negative-going signals.

A modified op-amp, which fits the same basic form as that of Fig. 7.12 but which minimizes some of its disadvantages, is shown in Fig. 7.13. The current source I, which may be some form of current mirror discussed in Sec. 5.12, minimizes the gain unbalance for v_{s1} and v_{s2}. Also, as discussed in connection with Fig. 5.24, the current mirror at the collectors of Q-1 and Q-2 simultaneously provides a low impedance at the collector of Q-1, which minimizes feedback capacitance loading for v_{s1}, and a current source for Q-2, which minimizes signal current being diverted from the input of Q-3. It can be structured to provide exact balanced load conditions for Q-1 and Q-2.

The simple biasing network for the base of Q-3 is replaced by the complementary emitter-follower combination, which reduces its base loading to a very low value and, in turn, minimizes the loading at the collector of Q-2, which reduces the feedback capacitance that loads the signal source v_{s2}.

The resistor R_{C3} in the collector circuit of Q-3 is eliminated by operating Q-1, Q-2, and Q-3 at sufficiently low current levels that not too much base current is supplied to Q-4. This mode of input circuit operation is also preferable for op-amps to minimize input circuit loading, which permits the input transistors Q-1 and Q-2 to be characterized by their transconductances.

The load resistance R_{C4} at the collector of Q-4 is replaced by the current source I_4

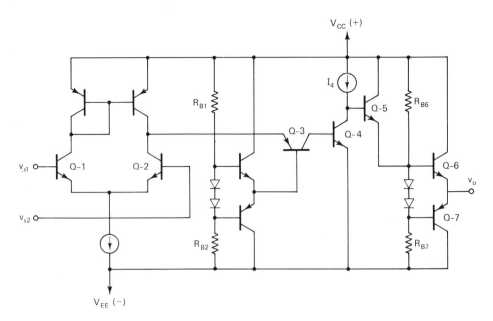

Figure 7.13 Basic op-amp structure.

implemented as a current mirror, which maximizes the signal current to be supplied to the base of Q-5.

At the output, the complementary transistors Q-6 and Q-7 are added to the emitter-follower output of the preceding circuit to provide a dynamically balanced source for the external load impedance to be supplied. These may be biased to operate as class A, class B, or class AB, as discussed in Chapter 5. The output impedance is much less than that of the single emitter follower because the input is also an emitter follower driven by a common-emitter stage unless r_{ceo} of Q-4 is abnormally high.

An op-amp incorporating a slightly different form of input circuit designed to reduce input circuit loading of the signal sources is suggested in Fig. 7.14 in somewhat skeleton form. Here the input stages Q-1 and Q-2 function essentially as emitter followers, each being loaded by a common-base stage driving a similar common-base stage on the opposite side, which functions as an emitter follower driving the emitter of the input transistor on the opposite side.

For example, the input resistance for Q-1 can be written approximately by inspection, using the relationships developed in Chapter 4, as

$$R_{i1} = r_{bes1} + (1 + \beta_1) \frac{r_{bes3} + r_{bes4} + (1 + \beta_4)[(r_{bes2} + R_{s2})/(1 + \beta_2)]}{1 + \beta_3} \quad (7.29)$$

As an approximation, assuming identical transistors all at the same operating point,

$$R_{i1} \simeq 4r_{bes} + R_{s2} \quad (7.30)$$

as compared with $2r_{bes} + R_s$ for the emitter-coupled stage.

The collector current of Q-3 controlled by its base current is the current source for the input transistor Q-5 of the Q-5, Q-6 current mirror, which forces the collector current

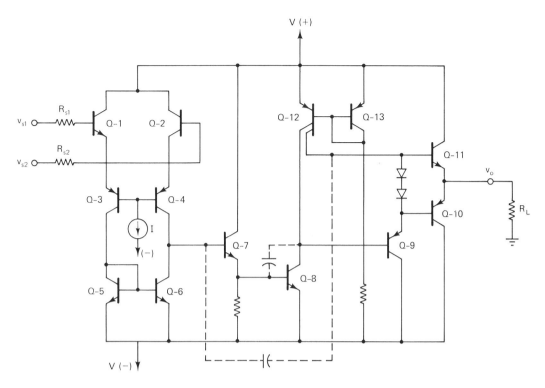

Figure 7.14 Complete op-amp with alternative input circuit.

of Q-6 to be essentially equal to it, which in turn creates balanced currents in Q-1 and Q-2 for $v_{s1} = v_{s2}$, except for the slight base current loading of Q-7, which is the input to the following stage. If a positive signal voltage, v_{s1} is applied to Q-1, the emitter current from Q-1 and into Q-3 is increased, which increases the current supplied to Q-5, which in turn forces the collector current in Q-6 to increase, thus decreasing the current supplied to the base Q-7, which is the same as a negative voltage increment applied to its base. On the other hand, if v_{s1} is held constant, and a positive signal voltage v_{s2} is applied to Q-2, the current at the collector of Q-4 increases, part of which causes an *increased* base current (positive signal) to the base of Q-7. In either case, the collector currents in Q-5 and Q-6 remain equal, and the net increase or decrease due to changes in v_{s1} and v_{s2} is applied to increase or decrease the base current of Q-7.

The signal current i_{b7} applied to the base of Q-7 due to one input (e.g., v_{s2}) is that of Q-2 acting as an emitter follower driving the common-base stage Q-4 and is approximately

$$i_{b7} \cong (1 + \beta_2) \frac{\beta_4}{1 + \beta_4} i_{b2} \tag{7.31}$$

The current due to a change in i_{b1} for balanced devices is the negative of the preceding value.

The emitter follower Q-7 in turn has an output signal current that supplies the base of Q-8 is given by

$$i_{b8} \cong (1 + \beta_7)i_{b7} \qquad (7.32)$$

for $R_{E7} \gg r_{bes8}$.

The current mirror Q-12, Q-13 supplies the dc collector current for Q-8 as well as the diode network, which supplies the correct bias current to the complementary output stage. The output of Q-8 supplies signal current to the emitter follower Q-9, whose base current is

$$i_{b9} = -\beta_8(1 + \beta_7)i_{b7} \qquad (7.33)$$

and whose output current supplies the base current of Q-10, given by

$$i_{b10} = -(1 + \beta_9)i_{b9} \qquad (7.34)$$

with the emitter current of Q-10 (assuming class B operation with Q-11 off) given by

$$i_{E10} = (1 + \beta_{10})i_{b10} \qquad (7.35)$$

with the output voltage being

$$v_o = -i_{e10}R_L \qquad (7.36)$$

The output voltage is given approximately, using the product of the preceding equations, by

$$v_o \cong -i_{b2} (1 + \beta_2) \frac{\beta_4}{1 + \beta_4} (1 + \beta_7)\beta_8(1 + \beta_9)(1 + \beta_{10})R_L \qquad (7.37)$$

If the input stages using Q-1 through Q-4 operate at very low current levels, as is the usual case for high-gain op-amps, and the resistances of the signal sources R_s are relatively low, from Eq. 4.30

$$i_{b2} \cong \frac{v_{s2}}{4r_{bes}} \qquad (7.38)$$

Also, for low collector currents, $r_{bes} \cong r_d$, so

$$i_{b2} \cong \frac{g_{m2}}{4\beta_2} v_{s2} \qquad (7.39)$$

and Eq. 7.37 can be replaced by

$$\frac{v_o}{v_{s2}} = -g_{m2} \frac{1 + \beta_2}{\beta_2} \frac{\beta_4}{1 + \beta_4} (1 + \beta_7)\beta_8(1 + \beta_9)(1 + \beta_{10})R_L \qquad (7.40)$$

which can be written approximately as

$$\frac{v_o}{v_{s2}} \cong \frac{g_{m2}}{4} \beta^4 R_L \qquad (7.41)$$

assuming that β's beyond the first stage are all much greater than 1 and are approximately equal.

Example

For an input stage collector current level of $I_C = 0.1$ mA,

$$g_m \cong \frac{I_C}{kT/q} = \frac{0.1}{0.026} = 3.85 \text{ mS}$$

Then, if $\beta \cong 50$ and $R_L = 2 \text{ k}\Omega$,

$$\frac{v_2}{v_s} \cong -\frac{3.85}{4}(50)^4(2) \cong 1 \times 10^7$$

The input resistance is

$$R_{i2} \cong 4r_d \cong 4\left(\frac{kT}{q}\right)\frac{\beta_2}{I_C} = 4(0.026)\frac{50}{0.1} = 52 \text{ k}\Omega$$

The frequency response of this amplifier is limited primarily by the Miller capacitance of Q-8, which is the primary gain stage. Its load impedance is approximately the input impedance of the emitter follower Q-9, which is

$$R_{i9} \cong r_{bes9} + (1 + \beta_9)[r_{bes10} + (1 + \beta_{10})R_L] \tag{7.42}$$

The resultant high-frequency input admittance loading of Q-8 through the collector–base feedback capacitance might be relatively unimportant for the usually low output impedance of an emitter follower. However, Q-7 is driven from a high-impedance source, which keeps its output impedance relatively high.

Op-amps of this general type were not intended as broad-band amplifiers, but primarily to have large open-loop gains. However, they may be modified, improved, and embellished to improve drift stability to provide overload protection and expedite monolithic processing. An example of a more complete circuit, which is basically that of Fig. 7.14, is shown in Fig. 7.15.

In this circuit, Q-13 and Q-14 are the common input current control circuits of two current-mirror current sources, one being Q-13, Q-12, which supplies current to the output stage and the stage driving it, (Q-12 being the same as that of the previous circuit), and the other Q-14, Q-15, which provides the current source for the bases of Q-3 and Q-4 in connection with Q-16 and Q-17 to provide additional drift stability. In this section, Q-16, supplied by the common-collector currents of Q-1 and Q-2, functions as the input portion of the current mirror, Q-16, Q-17, whose output is connected to that of Q-15. Thus the base current of Q-3 and Q-4 is not simply the output current of Q-15 but is

$$I_{B(3,4)} = I_{C15} - I_{C17}$$

This is a way of deriving the very small current required for the bases without having to control an equally small current in the collector of Q-15 if used alone. At the same time, the base current of Q-3 controls its collector current, which is the current supply for current mirror Q-5, Q-6. This interlocking feedback bias network improves the overall drift stability of the amplifier.

The small resistors in the emitters of Q-5 and Q-6 allow an external differential voltage to be applied, which can supply an offset current that can propagate through the amplifier to null out any residual offset voltage at the output relative to the input.

In this particular circuit in the signal path, the emitter–follower, common-emitter stages, Q-7 and Q-8, are combined as a Darlington amplifier. This simplifies integrated design by making the two collectors common, but degrades the frequency response to some extent because of the increased Miller capacitance loading at the collector of Q-6.

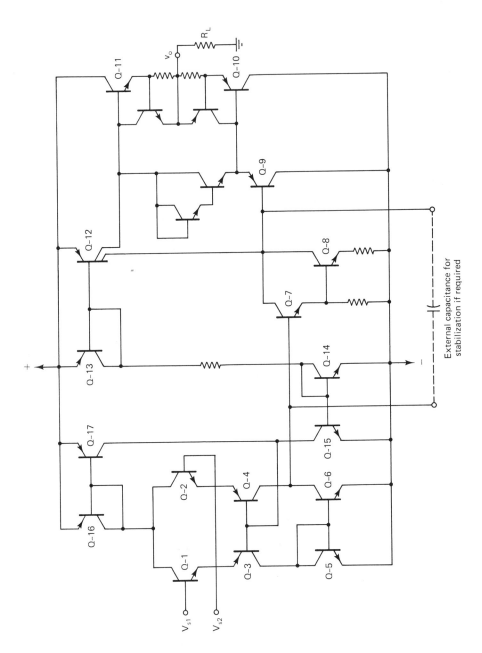

Figure 7.15 Op-amp with improved stability and drift characteristics.

Also, the biasing diodes for Q-10 and Q-11 in the previous circuit are replaced by the two transistors having basically the two base–emitter circuits in series. This particular connection simplifies processing by having three collectors common to one another. The additional transistors in the base–emitter circuits of Q-10 and Q-11 provide short-circuit protection in the manner discussed in Chapter 5.

An alternative version of a high-gain op-amp with characteristics similar to those of Figs. 7.14 and 7.15 utilizes the more conventional emitter-coupled input stage, as shown in Fig. 7.16. This circuit also eliminates the emitter follower Q-9 of the previous circuits and makes use of a slightly different form of constant-current source for the various stages, which replaces the large resistance normally at the input of the two current mirrors consisting of Q-6, Q-5, and Q-9 with the collector current of Q-12, whose base current is controlled by the source current of the n-channel JFET, Q-13, and whose base voltage is established by the zener diode. This circuit has improved temperature-compensation characteristics.

Many operational amplifiers using the circuits of Figs. 7.14 through 7.16 and minor modifications thereof have long been used as a more or less standard operational amplifier, referred to as the type 741 and available in single or quad matched form. Regardless of the specific circuit configuration, they have roughly equivalent specifications, with all being high gain ($>2 \times 10^5$) and narrow bandwidth ($<$60-Hz open-loop corner frequency) limited by the feedback capacitance of the primary common-emitter gain stage, sometimes

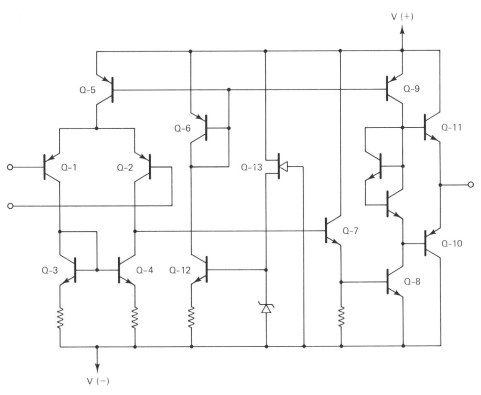

Figure 7.16 Op-amp with *pnp* differential amplifier input stage.

requiring additional external capacitance to allow stabilization in the voltage follower mode.

7.4 APPROACHES TO HIGH-GAIN EXTENDED BANDWIDTH OP-AMPS

It has been pointed out that in the conventional high-gain op-amp the bandwidth is controlled by the feedback capacitance of a high-gain inverting stage, which creates a dominant pole in the response function intended to be at a frequency orders of magnitude lower than that of other poles in order that the amplifier can be stabilized at low (preferably unity) gain. In some amplifiers this requires an external feedback capacitance, as indicated in the respective figures. The alternative is to build in the required excess feedback capacitance as is done in some designs. This restricts the flexibility of design applications, where high gain and maximum bandwidths are simultaneously required where stabilization for low gains is not required. The amount of frequency degradation of the dominant pole required depends on the extent to which the bandwidth of the remaining stages, especially the input stages, is limited. Therefore, a first approach to improving the overall bandwidth is to improve the bandwidth and at the same time the gain–bandwidth product of the input stages. First, it is important to understand the restrictions present in the conventional op-amps partly due to the high input resistance requirement, which can be done by examining the various relationships summarized in Table 7.1. For various reasons, including level shifting considerations, most conventional bipolar op-amp designs involve combinations of *npn* and *pnp* transitors as gain stages in the input differential amplifier section. This is inherent in the designs of both Figs. 7.15 and 7.16.

Examination of lines 1 and 2 of Table 7.1 shows that, with comparable structures and doping levels, the h_{FE} or β and the base transit time of the *pnp* structure will be inferior because of the larger diffusion constant for holes. Furthermore, in an integrated-circuit structure, *pnp* transistors are usually realized as some form of lateral structure or by some other process by which it is usually difficult to achieve very thin base regions. This further decreased β and increases the transit time.

Aside from consideration of the inferiority of *pnp* transistors in terms of gain–bandwidth product, operation at extremely low collector current levels in order to increase the input resistance has some favorable and some unfavorable implications. Examination of line 3 of Table 7.1 shows that the base diffusion capacitance is reduced by the same factor that the input resistance is increased. This would tend to have no effect on ω_β, as given by line 6, were it not for the other components of C_d, as indicated in line 4. At very low currents (high input impedance), where it is preferable to express gain in terms of g_m as the primary defining parameter, it is seen from line 7 that g_m is also reduced by approximately the same factor that r_d is increased. Then, if C_b were the dominant capacitance, a figure of merit as defined by line 8 would indicate that the lowering of g_m at low current would not deteriorate the overall gain–bandwidth product. However, at low currents, C_b becomes smaller relative to C_{je} and $C_{cb'}$, and thus the overall gain–bandwidth product, taking all capacitances into account, increases with increasing g_m. Hence the increase in input resistance with its corresponding decrease in g_m carries some penalty in frequency response.

TABLE 7-1 APPROXIMATE RELATIONSHIPS BASED ON CIRCUIT MODELS OF SEC. 2.8, ASSUMING UNIFORM IMPURITY DISTRIBUTIONS AND NEGLECTING DEPLETION LAYER RECOMBINATIONS (DLR)

Line	Parameter	*npn* Transistor	*pnp* Transistor
1	$h_{FE} \cong \beta$	$\dfrac{1}{\dfrac{N_{AB}W_B D_{pE}}{N_{DE}W_E D_{nB}} + \dfrac{W_B^2}{2\tau_n D_{nB}} + \text{DLR}}$	$\dfrac{1}{\dfrac{N_{DB}W_B D_{nE}}{N_{AE}W_E D_{pB}} + \dfrac{W_B^2}{2\tau_p D_{pB}} + \text{DLR}}$
2	τ_{FB}	$\dfrac{W_B^2}{2D_{nB}}$	$\dfrac{W_B^2}{2D_{pB}}$
3	C_b	$\dfrac{q}{kT}\dfrac{W_B^2}{2D_{nB}}\lvert I_E \rvert$	$\dfrac{q}{kT}\dfrac{W_B^2}{2D_{pB}}\lvert I_E \rvert$
4	C_d	$C_b + C_{je} + C_{cb'}$	$C_b + C_{je} + C_{cb'}$
5	r_d	$\dfrac{kT}{q}\dfrac{1+\beta}{\lvert I_E \rvert}$	$\dfrac{kT}{q}\dfrac{1+\beta}{\lvert I_E \rvert}$
6	ω_β	$\dfrac{1}{r_d C_d}$	$\dfrac{1}{r_d C_d}$
7	$g_m = \dfrac{\beta}{r_d}$	$\dfrac{q}{kT}\dfrac{\beta}{1+\beta}\lvert I_E \rvert$	$\dfrac{q}{kT}\dfrac{\beta}{1+\beta}\lvert I_E \rvert$
8	$\dfrac{g_m}{C_b}$	$\dfrac{2D_{nB}}{W_B^2} \cong \dfrac{1}{\tau_{FB}}$	$\dfrac{2D_{pB}}{W_B^2} \cong \dfrac{1}{\tau_{FB}}$

One approach to a possible improvement in bandwidth is to use FETs as the differential input stage, which permits a high input impedance independent of operating current levels. One such complete amplifier, which duplicates functionally the amplifier of Fig. 7.16, is shown in Fig. 7.17. In this circuit, the input differential amplifier uses *p*-channel JFETs utilizing a process technology compatible wth bipolar technology and requiring no offsetting *pnp* structure. At the expense of higher drain currents, g_m comparable to that of *npn* transistors at comparable collector currents can be achieved without a corresponding problem of lowered input resistance.

For the spike channel JFET at saturation, the maximum value of transconductance (for $V_{GS}' = 0$) from Eq. 3.51, rewritten for *p*-channel transistors, is

$$g_{mM} = \frac{q\mu_p N_A W d}{L} \tag{7.43}$$

and

$$C_o = \frac{\varepsilon_o \varepsilon_r W L}{d}$$

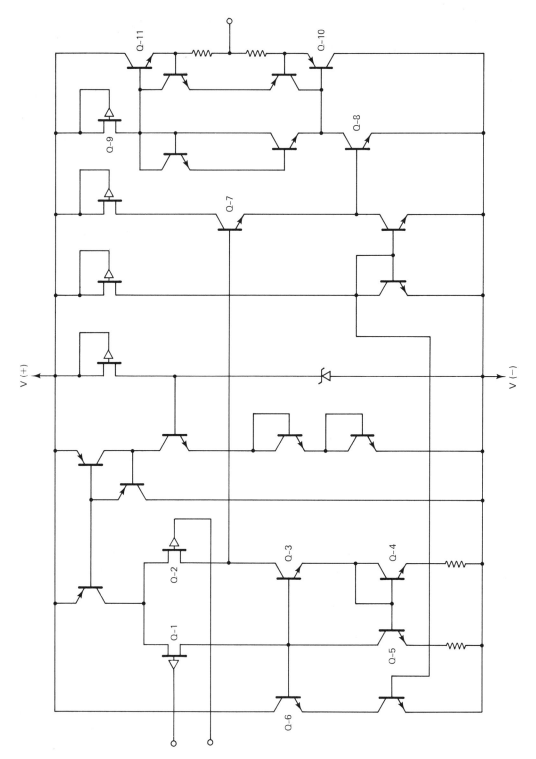

Figure 7.17 High-gain op-amp with JFET input stage.

with the gain–bandwidth product proportional to

$$\frac{g_{mM}}{C_o} = \frac{qN_A\mu_p}{\varepsilon_o\varepsilon_r}\frac{d^2}{L^2}$$

With d and N_A being used to control V_p, the figure of merit or gain–bandwidth product is inversely proportional to the square of the channel length; hence continued improvement of the frequency response of FETs depends on continued progress in fabrication of shorter channel devices. For values of $V_{gs} < |V_p|$, it can be shown that

$$g_m = \sqrt{\frac{2\mu_pW\varepsilon_o\varepsilon_r}{dL}i_{D\text{sat}}}, \qquad i_{D\text{sat}} < I_{DSS} \tag{7.44}$$

and

$$\frac{g_m}{C_o} = \sqrt{\frac{2\mu_pd}{\varepsilon_o\varepsilon_rWL^3}\,i_{D\text{sat}}} \tag{7.45}$$

These relationships demonstrate that both g_m and g_m/C_o increase with current up to the maximum value of I_{DSS}. Basically, increasing the g_m of the FET relative to that of the BJT depends on higher operating current levels, which are not normally feasible for the BJT because of input resistance requirements.

The improvement in first-stage bandwidth allows the dominant pole to be moved to a correspondingly higher frequency by using less capacitance degradation, while maintaining the same degree of stability with feedback. The output from Q-3 is compatible with the design of any of the amplifiers previously discussed. However, the complete amplifier of Fig. 7.17 has some additional unique features. The current source, Q-5, Q-3, Q-4, is the improved current mirror shown first in Fig. 5.27 to provide improved current balance to the drains of Q-1 and Q-2. The output from Q-2 drives the base of an emitter follower Q-7, while at the same time an output from Q-1 is applied to the dummy load Q-6 to equalize the loads on Q-1 and Q-2. Rather than resistor loads, current source loads are in the emitters of both Q-6 and Q-7 supplied by a common current-mirror source whose current is supplied through the drain circuit of a p-channel FET designed to be in saturation when its gate and source are connected to supply the required drain current. The current mirror in the source leads of Q-1 and Q-2 is the improved form shown first in Fig. 5.21; its current source is similar to that shown in Fig. 7.16 except for the use of a p-channel rather than an n-channel JFET and series diodes replacing part of the resistance in the emitter circuit of the transistor. This combination provides a current source with a very high degree of temperature stabilization. The collector for the emitter follower, rather than being connected directly to the positive supply voltage, is supplied through a p-channel JFET having $I_{DSS} > I_{C7}$, which puts its operation in its triode region for $v_{GS} = 0$. This makes its r_d quite low, so there is very little Miller capacitance loading at the input of Q-7. Its primary purpose is to limit the maximum base current that can be supplied to Q-8 from the emitter of Q-7 to prevent it from saturating. Bias current for the inputs of the output transistors Q-10 and Q-11 and their diode voltage-dropping networks is supplied through the drain of Q-9, which is designed to be in saturation for $v_{GS} = v_{DS}$. Its pinchoff voltage V_p is designed partly on the basis of considerations of temperature stabilization.

Further improvement in the bandwidth of op-amps, beyond reduction of the deliberate degradation of the dominant pole frequency made possible by improvement of the response of the first stage, lies in the structuring of the entire amplifier to improve all stages as much as possible, while still keeping track of and maintaining control over all significant pole locations. Many design variations are possible, some of which are intended to preserve the inherent very high gain characteristics of the conventional op-amp, but at extended bandwidth, and some of which are intended to achieve very large bandwidths with less gain and sacrifice of the unity-gain-stable feedback characteristics. Such extended bandwidth amplifiers designed for specific frequency characteristic (e.g., maximally flat response) are often referred to as *video amplifiers* because of the historic application to real-time visual data, such as required in television and other broad-band systems.

An example of a skeleton prototype of a broad-band amplifier can be derived as an extension of the amplifier suggested in Fig. 7.13, where the Miller capacitance loading of the v_{s2} input is reduced by driving the common-base stage, Q-3, in the differential cascode arrangement, along with some restructuring of the primary gain stage Q-4. One such suggested modification and extension is shown in Fig. 7.18.

In this circuit, the input transistors Q-1 and Q-2 are shown as n-channel JFETs operating at sufficient current levels to increase their g_m substantially, compared with that of BJTs operating at sufficiently low current levels to maintain their input resistance very high. The *pnp* common-base stage Q-3 simply transfers the signal current from the drain

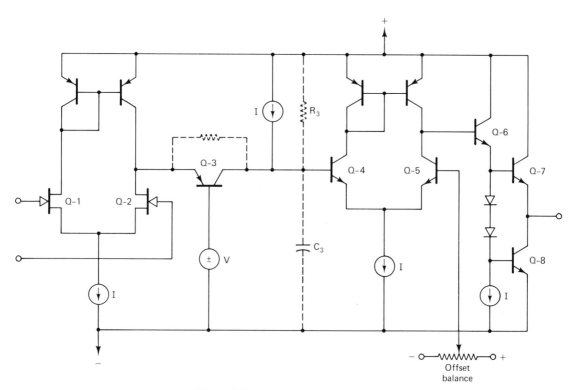

Figure 7.18 JFET input op-amp with improved bandwidth.

of Q-2 to the collector of Q-3 with substantially no loss, even with relatively low β and ω_β *pnp* transistors (the common-base cutoff frequency is given approximately by $\omega_\alpha = \beta\omega_\beta$). The inverting gain stage Q-4 of the circuit of Fig. 7.13 with its large feedback capacitance is replaced by the noninverting emitter-coupled stage, Q-4, Q-5, which in turn drives the emitter follower Q-6, which supplies the complementary output stage as before. The form of the overall frequency characteristic of the amplifier may be deduced from consideration of single-stage inverting high- and low-gain amplifiers analyzed in Sec. 4.12, noninverting amplifiers analyzed in Sec. 4.13, the cascade amplifier analyzed in Sec. 4.13, the cascode amplifier analyzed in Sec. 5.2, and the voltage follower in Sec. 5.12.

The inverting stage Q-2 being loaded by the common-base stage Q-3 has very little capacitance feedback loading. The input impedance of the common-base stage can, by appropriate design, be an almost parallel RC combination, as indicated in Sec. 4.13. With such a load, Q-3 is close to a unity-gain inverting voltage follower and, as such, has a gain almost independent of frequency in magnitude, but an excess phase shift at very high frequencies, as discussed in Sec. 4.12.

The overall response of the Q-2, Q-3 cascode amplifier is of the general form discussed in Sec. 5.2, even though the equations were derived specifically for FET amplifiers and even though the small amount of FET source loading was not considered. For example, Eq. 5.22 indicates domination by a two-pole response with $s = 1/[\omega R_L(C_2 + C_{ds2})]$ as the dominant pole for relatively low R_s. The remaining terms, containing a high-frequency pole, a left half-plane zero, and a right half-plane zero, would have to be taken into account if the feedback loop in a negative feedback configuration were to be closed for its low-gain mode. It might be observed, however, that the excess phase shift relating to the right half-plane zero is partially compensated by the left half-plane zero, and that with appropriate C/g_m ratios the net phase change due to the two zeros can be eliminated, although there will be a resultant increase in amplitude.

Thus, applying these concepts to the cascode stage in question, it would appear that the effective RC time constant at the output of Q-3 would be the controlling time constant and would not be dependent on Miller capacitance feedback. The frequency of this dominant pole can be lowered by increasing C or raised by a resistance R shunting the input resistance of Q-4. Substantial gain can be supplied by this source-coupled cascode combination.

Additional gain is supplied by the emitter-coupled stage Q-4, Q-5, which has freqeuncy characteristics described in Sec. 5.12, even though the equations were derived specifically for FETs. With proper device technology and selection of current levels, the input impedance of Q-4 is essentially a parallel RC, which is the load supplied by Q-3. The voltage gain of Q-4 is approximately $\frac{1}{2}$ and almost independent of frequency, while the gain of Q-5 is in magnitude approximately the same as the gain of a common-emitter amplifier, but noninverting with the dominant time constant being the effective RC at the output of Q-5. Then, to a fair approximation, the overall frequency response is that of a two-pole response controlled by the two RC time constants if all other poles are at much higher frequencies. Either RC time constant can be established to create *one dominant pole* if such a limitation is required for closed-loop feedback stability considerations.

The input to Q-5 can be used as a dc offset adjustment as indicated, or it can be used as another active input with an externally applied signal, or, as indicated in Fig. 7.19, it can be supplied through a separate common-base stage from the output of Q-1

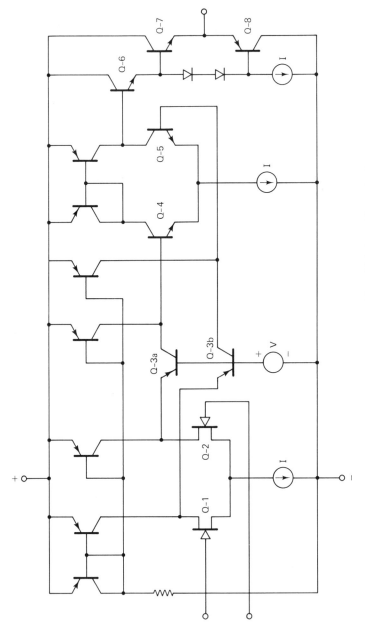

Figure 7.19 Broadband JFET input amplifier with balanced input circuit output.

with separate collector current sources for Q-1 and Q-2. In this modification, balanced outputs are carried through to the collector of Q-5, with the overall gain increased by a factor of 2.

In these video amplifiers, the pole-zero locations (by making use of the high-frequency poles) can be controlled to achieve specific response forms in both feedback and nonfeedback configurations (e.g., an n-pole maximally flat response).

There are many other possible configurations for wide-band amplifiers using all FETs, combined FETs–BJTs and all-FET structures, using JFET or IGFET technologies.

Development of MOSFET technology as a basis for integrated-circuit op-amps compatible with related digital circuits for systems requiring both digital and analog functions within a single structure offers particular challenges, which will be discussed separately in the next section.

7.5 IGFET (MOSFET) MONOLITHIC OPERATIONAL AMPLIFIERS

There are attractive reasons for considering MOS integrated technology as a vehicle for monolithic op-amp realization, particularly in LSI and VLSI systems, partly because of simplicity in internal level shifting and biasing and partly because development of short channel structures ($L < 1$ μm) offers opportunities for vastly improved bandwidths. However, it is unlikely that all-MOS structures will supplant all-bipolar or FET input-bipolar structures as readily available, off-the-shelf standard op amps; rather they will play a role as part of overall integrated structures whereby, in some cases, analog and digital components are combined in monolithic electronics subsystems and in special-purpose applications.

Historically, NMOS was first used in op-amp structures, but the advances in CMOS technology permit greater simplicity of design *because* of the availability of both n- and p-channel enhancement-mode devices, which makes for simpler level shifting and voltage and current sources without the necessity of using resistors of large ohmic value.

Some insight into the utilization of CMOS in op-amps and the design factors that are important can be gained by considering a functional duplication of basic bipolar prototypes and examination of what similarities and differences exist. For example, the circuit of Fig. 7.20 is functionally similar to that of Fig. 7.16.

In Fig. 7.20, the differential p-channel amplifier, Q-1, Q-2, drives the source follower Q-7, which in turn drives the common-source amplifier Q-8, with the output stages consisting of the complementary source followers Q-10 and Q-11. Instead of diode voltage drops to provide the bias for Q-10 and Q-11, as was indicated in Fig. 7.16, the source follower Q-9 itself is structured to provide the correct voltage separation by connecting one input directly to the output of Q-8 and the other to the source of Q-9. This voltage separation can be designed to allow the output stage to operate in class A, class B, or class AB modes. In this type of circuit, for simplicity in processing, all devices might be enhancement-mode structures with approximately the same magnitude of threshold voltage, V_T. Then the correct bias voltage for each stage can be obtained by proper proportioning of the K factors, as derived in Chapter 3 and applied to sources and loads as discussed in Sec. 5.13, with all devices operating in their respective saturation regions and $v_{GS} > (v_{DS} + V_T)$. In this particular circuit, Q-5 is the current source of the

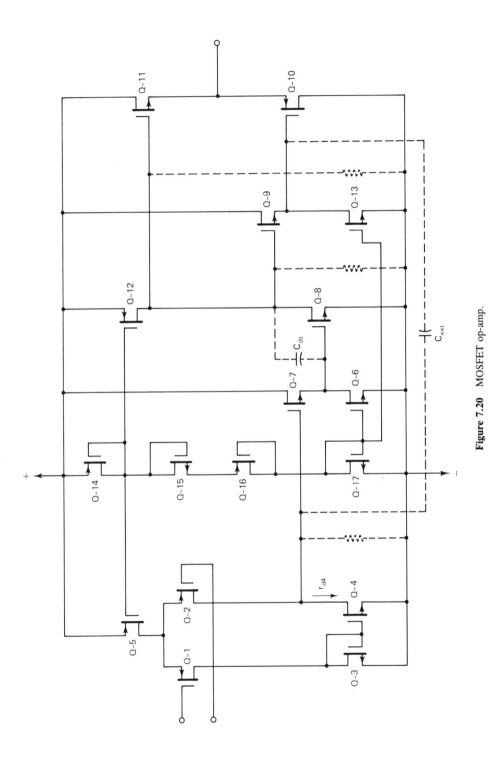

Figure 7.20 MOSFET op-amp.

input differential amplifier, Q-6 is the current source for the source follower Q-7, Q-12 is that for the inverting amplifier Q-8, and Q-13 is that for the source follower Q-9. All these current sources are provided the required gate voltages by the voltage partitioning array consisting of Q-14 through Q-17 all in, but near the edge of saturation with $v_{GS} = v_{DS}$ and K factors designed to proportion the voltage drops as indicated in Sec. 5.13. At the same time, Q-12 is structured to provide the required gate voltage for Q-11 and Q-9 to provide the required gate voltage separation for Q-10 and Q-11. The resistors shown dashed in Fig. 7.20 are loads for the output signal currents of the respective gain stages, which, along with effective shunt capacitances, control the gain and bandwidth of the respective stages. Internally, the dominant pole within the amplifier would be controlled by the feedback capacitance C_{ds8} and the output resistance of the source follower Q-7. This effective time constant can be overridden by the additional feedback capacitance C_{ext} to create a dominant single pole sufficiently low in frequency to permit unity-gain stability. For extremely high gain narrow-bandwidth amplifiers, the R's may be eliminated with the loads, then becoming the very large r_d's of the respective current sources for Q-3, Q-7, Q-8, and Q-9, as determined from the various small-signal saturation models discussed in Chapter 3. These load resistances can be made even larger by using more complex current sources. For example, Q-3, Q-4 can be replaced by the cascode current source shown in Fig. 5.28, with r_{d4} as indicated in Fig. 7.20 replaced by a value indicated in Eq. 5.83.

Many variations of the complete amplifier of Fig. 7.20 are possible. For example, often a current output may be utilized directly, particularly in a large-scale array where an output circuit suggested by Fig. 7.21 may be used. In this circuit, Q-8 of the previous

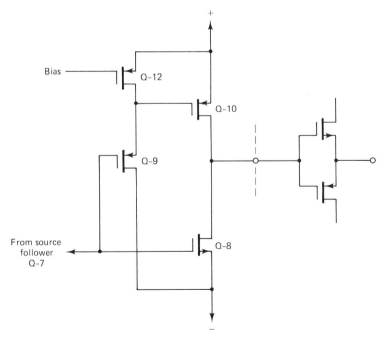

Figure 7.21 Alternative output stage for MOSFET op-amp.

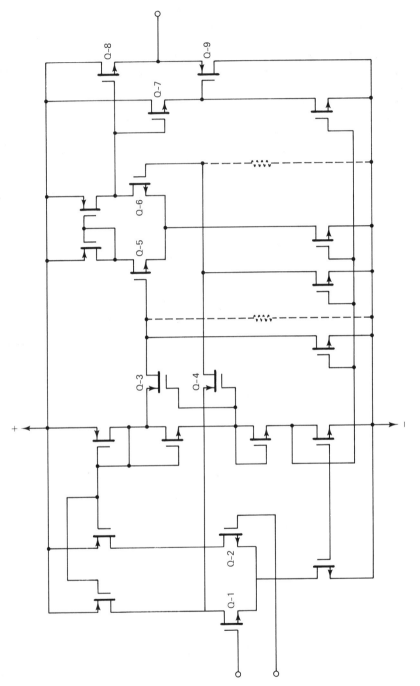

Figure 7.22 MOSFET op-amp with extended bandwidth.

circuit is the lower half of a complementary stage, with Q-10 as the upper half and Q-9 a p-channel source follower used to provide the required gate voltage separation. The current output stage, Q-8, Q-10, may be operated in class A or class AB modes. If required, an additional low-impedance output stage, Q-11, Q-12, may be provided as indicated, using FETs with lightly doped implanted channels to allow the two gates to be connected for a class AB mode of operation. Simpler versions of these op-amps for arrays are possible. For example, the source follower Q-7 might be eliminated, with the output of Q-2 driving the inverting stage Q-8 directly.

Wide-band (video) amplifiers using MOSFETS can be carefully crafted to maximize the bandwidth of each stage in the same manner as suggested in Sec. 7.4 for bipolar structures. In addition to using FET input stages, a circuit of the general form of Fig. 7.19 is suggested in Fig. 7.22, where the n-channel differential-cascode balanced amplifier Q-1, Q-2, Q-3, Q-4 drives a differential gain stage Q-5, Q-6 in a balanced fashion. The output of Q-6 drives the source-follower output stage, Q-8, Q-9, where gate voltage separation is provided by the source follower, Q-7. The respective current sources and voltage levels for the various stages are derived from a common voltage-divider network similar to that suggested in Fig. 7.20. The differential gain stages are all n-channel structures designed dimensionally with current levels to maximize their gain–bandwidth products. The RC time constants shown are intended to specify the locations of the dominant pole(s) for a prescribed overall frequency response. The two C's should, in general, be unequal because of the effective feedback capacitance of the inverting amplifier, Q-6.

The bandwidth of circuits of this type can be improved in various ways. For example, a common-base stage can be interposed between the output of Q-6 and the output stages in the usual cascode arrangement to lessen the feedback capacitance loading at the gate of Q-6. This would confine the amplifier to extremely small signal inputs to prevent overloading of the later stages. Alternatively, Q-6 may be operated at fixed bias, reducing the internal gain by a factor of 2, but eliminating feedback capacitance loading. For this type of operation, Q-4 would not be needed, but it might still be retained as a load for Q-1 in order to retain balanced loads for Q-1 and Q-2.

In all the all-MOS structures, there is a basic trade-off between operating current level and transconductance as compared with the bipolar amplifier.

7.6 OPERATIONAL TRANSCONDUCTANCE AMPLIFIER

Although the name operational transconductance amplifier (OTA) conjures up several possible meanings, each having to do with utilizing directly the basic property of the g_m of the input device, it is generally understood to refer to a very high gain operational amplifier with both high input and output impedances such that the overall voltage-to-current transfer characteristic, referred to here as effective transconductance g_{mE}, is the dominant property, and having the additional property that such effective transconductance is controllable over a very wide range (usually several decades) by an external bias voltage, as indicated symbolically in Fig. 7.23.

The effective transconductance is defined by

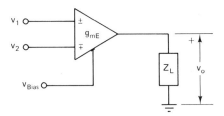

Figure 7.23 Operational transconductance amplifier symbol.

$$g_{mE} \cong \pm \frac{\partial i_o}{\partial(v_1 - v_2)}\bigg]_{v_o = 0} \qquad (7.46)$$

and whose value is also

$$g_{mE} = Kv_{Bias} \qquad (7.47)$$

where K is a constant of proportionality.

As a practical matter, the variable transconductance feature is most simply implemented by controlling the operating current of an input differential amplifier and utilizing an output stage with a very high output impedance. Such an implementation is shown schematically in Fig. 7.24. The intermediate amplifier with current gain A_i is supplied differentially from the output of the first stage in order that variations in i_E of the first stage do not result in output dc level shifts.

The intermediate amplifier should have low input impedances (consisting, perhaps,

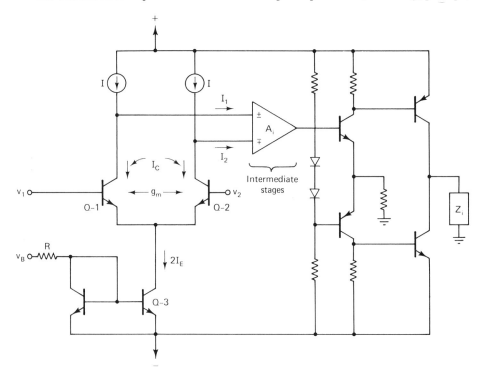

Figure 7.24 Basic operational transconductance amplifier indicating required properties.

of common-base stages). At a specific operating point, $I_{C1} = I_{C2} = I_C$, and the current supplied to each input of the intermediate amplifier is $I_1 = I_2 = I - I_C$. The intermediate stage is intended as a linear current amplifier whose effective current gain A_i is relatively independent of bias current. The dual complementary output stage drives the load from the collectors and thus presents a very high output impedance to the load.

For the circuit shown in Fig. 7.24, the transconductance for each input transistor is given by

$$g_m = \frac{1}{kT/q} \frac{h_{FE}i_E}{1 + h_{FE}} \tag{7.48}$$

If this g_m is in turn to be proportional to a bias control voltage v_B, then the current sources must include temperature-compensating elements. If this is done, we may state that

$$g_m = K_3 v_B \tag{7.49}$$

where K_3 is the constant of proportionality that includes temperature compensation.

We now define that total overall effective transconductance as

$$g_{mE} = \frac{g_m}{2} A_{iT} \tag{7.50}$$

where A_{iT} is the current gain of the rest of the amplifier following the input differential stage and including the complementary output stages.

Thus the output voltage across R_L is

$$v_o = g_{mE}(v_1 - v_2)R_L \tag{7.51}$$

where g_{mE} can be written in terms of v_B as $K_3 v_B A_{iT}$. Hence v_o in terms of bias voltage is

$$v_o = (K_3 A_{iT})v_B(v_1 - v_2)R_L \tag{7.52}$$

This equation defines a voltage-controlled amplifier whereby the overall gain, $v_o/(v_1 - v_2)$, is directly proportional to the bias control voltage, v_B.

Other implications are inherent in Eq. 7.52. For example, if v_B is a variable voltage $V_B + v_b$, where v_b is a signal variable,

$$v_o = (K_3 A_{iT})(V_B + v_b)(v_1 - v_2)R_L \tag{7.53}$$

which may be expressed as

$$v_o = \underbrace{(K_3 A_{iT} V_B)(v_1 - v_2)}_{\text{linear term}} + \underbrace{K_3 A_{iT} v_b(v_1 - v_2)}_{\text{product term}} \tag{7.54}$$

The first term is that of a linear amplifier whose gain is proportional to the average value of the bias voltage that controls the emitter current of the input stage, while the second term is proportional to the product of the differential input signal and the variable signal component of the bias voltage.

If two identical OTAs are connected as shown in Fig. 7.25, the voltage across R_L is given by

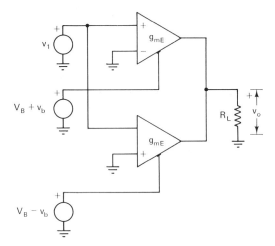

Figure 7.25   Two OTAs used as multiplier.

$$v_o = 2K_3 A_{iT}(v_1 v_b)R_L \qquad (7.55)$$

which is simply proportional to the product of the signal voltage and the variable bias voltage with the linear term canceled.

Two OTAs connected in this manner function as a multiplier. The inputs could have been driven differentially rather than one side grounded. Then v_1 in Eq. 7.55 is simply replaced by $(v_1 - v_2)$, where v_2 is the other input.

For FET input stages as indicated in Fig. 7.26(a), the situation is similar but slightly different. If it is assumed that the FETs are operated in their relatively low voltage saturation region, the approximate equation for each device is given by

$$i_D = \frac{K}{2}(v_{GS} - V_T)^2 \qquad (7.56)$$

where K is the structurally related coefficient. Also

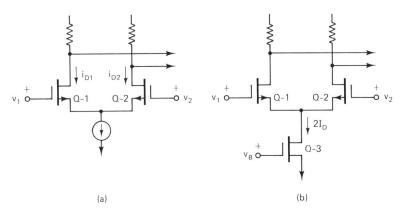

(a)

(b)

Figure 7.26   OTA input stage using FETs. (a) Differential amplifier with current source. (b) Current source for variable g_m requirement.

$$g_m\Big]_{I_D} = K(V_{GS} - V_T) \qquad (7.57)$$

At the operating point,

$$(V_{GS} - V_T) = \sqrt{\frac{2I_D}{K}}$$

so the transconductance can be written as

$$g_m\Big]_{I_D} = \sqrt{2K}\sqrt{I_D} \qquad (7.58)$$

which is not linearly proportional to bias current, as is the case for BJTs.

If, however, the current source is derived from another MOSFET, as indicated in Fig. 7.26(b), the equation for the bias current can be written as

$$2I_D = \frac{K_3}{2}(V_{GS3} - V_T) \qquad (7.59)$$

Thus the transconductance in terms of the gate–source voltage of Q-3D is

$$g_m = 2K\frac{K_3}{4}(V_{GS3} - V_T) \qquad (7.60)$$

which is linearly proportional to the bias voltage applied to the current source transistor, Q-3.

7.7 GAIN-STABILIZED TRANSCONDUCTANCE AMPLIFIER

The gain-stabilized transconductance amplifier (GSTA) is a particular form of highly stabilized operational amplifier. It is designed to possess the overall characteristics of the OTA, that is, the current–voltage transfer characteristic, but without the variable g_m feature; it includes a shunt feedback element, as symbolized in Fig. 7.27, and ideally has the overall transfer function

$$\frac{V_2}{V_1} = \pm\frac{Z_L}{Z_F} \qquad (7.61)$$

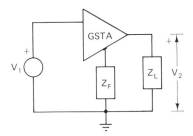

Figure 7.27 Symbolic representation of gain-stabilized transconductance amplifier.

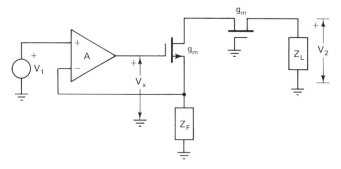

Figure 7.28 Simple GSTA structure.

The fact that Z_L and Z_F are both shunt elements yields a flexibility for certain applications that conventional op-amps do not have.

There are many possible implementations of the GSTA; one elementary form is suggested in Fig. 7.28, which consists of an ideal high-gain conventional op-amp with low impedance output, followed by a voltage-to-current converter, which is shown in the form of a cascode stage to minimize capacitance loading of its input. With $A \equiv V_x/V_i$, the overall response is

$$\frac{V_2}{V_1} \cong \frac{Z_L}{Z_F} \frac{1}{1 + (1/A) [1 + (1/g_m Z_F)]} \tag{7.62}$$

Ideally, A will be sufficiently large at all design frequencies such that

$$\frac{V_2}{V_1} \cong -\frac{Z_L}{Z_F} \tag{7.63}$$

7.8 MICROPOWER OPERATIONAL AMPLIFIERS

A micropower operational amplifier is generally considered to have the general characteristics of a conventional op amp but is designed to operate at extremely low voltage and current and hence at extremely low power levels. Transistors for such amplifiers are generally pushed to the minimum voltage limits, as indicated in Fig. 7.29 for a low-voltage *npn* transistor. Such a transistor may be biased in the vicinity of the $v_{CB} = 0$ contour with the collector–base junction allowed to extend dynamically into the forward-biased region as indicated. The particular characteristic shown is an extension of the discrete transistor shown in Fig. 2.12 into the micropower region.

Integrated-circuit transistors can be processed with structural parameters with higher β's at even lower values for v_{BE} at $v_{CB} = 0$. Collector currents of around 10 μA with $v_{BE} \cong 0.5$ V and $h_{FE} > 150$ are not unreasonable, with $v_{CEmin} \cong 0.2$ V without heavy saturation.

A skeleton prototype of a micropower amplifier that still preserves ± symmetry, at the same time maintaining the lowest possible total supply voltage, is suggested in Fig. 7.30, with the input–output levels referenced in the center of the supply-voltage range.

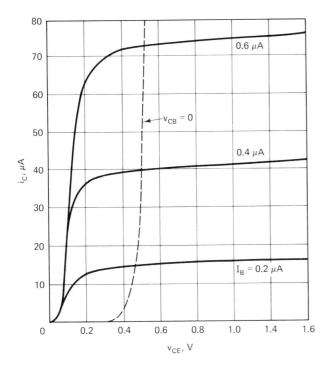

Figure 7.29 Low-current transistor characteristic.

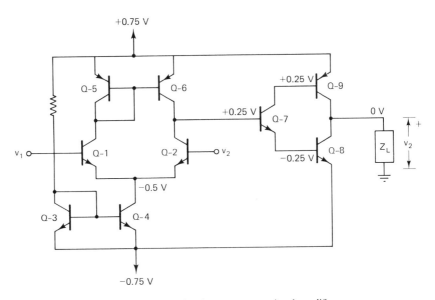

Figure 7.30 Basic micropower operational amplifier.

Sec. 7.8   Micropower Operational Amplifiers

315

The input stage is similar to that first suggested in Fig. 5.24. The approximate operating voltages are as indicated. All transistors except Q-3 and Q-4 are designed to operate in the vicinity of $v_{CB} \cong 0$. The transistors Q-3 and Q-4 are in the vicinity of $v_{CE} \cong 0.25$ V, since they operate at a constant current that permits an operating point closer to the edge of heavy saturation. In this design, the total supply voltage is minimized by deriving the output from the collectors of the complementary pair. In this respect, this, as well as most micropower implementations, tends more toward the characteristics of the OTA rather than those of a conventional op-amp. It is of course possible to design amplifiers of even lower supply voltage levels at the expense of preserving balance and symmetry.

7.9 OPERATIONAL AMPLIFIERS AS VOLTAGE COMPARATORS

A very high gain operational amplifier whose output is limited at upper and lower levels (usually at voltages near the $\pm V$ supply-voltage levels) as its input signal changes through its entire dynamic range referenced about a fixed voltage reference input is defined as a voltage comparator, as indicated in Fig. 7.31. The upper and lower output limits are designated as V_{OH} and V_{OL}, respectively. When the input signal $v_i = V_{REF}$, the output is approximately in the center of this range, and the change in input voltage required to overdrive the amplifier in both direction is

$$\Delta v_i = \frac{V_{OH} - V_{OL}}{A} \tag{7.64}$$

If A is very large, $\Delta v_i \to O$, and as $v_i \to V_{REF}$ from the negative side, the output switches abruptly to V_{OH}, and as it approaches V_{REF} from the positive side, the output switches abruptly to V_{OL}. This function defines a voltage comparator because the output signal is compared with another signal designated as V_{REF}. The direction of the output shift also identifies the direction of the input signal change.

An inverting comparator, with the v_i and V_{REF} voltages interchanged, functions in the same manner except that the output polarity changes are reversed.

Comparators are sometimes designed for a "built-in" offset voltage; that is, V_{OH} and V_{OL} need not be symmetrical about V_{REF}. For example, for $v_i \to V_{REF}$, output levels

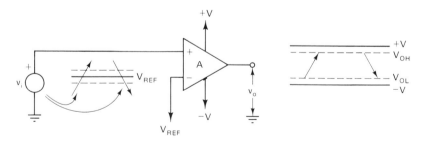

Figure 7.31 Op-amp as voltage comparator.

may be designed for $V_{OL} = 0$ and $V_{OH} = +5$ V, which would be compatible with standard saturated logic levels, or for $V_{OH} = 0$ and $V_{OL} = -5$ the outputs would be compatible with standard emitter-coupled logic levels.

Operational amplifiers designed to function as comparators should be designed to be particularly stable with respect to drift, and should be designed for large bandwidths and high slew rates, since they are always operated in their high gain mode, unprotected by negative external feedback.

Such comparators used in the open-loop mode may be referred to as *nonregenerative* comparators because there is no inverse functional relationship between output and input.

Regenerative Comparators

A comparator may be connected in a positive feedback mode as indicated in Fig. 7.32, with the input signal connected to the inverting input and the output signal and voltage reference connected to the noninverting terminal as shown.

When the output is $v_O = V_{OH}$, the feedback voltage is given by

$$v_N = V_{NH} = \frac{R_2 V_{OH} + R_1 V_{REF}}{R_1 + R_2} \tag{7.65}$$

and when $v_O = V_{OL}$, the feedback voltage is

$$v_N = V_{NL} = \frac{R_2 V_{OL} + R_1 V_{REF}}{R_1 + R_2} \tag{7.66}$$

Thus the difference between the two feedback levels is

$$V_{NH} - V_{NL} = \frac{R_2}{R_1 + R_2}(V_{OH} - V_{OL}) \tag{7.67}$$

As illustrated in Fig. 7.32, if the input signal starts from a low value and progresses upward, the output will switch from V_{OH} to V_{OL} when the input signal reaches a level corresponding to V_{NH}, and when it starts from a high level and moves downward, the output will shift from V_{OL} to V_{OH} when the input signal reaches V_{NL}. The difference between V_{OH} and V_{OL} is referred to as the *hysteresis* of the circuit.

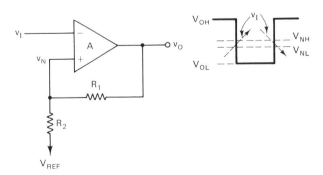

Figure 7.32 Op-amp regenerative comparator with hysteresis.

7.10 PROGRAMMABLE OPERATIONAL AMPLIFIERS

An operational amplifier designed to function in a variety of modes with only minor external changes is sometimes designated as a *programmable operational amplifier*. For example, an amplifier that has both low and high impedance outputs and is designed as an operational transconductance amplifier, but whose bias can be stabilized at a fixed value, can be programmed as a gain-stabilized transconductance amplifier or as a conventional operational amplifier. Also, an amplifier may be designed as a conventional narrow-band amplifier with one dominant pole; but by making available some interstage terminals, load resistances may be added to program it as a wide-band amplifier and/or a high-speed voltage comparator.

There are no specific minimum number of operational modes that will identify an amplifier as programmable. It is a term simply intended to indicate some degree of versatility for some amplifier designs.

7.11 ACTIVE FILTERS INCORPORATING OPERATIONAL AMPLIFIERS

Circuits composed of resistors, capacitors, and active elements such as operational amplifiers in a feedback mode have for many years been used as alternatives to traditional filters employing resistors, capacitors, and *inductors* at low (audio) frequencies, permitting the elimination of large and bulky inductive elements. In more recent times, integrated-circuit technology has led to the extension of the frequency range upward into the megahertz region. This section will present only a very brief introduction to this most extensive subject, which will illustrate some alternative techniques that are available.

Low-Pass Filters Using a Single Amplifier

The low-pass filter is specified as a two-port network that passes a range of frequencies with reasonable uniformity from zero to some arbitrarily defined maximum value. The single-pole response obtained with the low-pass RC network shown in Fig. 4.15 is a simple illustration of a low-pass filter, although a very poor one because the high-frequency asymptotic roll-off is only 6 dB/octave, while a two-pole maximally flat response decreases 12 dB/octave. The sharpness of the cutoff with frequency increases with the number of poles of the transfer function.

The relative magnitude of the n-pole maximally flat response as shown in Fig. 7.33 is given by an equation of the form

$$H(\omega) = \frac{1}{\sqrt{1 + (\omega/\omega_2)^{2n}}} \tag{7.68}$$

or in decibels as

$$H(\omega)]_{dB} = -10 \log \left[1 + \left(\frac{\omega}{\omega_2} \right)^{2n} \right] \tag{7.69}$$

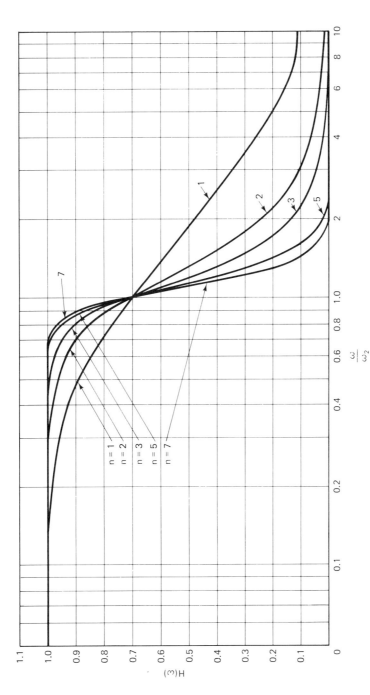

Figure 7.33 Frequency characteristics of *n*-pole, low-pass, maximally flat responses.

showing an asymptotic roll-off of -6 dB/octave for $n = 1$, with an additional -6 dB/octave for each additional pole. The corresponding transient responses are shown in Fig. 7.34.

Active filters incorporating the two-pole response arising from a quadratic equation in s are reasonably tractable analytically, and hence the most exhaustively studied; they are referred to as *biquadratic filters,* or *biquads.* Most low-pass biquad filter analysis and design assumes ideal operational amplifiers, with the response controlled by the passive network elements alone. One such filter using a noninverting ideal amplifier is shown in Fig. 7.35.

The overall voltage gain, assuming that A is nonfrequency dependent, may be put in the form

$$\frac{V_o}{V_s} = \frac{A\omega_2}{s^2 + k\omega_2 s + \omega_2^2} \tag{7.70}$$

where

$$\omega_2 = \frac{1}{\sqrt{R_1 C_1 R_2 C_2}}$$

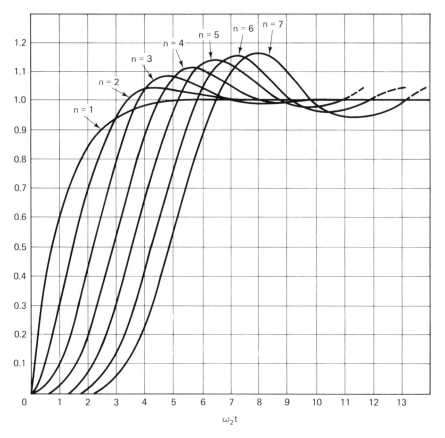

Figure 7.34   Step function response of maximally flat networks.

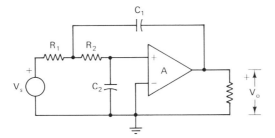

Figure 7.35 Two-pole, single-amplifier, low-pass filter.

and

$$k = \sqrt{R_1 C_1 R_2 C_2} \left[\frac{1}{R_1 C_1} + \frac{1}{R_2 C_1} + \frac{1}{R_2 C_2} (1 - A) \right]$$

Because of the nature of the coefficients of the first-power s term, which is highly dependent on A, there are possibilities of obtaining a number of specific response forms, one of which is the maximally flat response.

If the amplifier is to be stable, the poles of Eq. 7.70 are in the left half-plane given by

$$s = -\frac{k\omega_2}{2} \pm j \frac{k\omega_2}{2} \sqrt{\frac{4}{k^2 - 1}}$$

and specifically for the maximally flat response the magnitudes of the real and imaginary parts are made equal with $k = \sqrt{2}$.

For $k = \sqrt{2}$ and the special case of $R_1 C_1 = R_2 C_2$, Eq. 7.70 may be solved for a relationship between A and the capacitances, with the result being

$$\frac{C_2}{C_1} = A + \sqrt{2} - 2$$

In the special case with A highly stabilized by the op-amp in its unity-gain mode, $C_2/C_1 = 0.414$.

The amplifier may be highly stabilized in its voltage-follower mode, but there is no amplification. If A is made larger, overall gain is achieved at the sacrifice of some deviation from the prescribed response form due to variations in A. However, ω_2 is not dependent on A.

A slightly different form of low-pass biquad using an inverting amplifier is shown in Fig. 7.36. Its response is also of the general form of Eq. 7.70.

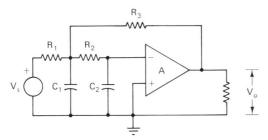

Figure 7.36 Another two-pole, single-amplifier, low-pass filter.

Sec. 7.11   Active Filters Incorporating Operational Amplifiers

321

Single-Amplifier Bandpass Filters

Before proceeding with the analysis and design of active RC bandpass filters, it is useful to review the properties of the RLC parallel resonant circuit and its terminology as a model. Such a parallel resonant circuit is shown in Fig. 7.37(a), with an applied current source or an equivalent voltage source as indicated in Fig. 7.37(b).

Usually, R_s is the lumped approximation of the distributed resistance of the inductor itself and is made as small as possible, while R_p, which may be part of the voltage or current source, is used to control the sharpness of the response.

The overall response is of the form

$$\frac{V_o}{V_s} = \frac{[Q_s\omega_oR_s/R_p][s + (\omega_o/Q_s)]}{s^2 + (\omega_o/Q_o)s + [(R_s + R_p)/R_p]\,\omega_o^2} \tag{7.71}$$

where

$$\omega_o^2 = \frac{L}{LC_o} = \text{resonant frequency of parallel } LC \text{ circuit with } R_s = 0$$

$$Q_s = \frac{\omega_oL}{R_s} = Q \text{ factor of a series } RL \text{ circuit}$$

$$Q_p = \frac{R_p}{\omega_oL} = Q \text{ factor of parallel } RL \text{ circuit}$$

$$Q_o = \frac{Q_sQ_p}{Q_s + Q_p} = \text{effective } Q \text{ factor of } R_s,\ R_p,\ L \text{ combination}$$

As an approximate form for $R_s \ll R_p$, the zero of the function is far removed from the pole locations, and

$$\frac{v_o}{v_s} \cong \frac{(\omega_o/Q_o)s}{s^2 + (\omega_o/Q_o)s + \omega_o^2} \tag{7.72}$$

which can be written as

$$\frac{V_o}{V_s} = \frac{1}{1 + jQ_o[(\omega/\omega_o) - (\omega_o/\omega)]} \tag{7.73}$$

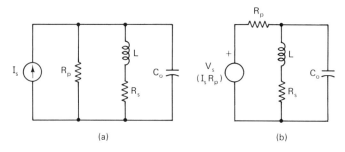

(a)   (b)

Figure 7.37 Single-tuned RLC bandpass filters. (a) Current source form. (b) Voltage source form.

for $Q_p \cong Q_o$. These equations represent an approximate form of the parallel resonant circuit.

If δ is defined as the fractional frequency deviation given by

$$\delta = \frac{\omega}{\omega_o} - 1$$

(7.74)

then

$$\frac{v_o}{v_s} = \frac{1}{1 + j[(2 + \delta)/(1 + \delta)]\delta Q_o}$$

The responses for various values of Q_o in terms of fractional frequency deviation are plotted in Fig. 7.38. For $\delta \ll 1$,

$$\frac{V_o}{V_s} \cong \frac{1}{1 + j2\delta Q_o} \tag{7.75}$$

The fractional bandwidth, β, is obtained for $\pm\delta$ for the 0.707 response points (i.e., for $2\delta Q_o = \pm 1$ or $\delta = \pm 1/Q_o$). Thus the fractional bandwidth β is

$$\beta = \frac{1}{Q_o} \tag{7.76}$$

and the actual bandwidth is

$$BW = f_o\beta = \frac{f_o}{Q_o} \tag{7.77}$$

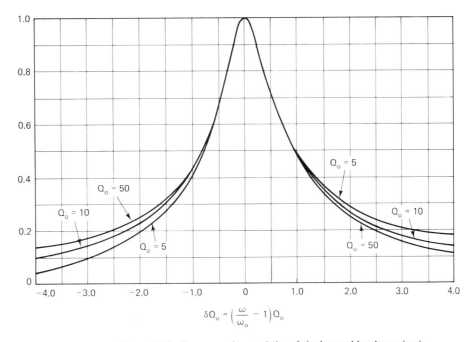

Figure 7.38 Frequency characteristics of single-tuned bandpass circuit.

Sec. 7.11   Active Filters Incorporating Operational Amplifiers

Therefore, an operational definition of circuit Q_o can be expressed as

$$Q_o = \frac{f_o}{BW} \tag{7.78}$$

which can be expressed in terms of the inductance of Q itself. Also,

$$BW \cong \frac{1}{2\pi R_o C_o} \tag{7.79}$$

where R_o is the total parallel shunt resistance for R_s sufficiently small.

Consider the parallel RLC tuned circuit being driven by a transconductance amplifier as shown in Fig. 7.39. The gain at the center frequency is

$$A_o = -g_m R_o \tag{7.80}$$

and, using Eq. 7.77, the gain–bandwidth product is defined as

$$A_o BW = \frac{g_m}{2\pi C_o} \tag{7.81}$$

If C_o is chosen as the minimum value possible for the device and circuit in question, then if R_o is decreased to increase the bandwidth, the gain is reduced by the same amount.

There are many ways that the approximate form of the parallel RLC response as given by Eq. 7.72 can be obtained using RC networks with single or multiple operational amplifiers with feedback. All of these are referred to as *bandpass biquads*. An example of a single-amplifier bandpass biquad is shown in Fig. 7.40. The overall gain can be expressed as

$$\frac{V_o}{V_s} = -\frac{A}{R_1 C_1 (1 + A)} \frac{s}{s^2 + (\omega_0/Q_o)s + \omega_o^2} \tag{7.82}$$

where

$$\omega_o^2 = \frac{1}{R_1 C_1 R_2 C_2 (1 + A)}$$

$$Q_o = \frac{R_1 C_1 R_2 C_2 (1 + A)}{R_1 C_1 + R_2 C_2 + R_1 C_2}$$

In this particular circuit, the effective Q and the center frequency are both dependent on A, which is subject to variation due to power supply and temperature variations. Also, inasmuch as the ideal op-amp is assumed, the use is limited to a very small fraction of its open-loop bandwidth for large values of Q, which in turn requires large gain.

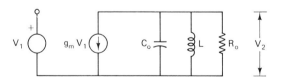

Figure 7.39 Approximate form of parallel *RLC* circuit.

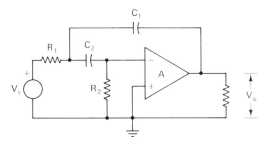

Figure 7.40 Active *RC* single inverting amplifier two-pole (biquad) bandpass filter.

An alternative single-amplifier bandpass biquad utilizing a noninverting amplifier is shown in Fig. 7.41, which requires one additional resistor, whose response is

$$\frac{V_o}{V_s} = \frac{As}{R_2 C_1 [s^2 + (\omega_o/Q_o)s + \omega_o^2]} \tag{7.83}$$

where

$$\omega_o^2 = \frac{1 + (R_2/R_1)}{R_2 C_1 R_2 C_2}$$

and

$$Q_o = \frac{\{[1 + (R_2/R_1)]R_2 C_1 R_3 C_2\}^{1/2}}{(R_2/R_1)R_3 C_2 + R_2 C_1 + R_2 C_2 + R_3 C_2(1 - A)}$$

In this particular circuit, the center frequency is not dependent on amplifier gain, but the Q_o is, in contrast to the circuit of Fig. 7.40, where both Q_o and ω_o are gain dependent.

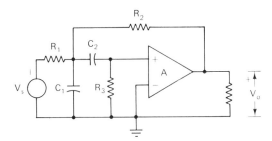

Figure 7.41 Active *RC* single noninverting amplifier biquad bandpass filter.

Single-Amplifier High-Pass Biquadratic Filter

A high-pass filter ideally has high attenuation up to some specified cutoff frequency and has relatively uniform response at higher frequencies. A high-pass biquadratic filter has the general response form of

$$\frac{V_o}{V_s} = K \frac{s^2}{s^2 + (\omega_o/Q_o)\,s + \omega_o^2} \tag{7.84}$$

which can be normalized in terms at real frequencies as

$$\frac{1}{K} \frac{V_o}{V_s} = \frac{1}{1 - (\omega_o/\omega)^2 - j(\omega_o/\omega)(1/Q_o)} \tag{7.85}$$

whose magnitude can be expressed as

$$\left| \frac{1}{K} \frac{V_o}{V_s} \right| = \frac{1}{\sqrt{1 - [2 - (1/Q_o^2)](\omega_o/\omega)^2 + (\omega_o/\omega)^4}} \tag{7.86}$$

If the middle term is made to vanish by setting $Q_o = 1/\sqrt{2}$, the high-pass maximal response is obtained as

$$\left| \frac{1}{K} \frac{V_o}{V_s} \right| = \frac{1}{\sqrt{1 + (\omega_o/\omega)^4}} \tag{7.87}$$

Or, by taking the derivative of Eq. 7.86 with respect to ω_o/ω and setting it equal to zero, the response can be shown to peak at a frequency given by

$$\left(\frac{\omega_o}{\omega} \right)^2 = 1 - \frac{1}{2Q_o^2} \tag{7.88}$$

with an amplitude at this peak frequency given by

$$\frac{1}{K} \frac{V_o}{V_s} = \frac{2Q_o^2}{\sqrt{4Q_o^2 - 1}} \tag{7.89}$$

As an example, for a value of $Q_o = 1$, the relative response is unity at $\omega = \omega_o$ and has a peak of 15.5% at $\sqrt{2}\omega_o$. The maximally flat response and this peaked response are plotted in Fig. 7.42.

These responses are compared with a single-pole, high-pass, single-time constant RC filter whose response is given by

$$\frac{V_o}{V_s} = \frac{1}{\sqrt{1 + (\omega_o/\omega)^2}} \tag{7.90}$$

where $\omega_o = 1/(RC)$.

In general, the high-pass maximally flat response is of the form

$$\left| \frac{1}{K} \frac{V_o}{V_s} \right| = \frac{1}{\sqrt{1 + (\omega_o/\omega)^{2n}}} \tag{7.91}$$

where n is the number of poles of the function. Such responses are observed to be simply frequency inversions of low-pass maximally flat responses with ω/ω_o being replaced by ω_o/ω.

An example of a single-amplifier RC high-pass biquad is shown in Fig. 7.43 using a noninverting amplifier. The overall response is of the form

$$\frac{V_o}{V_s} = \frac{As^2}{s^2 + (\omega_o/Q_o)\, s + \omega_o^2} \tag{7.92}$$

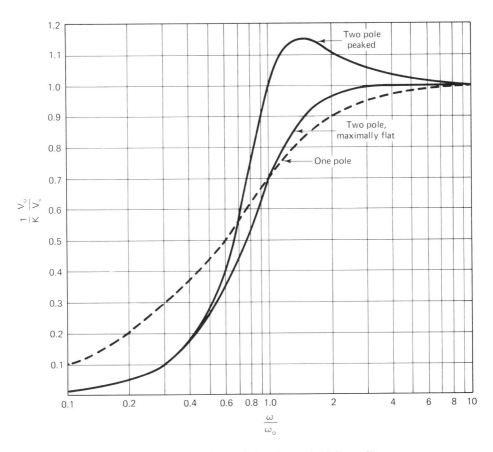

Figure 7.42 Frequency characteristics of two-pole high-pass filter.

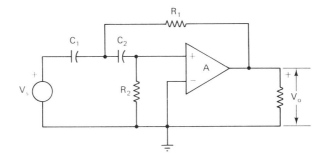

Figure 7.43 Single noninverting amplifier high-pass RC biquad.

where

$$\omega_o^2 = \frac{1}{R_1C_1R_2C_2}$$

and

$$Q_o = \frac{\sqrt{R_2C_2R_1C_1}}{R_1C_1 + R_1C_2 + R_2C_1\,(1 - A)}$$

An inverting amplifier may be used in the circuit of Fig. 7.44 to yield a response of the form

$$\frac{V_o}{V_s} = K\,\frac{s^2}{s^2 + (\omega_o/Q_o)\,s + \omega_o^2} \tag{7.93}$$

where again Q_o and ω_o can be determined in terms of the circuit elements.

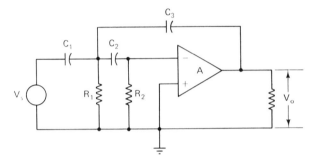

Figure 7.44 Single inverting amplifier high-pass *RC* biquad.

Multiple-Amplifier Biquadratic RC Filters

There are many ways in which multiple amplifiers with multiple feedback loops have been used in attempts to reduce the sensitivity of Q and ω_o to amplifier characteristics. A classic example is the circuit of Fig. 7.45, where op-amps are used in their high-again mode as summing amplifiers and integrators.

Outputs can be taken at a, b, and c to yield low-pass, bandpass, and high-pass characteristics, respectively, with equations

$$\frac{V_a}{V_s} = \frac{R_2(R_4 + R_3)}{R_3(R_1 + R_2)R_5R_6C_1C_2}\,\frac{1}{s^2 + (\omega_o/Q_o)\,s + \omega_o^2} \tag{7.94}$$

$$\frac{V_b}{V_s} = -\frac{R_2(R_4 + R_3)R_6C_2}{R_3(R_1 + R_2)R_5R_6C_1C_2}\,\frac{s}{s^2 + (\omega_o/Q_o) + \omega_o^2} \tag{7.95}$$

$$\frac{V_c}{V_s} = \frac{R_2(R_4 + R_3)C_2C_1R_5R_6}{R_3(R_1 + R_2)R_5R_6C_1C_2}\,\frac{s^2}{s^2 + (\omega_o/Q_o)\,s + \omega_o^2} \tag{7.96}$$

where

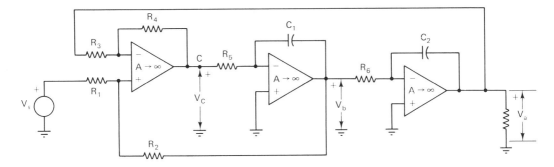

Figure 7.45 Multiple amplifier filter with high-low and bandpass outputs.

$$\omega_o^2 = \frac{R_4/R_3}{R_5 R_6 C_1 C_2} \tag{7.97}$$

and

$$Q_o = \frac{R_1 + R_2}{R_1(R_4 + R_3)} \sqrt{\frac{R_4 R_3 R_5 C_1}{R_6 C_2}} \tag{7.98}$$

With extremely high gain op-amps used as summing amplifiers and integrators, both the reference frequency, ω_o, and the effective Q are independent of amplifier gains. However, applications of such filters are confined to extremely low frequencies unless expensive op-amps of exceptionally high gain–bandwidth products are used, inasmuch as the response depends on the ideal amplifier in their high-gain modes.

There are more direct and analytically simple ways of using multiple amplifiers to achieve biquadratic responses, which have the potential of realizing filters at much higher frequencies. One particular form, known as the *amplifier-isolated positive feedback* (AIPF) filter, employs operational amplifiers stabilized in their low-gain mode used mainly as isolating elements and summing amplifiers between cascaded low-pass, bandpass, and high-pass simple RC filter sections in a positive feedback configuration. A simple bandpass biquad incorporating this principle is shown in Fig. 7.46. This particular configuration incorporates low-pass filters at the summing inputs whereby the source resistance and the parasitic shunt capacitance become part of the response. Then the only high-frequency limitation is the parasitic capacitance at the input of A_2 and the high-frequency characteristics of the amplifiers themselves.

The overall response of this particular AIPF filter may be written as

$$\frac{V_o}{V_1} = \frac{s + \omega_3}{s + \omega_1} \frac{A_2}{2} \frac{\omega_1 s}{s^2 + (\omega_2 + \omega_3 - A_2 A_3 A_3)s + \omega_2 \omega_3} \tag{7.99}$$

where $\omega_1 = 1/(R_1 C_1)$, $\omega_2 = 1/(R_2 C_2)$, and $\omega_3 = 1/(R_3 C_3)$.

For the special case of equal time constants, $\omega_1 = \omega_2 = \omega_3 \cong \omega_o$,

$$\frac{V_o}{V_1} = \frac{A_2}{2} \frac{\omega_o s}{s^2 + (2 - A_2 A_3)\omega_o s + \omega_o^2} \tag{7.100}$$

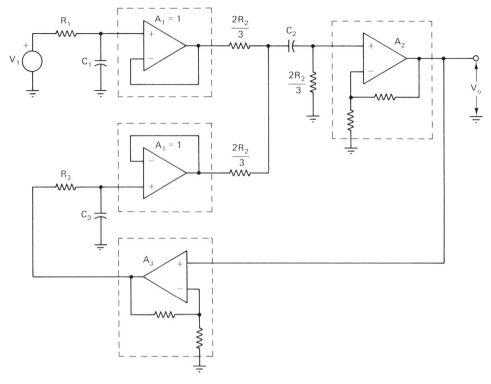

Figure 7.46 Simple amplifier isolated positive feedback (AIPF) active filter.

with the effective Q given by

$$Q_o = \frac{1}{2 - A_2 A_3} \tag{7.101}$$

In such an amplifier, relatively high Q's can be achieved because A_2 and A_3 are in their highly stabilized low-gain modes ($A_2 A_3 < 2$) and the operating frequency can be extended to a substantial fraction of the unity gain–bandwith products of the amplifiers.

At an expense of some additional complexity, readily controllable n-pole responses can be obtained using multiple feedback incorporating two-pole RC low-pass, bandpass, and high-pass RC filter sections, as illustrated by the example of Fig. 7.47.

For the special case of $R_1 C_1 = R_2 C_2$ and with all C's identical, the overall response reduces to

$$\frac{V_O}{V_1} = \frac{(A_1 A_2/4)\omega_o s}{s^4 + P_3 \omega_o s^3 + P_2 \omega_o^2 s^2 + P_1 \omega_o s + \omega_o 4} \tag{7.102}$$

where

$$P_3 = 6 - \frac{A_2}{4} A_{f3}, \qquad P_2 = 4 - \frac{A_2}{4} A_{f2}, \qquad P_1 = 9 - \frac{A_2}{4} A_{f1}$$

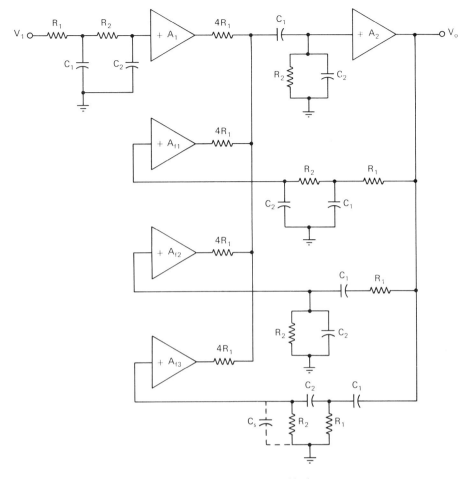

Figure 7.47 AIPF active filter with multipole response.

and

$$\omega_o^2 = \frac{1}{R_0 C_0} = \frac{1}{R_1 C_1} = \frac{1}{R_2 C_2}$$

except for the high pass network where

$$\omega_o^2 = \frac{1}{R_0 C_0 [1 + (C_s/C_2) + (C_s/C_1)]}$$

C_0 can be adjusted to make all ω_o's identical.

Thus all the pole polynomial coefficients can be established independently by varying A_f, A_{f2} and A_{f3} separately for any desired form of four-pole bandpass characteristics, including the maximally flat form.

It may also be observed that all parasitic shunt capacitances of the various amplifier

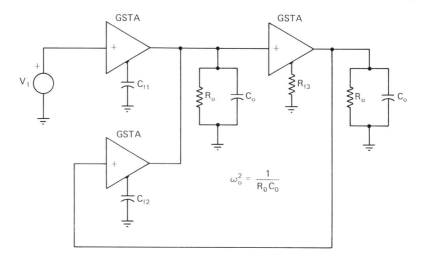

Figure 7.48 AIPF active filters using gain-stabilized transconductance amplifiers.

inputs may be incorporated as part of the structure, rather than the response being degraded or skewed by them.

In general, the *gain-stabilized transconductance amplifier* represents a more ideal element for the high-frequency bandpass biquad because both low- and high-pass *RC* time constants can be realized using shunt capacitances, rather than being limited by them. A very simple example is shown in Fig. 7.48.

Using GSTA of sufficiently high gain, the overall response is

$$\frac{V_O}{V_1} = \frac{C_{f1}/C_O\omega_o s}{s^2 + [2 - (C_{f2}/C_O)(R_2/R_{f3})]\,\omega_o s + \omega_o^2} \tag{7.103}$$

and it may be observed that this overall response depends on the passive elements only.

7.12 SYNTHESIS OF IMPEDANCE ELEMENTS

Amplifiers with feedback can be incorporated in conjunction with real impedance elements to create driving-point impedance functions that are related to some impedance elsewhere in the circuit. One such synthesized network element, known as a *gyrator,* is defined in Fig. 7.49. Such an element has a driving-point impedance that is the inverse of the load impedance Z_L.

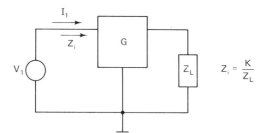

Figure 7.49 Symbolic representation of a gyrator.

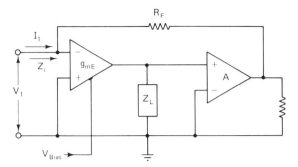

Figure 7.50 Simple gyrator circuit.

One simple implementation of the gyrator is proposed in Fig. 7.50, which incorporates a normal op-amp and an OTA as indicated. Assuming infinite input impedance for the OTA and negligible output impedance for the high-gain op-amp, the input impedance may be written as

$$Z_i = \frac{R_F}{1 + g_{mE}Z_L A} \tag{7.104}$$

which for $g_{mE}Z_L A \gg 1$ is

$$Z_i \cong \frac{R_L}{g_{mE}A Z_L} \tag{7.105}$$

Another useful network element is the negative impedance converter (NIC) defined by Fig. 7.51. A simple operational amplifier implementation of the gyrator is shown in Fig. 7.52. The input impedance may be written as

$$Z_i = \frac{R_1[R_2 + Z_L(1 - A)]}{Z_L + R_2(1 + A)} \tag{7.106}$$

which can be expressed approximately as

$$Z_i \cong -\frac{R_1}{R_2} Z_L \tag{7.107}$$

for $R_2(1 + A) \gg |Z_L|$ and $|Z_L(1 - A)| \gg R_2$.

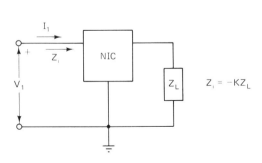

$$Z_i = -KZ_L$$

Figure 7.51 Symbolic representation of a negative impedance converter.

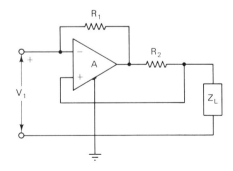

Figure 7.52 Negative impedance converter implementation.

Sec. 7.12   Synthesis of Impedance Elements

333

PROBLEMS

7.1 The circuit shown is that of a very simple low-input current operational amplifier and it is intended that $V_o = 0$ for $V_{i1} = V_{i2} = 0$. The two input transistors are nominally balanced for $I_{B1} = I_{B2} = 0.2$ μA. The dc current gain of Q-3 is approximately $h_{FE3} \cong 100$ and that of Q-4, $h_{FE4} \cong 150$. I_3 is set at 2.0 mA.

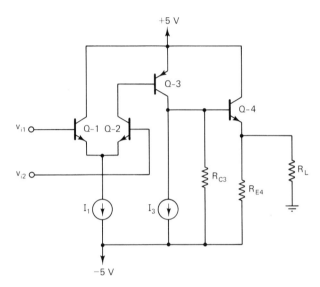

Prob. 7.1

(a) Determine R_{C3} to set the base current of Q-3 at 0.01 mA, and determine R_{E4} to fulfill the zero input–output offset voltage condition.
(b) Determine the open-loop gain of the amplifier for $R_L = 10$ kΩ and $v_{i2} = 0$.
(c) What is the output impedance looking back from R_L?
(d) As defined, what is the input bias current?
(e) Because of unbalances in Q-1 and Q-2, it is found that, for $V_{i1} = V_{i2} = 0$, $I_{B1} = 0.21$ μA and $I_{B2} = 0.19$ μA. What is the input offset current by definition?
(f) Assuming all h_{FE}'s unchanged, what is the output voltage offset due to I_{OS} and how much should I_3 be changed to rebalance the circuit?

7.2 In the circuit shown, V_{Bias} is initially set to make $V_2 = 0$ for $V_1 = 0$. All dc currents are as indicated, and Q-1 and Q-2 are assumed to be perfectly balanced, with $g_m = 5$mS. Also $\beta = 100$ for Q-3 and Q-4 with $|V_{BE}| \approx 0.8$ V.
(a) Calculate values for R_3 and R_4.
(b) Determine approximately the open-loop gain, carefully noting what approximations you think are valid.
(c) Now assume that Q-1 and Q-2 are structurally unbalanced such that $I_{D1} = 0.9$ mA and $I_{D2} = 1.1$ mA when the bias current is 2 mA. Assuming that the g_m's and β's are unchanged, determine the input offset voltage.
(d) With SW closed, determine approximately the overall voltage gain for $R_{F2}/(R_{F1} + R_{F2}) = 0.1$.
(e) With $V_i = 0$, determine the change in bias current required to bring the offset output voltage back to zero.

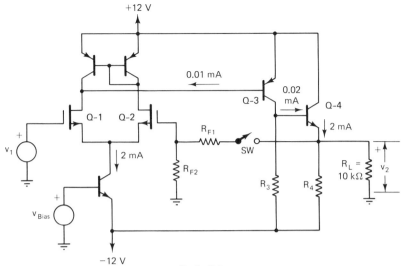

Prob. 7.2

7.3 A basic differential amplifier having the characteristic given is used in the circuit shown as the input section of an operational amplifier. Assume that, for $V_1 = V_2 = 0$, equal base currents of $I_{B1} = I_{B2} = 0.06$ μA flow. (Neglect voltage drops across R_{B1} and R_{B2}.)

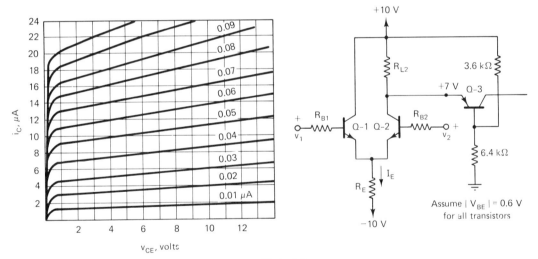

Prob. 7.3

(a) What are the collector currents I_{C1} and I_{C2}?

(b) The emitter current for Q-3 is specified as $I_{E3} = 20$ μA. Determine nominal values for R_E and R_{L2}.

(c) Determine A_{21} and A_{22} as defined by the approximate differential amplifier equations given in Chapter 5. What is the common-mode rejection ratio based on the approximations used?

(d) For integrated-circuit implementation, why is it undesirable to use resistors in the positions of R_E, R_{L2}, and R_{B3}?

7.4 The input section of the op-amp shown in Fig. 7.13 is an improved version of the basic amplifier of Prob. 7.3, using Q-1 and Q-2 with the characteristics of those of Prob. 7.3. The current source $I_E = 28$ μA. The upper current mirror is structured such that the mirrored current is twice that produced at the collector of Q-1. As in Prob. 7.3, the dc voltage $V_{C3} = 7.0$ V.

(a) For $V_{S1} = V_{S2} = 0$, make estimates for I_{C1}, I_{C2}, and I_{E3} and state a value for input offset current assuming $I_{C1} = I_{C2}$.

(b) On the other hand, if $I_{B1} = I_{B2}$, determine the value for I_{E3}.

7.5 Consider a high-gain operational amplifier whose dominant frequency-dependent stage has excess phase shift, with its overall response given by

$$A = \frac{A_{0M}[1 - jk(\omega/\omega_2)]}{1 + j(\omega/\omega_2)}$$

and is used in the voltage-follower mode.

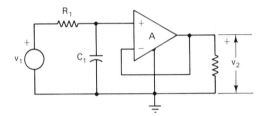

Prob. 7.5

(a) In the circuit shown, determine K_1 and B in the conventional feedback equations.

(b) Write the overall gain equation V_1/V_2 and, making $R_1C_1 = k/\omega_2$, what should be the value kA_{0M} to yield a response whose magnitude is independent of frequency?

(c) At what relative frequency is the overall phase shift equal to $-90°$?

7.6 For the very simple operational amplifier shown:

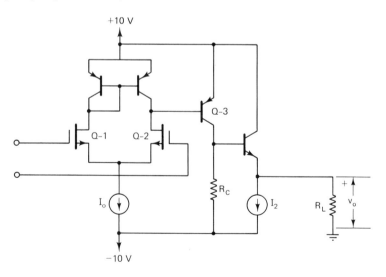

Prob. 7.6

Properties of Amplifiers with Feedback   Chap. 7

(a) Label the input terminals $(+)$ or $(-)$ to indicate which is inverting and which is non-inverting with respect to the output.

(b) Write approximate equations for v_o/v_i using alternately inverting and noninverting inputs. (Neglect the effect of μ_{re} and r_{ceo} of the BJTs and r_d of the FETs, retaining only the parameters β, r_{bes}, and g_m.)

(c) The amplifier is to be used in its negative voltage feedback mode using the terminal not used for the input signal. Show R_F placed properly in the circuit, and determine its value to make $v_o/v_i = 1$ for $g_{m1} = g_{m2} = 5$ mS, $\beta_2 = \beta_4 = 100$, $R_c = 5$ kΩ, and $R_L = 1$ kΩ. (Assume that I_2 biases Q-4 properly to make $v_o = 0$ for $v_i = 0$.)

(d) Assuming that the frequency response is determined primarily by a dominant pole controlled by the characteristic of Q-3, determine f_2 (without R_F) in the overall gain expression

$$\frac{v_o}{v_i} = \frac{\pm A_{0M}}{1 + j\dfrac{f}{f_2}}$$

Hint: First make an estimate of $C_{cb'}$ and $C_{b'c}$ for Q-3 based on its dc current.

7.7 For the low-pass filter shown in Fig. 7.36, write an equation for V_o/V_s in the s-plane separating out ω_2 and k. Determine the relationships for the maximally flat form of response.

7.8 For the bandpass filter of Fig. 7.44, verify the form of Eq. 7.93 and obtain expressions for Q_o and ω_o.

7.9 The amplifier shown consists of an OTA followed by a conventional op-amp, which has sufficient feedback that its effective gain $A' = V_3/V_2$ is still very high, but the output resistance, R_o' is negligible.

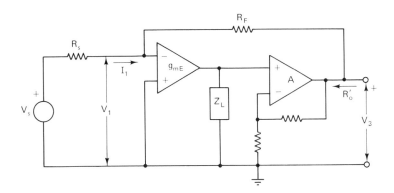

Prob. 7.9

(a) Obtain an expression for overall gain V_3/V_S in terms of g_{mE} and A'.

(b) Determine the input impedance $Z_i = V_1/I_1$.

(c) Under what conditions will Z_i be of the form $Z_i \cong K/Z_L$, and determine K.

7.10 A very high gain differential input comparator ($A >>> 1$) designed to have zero input–output offset with output limiting levels, $V_{2H} = 5.0$ V and $V_{2L} = 0.0$ V, is connected in the feedback circuit as shown. Input pulses having approximately equal slow rise and fall times are applied at the input, and it is intended that output transitions between V_{OL} and V_{OH} occur when $V_i = +2.0$ and $V_i = +3.0$.

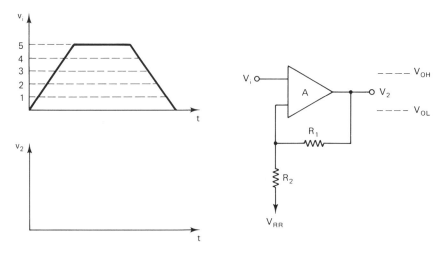

Prob. 7.10

(a) Use ($\pm$) signs at the input to indicate inverting and noninverting conditions.

(b) Determine V_{RR} and $R_2/(R_1 + R_2)$ to meet the conditions specified.

(c) Carefully sketch the output waveform in the space indicated, being sure to show correctly the transition times involved.

REFERENCES

AHUJA, B. K. "An Improved Frequency Compensation Technique for CMOS Operational Amplifiers," *IEEE J. Solid State Circuits,* vol. SC-18, no. 6, Dec. 1983, pp. 629–644.

BRUTON, L. T. *RC-Active Circuits,* Prentice-Hall, Englewood Cliffs, N.J., 1980.

CAVE, D. L., AND W. R. DAVIS. "A Quad JFET Operational Amplifier Integrated Circuit Featuring Temperature Compensated Bandwidth," *IEEE J. Solid State Circuits,* vol. SC-12, no. 4, Aug. 1977, pp. 382–388.

CONNELLY, J. A., ED. *Analog Integrated Circuits,* Wiley, New York, 1975.

GHAUSI, M. S. *Electronic Devices and Circuits, Discrete and Integrated,* Holt, Rinehart, and Winston, New York, 1985.

GLASFORD, G. M. "Direct Analytical Comparisons of Video Preamplifier Configuration for Optimum Signal-to-Noise Ratio Performance," *IEEE Trans. Broadcasting,* vol. BC-15, no. 2, June 1969, pp. 44–54.

————. "Nonlinear Device Models Applied to Micropower and Operational Transconductance Amplifiers," *Conf. Record, Tenth Asilomar Conference on Circuits, Systems, and Computers,* pp. 275–340. Pacific Grove, Calif., Nov. 1976, Western Periodicals, North Hollywood, Calif.

————. "Design Aspects and Performance Limitations of Amplifier Isolated Active R-C Filters," *Conf. Record, Twelfth Asilomar Conference on Circuits, Systems, and Computers,* Nov. 1978, pp. 44–48, IEEE Computer Society, IEEE Service Center, Piscataway, N.J.

————. "Extending the Performance and Frequency Range of AIPF Active Filters," *Proc. IEEE 1980 International Symposium on Circuits and Systems,* April 1980, IEEE Service Center, Piscataway, N.J.

————. "Controlling the Pole Locations in CMOS Gain Stabilized Transconductance Amplifiers," *Conf. Record, 16th Asilomar Conference on Circuits, Systems, and Computers,* Nov. 1982, IEEE Computer Society, IEEE Service Center, Piscataway, N.J.

————. "Current Developments in CMOS Gain-Stabilized Transconductance Amplifiers for High Frequency Bandpass Filter Application," *Proc. IEEE 1983 International Symposium on Circuits and Systems,* pp. 1282–1285, IEEE Service Center, Piscataway, N.J.

————. "Optimizing the Architecture of Extended Bandwidth Programmable CMOS Operational Amplifiers," *Proc. 27th Midwest Symposium on Circuits and Systems,* June 11–12, 1984, Morgantown, W. Va., Western Periodicals, North Hollywood, Calif.

GRAY, P. R., AND R. G. MEYER. "MOS Operational Amplifier Design—A Tutorial Overview," *IEEE J. Solid State Circuits,* vol. SC-17, no. 6, Dec. 1982.

GRAY, P. R., AND R. G. MEYER. *Analog Integrated Circuits,* Wiley, New York, 1984.

PRENSKY, S. D. *Manual of Linear Integrated Circuits, Operational Amplifiers and Analog IC's,* Reston, Reston, Va., 1974.

SOCLOF, S. *Analog Integrated Circuits,* Prentice-Hall, Englewood Cliffs, N.J., 1985.

SOLOMAN, J. E. "The Monolithic Op Amp: A Tutorial Study," *IEEE J. Solid State Circuits,* vol. SC-9, no. 6, Dec. 1974, pp. 314–332.

TSIVIDIS, Y. P., D. L. FRASER, JR., AND J. E. DZIAK. "A Process-insensitive High-performance NMOS Operational Amplifier," *IEEE J. Solid State Circuits,* vol. SC-15, no. 6, Dec. 1980, pp. 921–928.

VAN VALKENBURG, M. E. *Analog Filter Design,* Holt, Rinehart and Winston, New York, 1982.

Chapter 8

Nonlinear Distortion in Devices and Applications of Nonlinear Characteristics

INTRODUCTION

Except for the development of complete device models carried out in Chapters 1 through 3, subsequent discussions have centered on incrementally linear characteristics of bipolar and field-effect transistors and have involved the use of appropriate linear circuit models for the various amplifier configurations. However, the device characteristics are only approximately linear, and when used over a substantial portion of their dynamic range in their normal active region, the result is distortion of output voltages or currents relative to the input signal.

Sometimes, in evaluating the transfer characteristics of single-device and multi-device amplifiers, particularly those involving substantial levels of power output, distortion must be taken into account and the performance degradation resulting from it evaluated. The output stages such as the class B power amplifiers of operational amplifiers represent particular examples; however, large amounts of negative feedback can be used to reduce large amounts of distortion occurring at the output of such amplifiers to negligible proportions.

On the other side of the coin of nonlinear characteristics, we find that the very property of nonlinearity that degrades the performance of linear systems is made use of in many other applications, such as logarithmic amplifiers, analog multipliers, and frequency converters. This subject has already been lightly touched on in connection with possible applications of the operational transconductance amplifier discussed in Chapter 7. The entire subject of nonlinearity will be explored in this chapter.

8.1 NONLINEAR CHARACTERISTICS OF FIELD-EFFECT TRANSISTORS

An analysis of the dominant nonlinearity component of FETs in their saturation (normal active) region can be carried out using the simplified semiempirical model for the FET. One such model for the FET, as given by Eq. 3.36 for $n = 2$, is

$$i_D = I_{DSS} \left[1 + \frac{1}{V_P} \left(v'_{GS} + \frac{V_{DS}}{\mu} \right) \right]^2 \tag{8.1}$$

where I_{DSS} and V_P are defined by Eqs. 3.19 and 3.20 for the spiked channel doping profile.

A comparable "universal" empirical equation for the MOSFET obtained from Eq. 3.66 with $v_{GS} + v_{DS}/\mu$ substituted for v_{GS} yields

$$i_D = \frac{K}{2} V_T^2 \left[1 + \frac{1}{|V_T|} \left(v_{GS} + \frac{v_{DS}}{\mu} \right) \right]^2 \tag{8.2}$$

for the n-channel depletion-mode FET, where K is obtained from Eq. 3.54 and V_T is given approximately by Eq. 3.56 or, more exactly, by Eq. 3.64. Similarly,

$$i_D = \frac{K}{2} V_T^2 \left[\frac{1}{V_T} \left(v_{GS} + \frac{V_{DS}}{\mu} \right) - 1 \right]^2 \tag{8.3}$$

for the enhancement-mode n-channel device, where V_T is given approximately by Eq. 3.63.

As an example, the nonlinearities as defined by Eq. 8.1 will be analyzed as being representative of both JFET and IGFET structures, although other models developed in Chapter 3 are more explicitly descriptive of structural parameters such as short-channel effects. These models are slightly more complicated analytically because μ and r_d are both voltage dependent.

Equation 8.1 may be written in terms of operating point (dc) values and incremental values by substituting

$$i_D = I_D + i_d, \qquad v_{GS} = V_{GS} + v_{gs}, \qquad v_{DS} = V_{DS} + v_{ds} \tag{8.4}$$

and expanding the resultant equation and segregating linear and product terms. The dc component is reconstituted as in the original form as

$$I_D = \left[1 + \frac{1}{V_P} \left(V'_{GS} + \frac{V_{DS}}{\mu} \right) \right]^2 \tag{8.5}$$

with the incremental components summed:

$$i_d = \frac{2I_{DSS}}{V_P} \left[1 + \frac{1}{V_P} \left(V'_{GS} + \frac{V_{DS}}{\mu} \right) \right] v_{gs} + \frac{2I_{DSS}}{\mu V_P} \left[1 + \frac{1}{V_P} \left(V_{GS} + \frac{V_{DS}}{\mu} \right) \right] v_{ds}$$

$$+ \frac{I_{DSS}}{V_P^2} v_{gs}^2 + \frac{I_{DSS}}{\mu^2 V_P^2} v_{ds}^2 + \frac{2I_{DSS}}{\mu V_P^2} v_{gs} v_{ds} \tag{8.6}$$

The coefficient associated with v_{gs} may be recognized as the transconductance at the operating point given by

$$g_{mo} = \frac{2I_{DSS}}{V_P}\left[1 + \frac{1}{V_P}\left(V'_{GS} + \frac{V_{DS}}{\mu}\right)\right] \tag{8.7}$$

and that associated with v_{ds} as the inverse of the drain resistance at the operating point:

$$\frac{1}{r_{do}} = \frac{2I_{DSS}}{\mu V_P}\left[1 + \frac{1}{V_P}\left(V'_{GS} + \frac{V_{DS}}{\mu}\right)\right] \tag{8.8}$$

Then Eq. 8.6 can be written as

$$i_d = g_{mo}v_{gs} + \frac{1}{r_{do}}v_{ds} + \frac{I_{DSS}}{V_P^2}v_{gs}^2 + \frac{I_{DSS}}{\mu^2 V_P^2}v_{ds}^2 + \frac{2I_{DSS}}{\mu V_P^2}v_{gs}v_{ds} \tag{8.9}$$

Observing that

$$\frac{\partial g_m}{\partial v_{GS}} = \frac{2I_{DSS}}{V_P^2}$$

Eq. 8.9 may be written as

$$i_d = g_{mo}v_{gs} + \frac{1}{r_{do}}v_{ds} + \frac{1}{2}\frac{\partial g_m}{\partial v_{GS}}v_{gs}^2 + \frac{1}{2}\frac{1}{\mu^2}\frac{\partial g_m}{\partial v_{GS}}v_{ds}^2 + \frac{1}{\mu}\frac{\partial g_m}{\partial v_{GS}}v_{gs}v_{ds} \tag{8.10}$$

The hypothetical characteristic shown in Fig. 8.1 is a plot of Eq. 8.1 for an FET having $I_{DSS} = 10$ mA, $V_P = 4$ V, and $\mu = 100$. If an operating point is selected (e.g., $V_{DS} = 14$ V and $V'_{GS} = -2.0$ V) and $I_D = 2.8$ mA as shown, Eqs. 8.9 and 8.10 are incremental variations about that point.

Since $i_d = f(v_{gs}, v_{ds})$ if the FET is used as an amplifier with load resistance R_L, the equation for output voltage depends on the operating path defined by

$$v_{DS} = V_{DD} - i_D R_L \tag{8.11}$$

Two specific values for R_L are indicated.

The output voltage (across R_L) as a function of v_{GS} may be determined graphically from Fig. 8.1 or analytically using Eq. 8.9 or 8.10. After substituting $i_d = -v_{ds}/R_L$ into Eq. 8.10, it can be rearranged as a quadratic equation in v_{ds}, which can be solved for v_{ds}. The result can be written in the form (using $\mu = g_m r_d$ and expanding the square root component into a power series)

$$v_{ds} = \frac{-g_m R_L}{1 + (R_L/r_d)}v_{gs} - \frac{1}{2}\frac{\partial g_m}{\partial v_{GS}}R_L\left[\frac{1}{1 + (R_L/r_d)}\right]^3 v_{gs}^2$$
$$+ \frac{R_L}{g_m r_d}\left(\frac{\partial g_m}{\partial v_{GS}}\right)^2 R_L\left[\frac{1}{1 + (R_L/r_d)}\right]^5 v_{gs}^3 + \cdots \tag{8.12}$$

For $R_L \ll r_d$, the equation becomes

$$v_{ds} - g_m R_L v_{gs} - \frac{1}{2}\frac{\partial g_m}{\partial v_{GS}}R_L v_{gs}^2 + \frac{1}{2\mu}\frac{R_L}{r_d}\left(\frac{\partial g_m}{\partial v_{GS}}\right)^2 R_L v_{gs}^3 \tag{8.13}$$

Furthermore, in the limit the v_{gs}^3 and higher-order terms vanish and

$$v_{ds} - g_m R_L v_{gs} - \frac{1}{2}\frac{\partial g_m}{\partial v_{GS}}R_L v_{gs}^2 \tag{8.14}$$

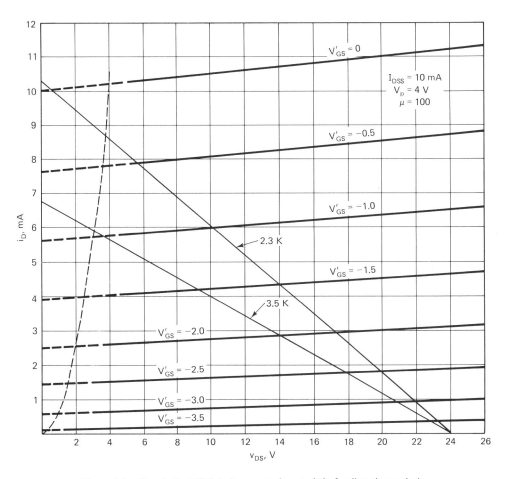

Figure 8.1 Hypothetical FET drain current characteristic for distortion analysis.

The nonlinearity of Eq. 8.12 for a steady-state single-frequency input of $v_{gs} = V_{gs} \cos \omega t$ after substituting $\partial g_m / \partial v_{GS} = 2I_{DSS}/V_P^2$ may be written as

$$v_{ds} = \frac{I_{DSS}R_L}{2V_P^2} \left[\frac{1}{1 + (R_L/r_d)} \right]^3 V_{gs}^3$$

$$- \left\{ \frac{\mu R_L}{r_d + R_L} - \frac{6}{4\mu} \left(\frac{I_{DSS}R_L}{V_P^2} \right)^2 \left[\frac{1}{1 + (R_L/r_d)} \right]^5 V_{gs}^2 \right\} V_{gs} \cos \omega t$$

$$- \left\{ \frac{1}{2} \frac{I_{DSS}R_L}{V_P^2} \left[\frac{1}{1 + (R_L r_d)} \right]^2 V_{gs} \right\} V_{gs} \cos 2\omega t$$

$$+ \left\{ \frac{2}{4\mu} \left(\frac{I_{DSS}R_L}{V_P^2} \right)^2 \left[\frac{1}{1 + (R_L/r_d)} \right]^3 V_{gs}^2 \right\} V_{gs} \cos 3\omega t$$

$$+ \text{ higher-order harmonics} \qquad (8.15)$$

This equation shows (1) a shift in the dc or average value of the output voltage, (2) a reduction of the fundamental component compared to its value without distortion for the same μ and r_d at the operating point, (3) a second-harmonic distortion component, and (4) a third-harmonic component that is very small relative to the second-harmonic component.

The nature of the harmonic components is illustrated in Fig. 8.2, where the separate harmonic components are shown. These have not been calculated for any particular value of R_L but are intended to illustrate their general nature, particularly with respect to phase. If the components were added, it is seen that the second harmonic stretches the positive peak and flattens the negative one, whereas the third harmonic tends to flatten both peaks. This illustrates the basic difference between even-order distortion and odd-order distortion.

The magnitudes of the harmonic coefficients are used to determine the fractional or percentage of each component relative to the fundamental. Examination of Eq. 8.15 shows that second-harmonic distortion is dominant and that normally third- and higher-order harmonics are small relative to it. The following example is used to illustrate the nature of problems that can be addressed using the material that is developed in this section.

Example

Using the theoretical FET illustrated in Fig. 8.1, with values of load resistances as indicated, the relationships among all parameters can be established and values of second- and third-harmonic components as a percentage of the fundamental determined for specified values of signal amplitude.

(a) Determine g_m and r_d from Eqs. 3.37 and 3.38.

$$g_m = \frac{20}{4}\left[1 + \frac{1}{4}\left(-1.5 + \frac{14}{100}\right)\right] = 3.30 \text{ mS}$$

$$\frac{1}{r_d} = \frac{20}{400}\left[1 + \frac{1}{4}\left(-1.5 + \frac{14}{100}\right)\right], \qquad r_d = 30.30 \text{ k}\Omega$$

Check: $g_m r_d = 3.30 \times 30.30 = 100$

(b) Determine the signal content from Eq. 8.15 for $V_{g\max} = 1.2$ V.

$$\text{DC level shift} = -\frac{I_{DSS}R_L}{2V_p^2}\left[\frac{1}{1 + (R_L/r_d)}\right]^3 V_{GS}^3 = -\frac{23}{32}\left(\frac{1}{1 + 2.3/30.30}\right)^3 (1.2)^3$$

$$= -1.00 \text{ V}$$

$$\text{Fundamental} = -\left[\frac{\mu R_L}{r_d + R_L} - \frac{6}{4\mu}\left(\frac{I_{DSS}R_L}{V_p^2}\right)^2\left[\frac{1}{1 + (R_L/r_d)}\right]^5 V_{gs}^2\right]V_{GS}$$

$$= -\left[\frac{230}{32.6} - \frac{6}{400}\left(\frac{23}{16}\right)^2 (0.9294)^5 (1.2)^2\right]1.2$$

$$= -8.43 \text{ V}$$

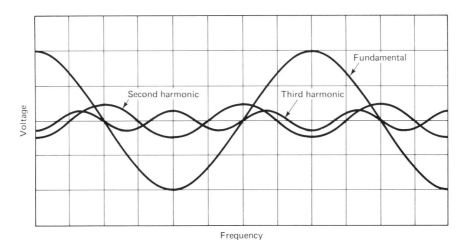

Figure 8.2 Harmonic distortion components in typical power output stage (mostly even harmonics).

$$\text{Second harmonic} = -\frac{1}{2}\frac{I_{DSS}R_L}{V_p^2}\left[\frac{1}{1+(R_L/r_d)}\right]^2 V_{gs}^2 = -\frac{1}{2}\frac{23}{16}(0.9294)1.44$$

$$= -0.962 \text{ V}$$

$$\text{Third harmonic} = \frac{2}{4\mu}\left(\frac{I_{DSS}R_L}{V_p^2}\right)^2\left(\frac{1}{1+(R_L/r_d)}\right)^3 V_{gs}^3 = \frac{2}{400}\left(\frac{23}{16}\right)^2(0.9294)^3(1.2)^3$$

$$= 0.0143$$

(c) Neglecting third- and higher-order harmonics, determine minimum and maximum values of v_{DS}.

$$v_{DS}(\text{min}) = 14 - (8.43 + 0.962) = 4.61 \text{ V}$$

$$v_{DS}(\text{max}) = 14 + (8.43 - 0.962) = 21.5 \text{ V}$$

General Comments on FET Distortion

The nonlinear equations for $i_d = f(v_{gs}, v_{ds})$ (Eqs. 8.9 and 8.10) were based on an assumed analytical model that permitted the determination of circuit parameters in terms of structural device parameters. However, if the model is not precisely that suggested by Eq. 8.1, Eq. 8.10 is still valid as an approximation over a limited operating range as long as g_m and r_d can be determined at the operating point and $\partial_{gm}/\partial v_{GS}$ evaluated. If an analytical expression that differs from Eq. 8.1 is available with a higher-order nonlinearity, a power series in v_{gs} and v_{ds} may be obtained as a two-dimensional Taylor series expansion, and the result will be an expanded version of Eq. 8.12 with high-order terms present, which will result in higher-order harmonics in an equation similar to Eq. 8.15.

However, for most FETs, Eqs. 8.9 and 8.15 represent reasonable approximations for reasonable excursions of v_{gs}.

8.2 NONLINEAR REPRESENTATION OF BIPOLAR TRANSISTORS

The nonlinearity of a bipolar transistor is basically a more complex function than that of the FET, because the exponential nonlinearity of the input voltage–current characteristic and the input–output current relationship exist simultaneously. Thus the nature of the overall nonlinearity depends highly on whether the amplifier is used basically as a current or a voltage amplifier. Considering the collector characteristic of the transistor of Fig. 8.3 as typical, it was suggested in Chapter 2 that a semiempirical equation for the collector current could be written in the form

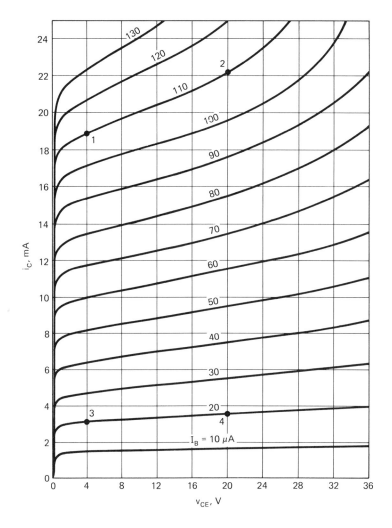

Figure 8.3 BJT nonlinear collector characteristics.

$$i_C = I_{OS} + g_o v_{CE} + \beta_o i_B + f_1(i_B, v_{CE}) \tag{2.47}$$

where $f_1(i_B, v_{CE})$ is a term of the appropriate form to model the nonlinearities involved. In the low-level injection region, it has been shown that this nonlinearity can be modeled reasonably well by an equation of the form

$$f_1(i_B, v_{CE}) = k_B i_B^x v_{CE}^y \tag{2.48}$$

The nonlinear collector current expression may be written as a two-dimensional Taylor series expansion about a prescribed operating point I_B, V_{CE}. It is convenient first to express the incremental collector current as

$$i_c = \beta_o i_b + g_o v_{ce} + f_1(i_B, v_{CE}) - f_1(I_B, V_{CE}) \tag{8.16}$$

rather than to expand $f_1(i_B, v_{CE})$ alone. The form of the result, irrespective of the form of $f_1(i_B, v_{CE})$, can be written in terms of h_{fe} and h_{oe} as

$$i_c = (h_{fe} i_b + h_{oe} v_{ce}) + \frac{2}{2!} \frac{\partial h_{fe}}{\partial v_{CE}} i_b v_{ce} + \frac{1}{2!} \frac{\partial h_{fe}}{\partial i_B} i_b^2$$

$$+ \frac{1}{2!} \frac{\partial h_{oe}}{\partial v_{CE}} v_{ce}^2 + \frac{1}{3!} \frac{\partial^2 h_{fe}}{\partial i_B^2} i_b^3 + \frac{1}{3!} \frac{\partial^2 h_{oe}}{\partial v_{CE}^2} v_{ce}^3$$

$$+ \frac{3}{3!} \frac{\partial^2 h_{oe}}{\partial i_B^2} i_b^2 v_{ce} + \frac{3}{3!} \frac{\partial^2 h_{oe}}{\partial v_{CE}^2} v_{ce}^2 i_b + \cdots \tag{8.17}$$

For the specific form of nonlinearity suggested by Eqs. 2.47 and 2.48,

$$h_{fe} = \beta_o + x k_B v_{CE}^y i_b^{x-1} \tag{8.18}$$

$$h_{oe} = g_o + y k_B i_B^x v_{CE}^{y-1} \tag{8.19}$$

and all the terms in Eq. 8.17 involving derivatives of h_{fe} and h_{oe} can be determined analytically.

When the transistor is used as a current amplifier, $R_S \gg r_{bes}$, and with a sufficiently low load resistance, $R_L \ll 1/g_o$, the overall nonlinear model suggested by Eq. 8.17 can be used to determine the overall linearity.

An approximate input voltage–current relationship can be established using the relationships developed in Sec. 2.9. Starting with

$$i_E \cong I_{ES} e^{a v_{B'E}} \tag{8.20}$$

expanded into a Taylor series about the operating point I_E, $V_{B'E}$ and written in terms of

$$r_e \cong \frac{kT}{q} \frac{1}{|I_E|} \tag{8.21}$$

and the result is

$$i_e = \frac{v_{b'e}}{r_e} + \frac{1}{|I_E|} \frac{v_{b'e}^2}{2r_e^2} + \frac{1}{I_E^2} \frac{v_{b'e}^3}{6r_e^3} + \cdots \tag{8.22}$$

Then using

$$i_b \cong \frac{i_e}{1 + h_{fe}} \quad \text{and} \quad r_d = (1 + h_{fe})r_e$$

$$i_d \cong \frac{v_{b'e}}{r_d} + \frac{1 + h_{fe}}{|I_E|} \frac{v_{b'e}^2}{2r_d^2} + \frac{1}{I_E^2} \frac{(1 + h_{fe})^2 v_{b'e}^3}{6r_d^3} + \cdots \qquad (8.23)$$

This relationship assumes negligible internal feedback through the $\mu_{re}v_{ce}$ term.

Using the linear h-parameter model in the amplifier shown in Fig. 8.4 as a reference, the overall nonlinear voltage relationship consists of Eq. 8.17 combined with Eq. 8.23 along with

$$i_b = \frac{v_s - v_{b'e}}{R_s + r_b'} \qquad (8.24)$$

and

$$i_c = -\frac{v_{ce}}{R_L} \qquad (8.25)$$

The resultant overall nonlinearity leads to an exceedingly complex nonlinear input–output relationship analytically, but can be determined numerically using appropriate computer programs. However, it is often the case that *either* the input or output nonlinearity dominate and Eq. 8.17 or 8.23 can be used alone.

For $R_s \gg r_{bes}$, the input–output current relationship dominates, and Eq. 8.17 can be used alone along with Eq. 8.25, which even so is extremely complicated unless Eq. 8.17 can be truncated to a small number of terms, in which case a solution similar to that suggested for the FET may be used.

When the transistor is used as a low-current voltage amplifier, that is, $(R_s + r_b')$ $\ll r_d$, the extreme nonlinearity of the input voltage–current relationship expressed by Eq. 8.23 dominates, and the h_{fe} and h_{oe} nonlinearity may be ignored.

In this case, using $i_c \cong h_{fe}i_b$ and $g_m = h_{fe}/r_d$ in Eq. 8.23,

$$i_c \cong g_m v_{b'e} + \frac{g_m^2 v_{b'e}^2}{2|I_E|} + \frac{g_m^3 v_{b'e}^3}{6|I_E|^2} + \cdots \qquad (8.26)$$

and for very small $R_S + r_d'$ relative to r_d, $v_1 \cong v_{b'e}$.

This equation is extremely nonlinear except for very small values of $v_{b'e}$. Often, when a very linear voltage amplifier is required, a compensating nonlinear input resistance consisting of a forward-biased diode with a resistance network can be used to improve

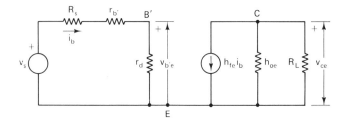

Figure 8.4 *H*-parameter BJT circuit model for nonlinear analysis.

Nonlinear Distortion in Devices   Chap. 8

the input-current, input-voltage linearity, or alternatively a shunting diode may be used as part of the load resistance.

When both input and output nonlinearities must be considered simultaneously, as well as the internal feedback term involving $\mu_{re} = r_e/r_{ceo}$, the complete two-generator nonlinear h-parameter model becomes exceedingly cumbersome because of the complexity of the interparameter nonlinearities. As an alternative, the T-circuit model has been suggested as a somewhat more manageable model to take into account both input and output nonlinearities, as well as the feedback component linking the two. Such a model is indicated in Fig. 8.5(a).

Techniques have been developed using a Volterra series in CAD programs to obtain complete numerical solutions for similar nonlinear models. The model can be further extended to include the effects of nonlinear capacitances in the frequency-dependent

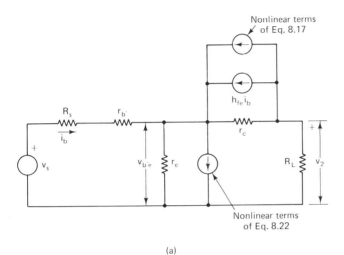

(a)

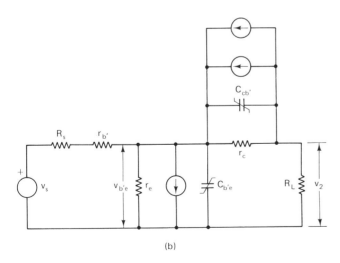

(b)

Figure 8.5 BJT T-equivalent circuit used in nonlinear analysis. (a) Low-frequency components. (b) Nonlinear capacitances included.

model indicated in Fig. 8.5(b), where the capacitance nonlinearities can be expanded as a Taylor series.

Volterra series techniques have been shown to be reasonably successful in modeling mild nonlinearities using the T-equivalent circuit.

8.3 TRANSISTOR POWER AMPLIFIER DISTORTIONS

Bipolar transistor amplifiers designed to deliver substantial amounts of power to a load will operate over a wide current and voltage range, from low currents where low-level injection is valid and where the model might be similar to that suggested by Eqs. 2.47 and 2.48 to currents where high-level injection effects are present, where an equation of the general form of Eq. 2.120 might be valid. At the same time, input voltage–current nonlinearities must be considered, and a determination of overall nonlinearity is extremely cumbersome and graphical techniques are often resorted to.

The characteristics shown in Fig. 8.6 are those of a medium power transistor driven from a current source. The dashed contour is a line of constant collector circuit power dissipation. Several load lines are selected, each an operating point to allow maximum voltage swing.

For the load line, with $R_L = 0.4k\Omega$ at $V_{CE} = 40$ V, $I_C = 100$ mA with $I_B = 2.5$ mA, a current swing at $i_D = \pm 2.5$ mA would effect an output current change from $i_c = 0$ to $i_C = 180$ mA and, as shown in Fig. 8.7(a), the output current and voltage would exhibit compression for both positive and negative excursions. The compressions at higher voltage as $v_{CE} \rightarrow V_{CC}$ is the nonlinearity due to decreasing h_{fe} at low currents and at low voltages due to the approach of saturation.

The flattening of both peaks indicates substantial odd-harmonic distortion, predominantly third, whereas the dissymmetry of the two halves predicts even-harmonic distortion, predominantly second.

If the input current deviation is reduced to $I_b = \pm 2.0$ mA, as indicated by the dashed curve, all distortion components are reduced substantially and the amplifier is almost linear. If the same load line is retained and the input operating current is reduced to $I_B = 2$ mA with $i_b = \pm 2$ mA, a plot would show compression at the low current end as cutoff is approached and expansion for increasing current, indicating even-harmonic distortion only.

If the load resistance is decreased to 270 Ω as indicated by the corresponding load line of Fig. 8.6, for the particular transistor shown, the peak compression at high currents is due to a combination of saturation and high-level injection, while at low currents the compression is reduced except at very low values of base current because of the increased current gain at high voltages due to carrier multiplication effects. The resultant plot is shown in Fig. 8.7(b), and it might be predicted that both even- and odd-harmonic components are increased. Furthermore, a larger input voltage deviation is required to provide essentially the same output (i.e., the overall gain is less).

A further decrease in load resistance to 200 Ω as indicated by the remaining load line which skirts the edge of allowable collector circuit dissipation, shows even greater compression at high currents as high-level injection effects dominate and still further current gains at low levels. As indicated in Fig. 8.7(c) for an equal input base current

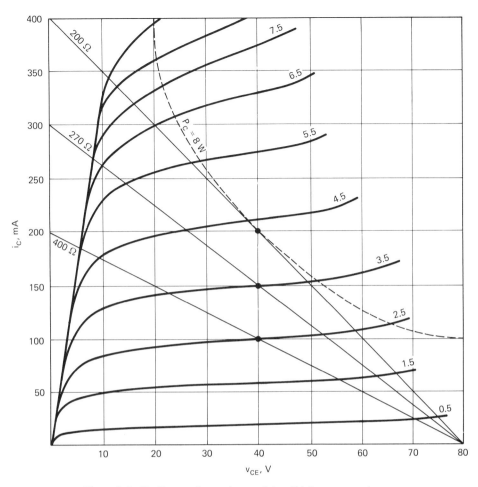

Figure 8.6 Nonlinear collector characteristic of high power transistor.

excursion of $i_D = \pm 4$ mA about $V_{CE} = 40$ V and $I_B = 4.3$ mA, the dissymmetry results in excessive even-harmonic distortion; whereas reducing the operating current to, for example, 3.5 mA, which increases the value of V_{CE} to approximately 50 V, again permits a better balance of negative and positive peak distortion if the input swing is held to approximately $i_D = \pm 3.0$ mA, which of course yields a reduced output signal voltage.

It can be seen that optimization of the load resistance for a particular application depends on a combination of factors involving minimization of distortion, along with maximizing the power delivered to the load and maximizing of efficiency in terms of allowable dissipation.

The collector circuit power dissipation at the operating point is $P_C = V_{CE}I_C$, while the *average* collector dissipation under signal conditions is

$$P_C(Av) = \frac{1}{T} \int_0^T V_{CE}I_C dt \qquad (8.27)$$

which may differ from the zero-signal value depending on the type of nonlinearity.

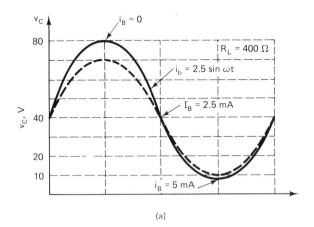

(a)

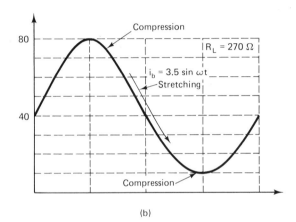

(b)

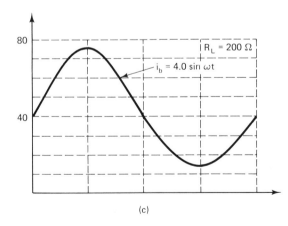

(c)

Figure 8.7 Distortion components in power-output stage.

The average *signal power* delivered to the load is given by

$$P_s = \frac{1}{T} \int i_{ce}^2 R_L \, dt \tag{8.28}$$

or

$$P_s = \frac{1}{T} \int \frac{V_{ce}^2}{R_L} \, dt \tag{8.29}$$

However, the useful signal power would exclude that resulting from distortion components. Suppose, for example, an analysis of the output current in response to a base current signal $i_b = I_b \cos \omega t$ would result in an output current expressed as

$$i_c = I_o + I_1 \cos \omega t + I_2 \cos 2\omega t + I_3 \cos 3\omega t + \cdots \tag{8.30}$$

where I_0 is a dc level shift from the established operating point (zero-order harmonic), I_1 is the magnitude of the fundamental component, and I_2, I_3, etc., are the magnitude of the harmonic distortion components. The components of power delivered to the load are

$$P_L = I_o^2 R_L + I_1^2 R_L \cos \omega t + I_2^2 R_L \cos 2\omega t + I_3^2 R_L \cos 3\omega t + \cdots \tag{8.31}$$

The useful component of Eq. 8.31 is only that generated by the fundamental component of current, or $I_1 R_L \cos \omega t$.

The average value of the fundamental component of power is

$$P_1 = \frac{R_L}{T} \int_0^T I_1^2 \cos^2 \omega t \tag{8.32}$$

or

$$P_1 = \frac{R_L I_1^2}{2} \tag{8.33}$$

or in terms of the rms value of current, $I_{RMS} = I_1/\sqrt{2}$,

$$P_1 = I_{RMS}^2 R_L \tag{8.34}$$

Likewise, the power in the various harmonics may be evaluated and the percentage of harmonic distortion expressed as

$$\% \text{ Distortion} = \frac{P_1 + P_3 + P_4 + \cdots}{P_1} \times 100 \tag{8.35}$$

The efficiency of a power amplifier is the useful power output given by Eq. 8.33 or 8.44 divided by Eq. 8.27, or

$$\% \text{ EFF} = \frac{P_1}{P_C(\text{Av})} \times 100 \tag{8.36}$$

Input Nonlinearities

The foregoing analysis has assumed an essentially current source input and hence has not taken into account the voltage–current nonlinear relationship at the input given approximately by Eq. 8.23. For essentially voltage source inputs (R_s not $>> r_{bes}$), the nonlinearity of the i_b, v_{be} relationship must be considered; this can be approximated by Eq. 8.23, which is essentially an expansion about the operating point of

$$i_B = \frac{I_{ES}}{1 + h_{FE}} e^{av_{B'E}}$$

For most of the output dynamic range, the exponential nonlinearity of the input and the resultant output current nonlinearity given by Eq. 8.26 will give rise to harmonics that dominate those suggested by the output–input current relationships alone. However, as high-level injection or saturation is approached, the input exponential nonlinearity will compensate in some degree the compression of the output current and reduce to some extent the specific odd-order harmonics that result from this compression.

Emitter-Follower Power Amplifier Linearity

When the load resistance R_L is supplied from the emitter as indicated in Fig. 8.8, the nonlinearity of the complete input–output voltage–current relationship and hence the voltage gain is reduced substantially. For convenience, the dc value of base-spreading resistance r_B' is separated out and incorporated into the resistance as indicated.

The nonlinear base–emitter voltage–current relationship is approximately

$$i_E = I_{ES}e^{v_{B'E}/V_T}$$

Also

$$v_{B'E} = V_1 - i_E \left(\frac{R_S}{1 + h_{FE}} + R_L \right)$$

and

$$v_2 = i_E R_L$$

(8.38)

Using these relationships, the output voltage can be written as

$$v_2 + \frac{V_T \ln (v_2/R_L I_{ES})}{1 + (R_S'/R_L)(1/1 + h_{FE})} = \frac{v_1}{1 + (R_s/R_L)[1/(1 + h_{FE})]}$$

(8.39)

If

$$\frac{R_S'}{R_L} \frac{1}{1 + h_{FE}} << 1,$$

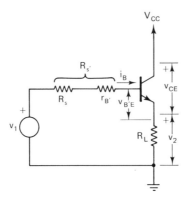

Figure 8.8 Emitter-follower power-output stage.

the output voltage is essentially independent of variations in r'_B and h_{FE}, and the incremental voltage gain $\Delta v_2/\Delta v_1 \to 1$. Even for higher values of

$$\frac{R'_S}{R_L} \frac{1}{1 + h_{FE}},$$

the gain is much less affected by current gain nonlinearity than is the common-emitter power amplifier.

The emitter follower may be viewed as a negative feedback amplifier whose output distortion is reduced by the feedback factor as discussed in Chapter 6. However, since the input voltage must change over a dynamic range slightly greater than the output voltage range, the linearity of the source transistor must be considered; but since it is loaded by relatively high input resistance of the source follower, it can be designed with a current source supply and operate over a very low range of current deviation. Since its effective load resistance, mostly the input resistance of the source follower is high, it is relatively linear over the required dynamic range.

8.4 COMPLEMENTARY-PAIR TRANSISTOR POWER AMPLIFIERS

Complementary *pnp–npn* transistors as the output stage of operational amplifiers were discussed in Chapter 7. Such a stage is extremely useful as a power output stage from the standpoint of maximizing efficiency and minimizing distortion. The complementary pair can be used as a composite emitter follower or a composite common-emitter amplifier and can operate in class A, class B, or class AB mode. The complementary emitter follower shown in Fig. 8.9 is similar in form to those shown in Chapter 7, with the signal at the appropriate dc level applied to A or B or a combination of both. The resistors R_{B1} and R_{B2} are selected to provide the correct voltage separation for Q-1 and Q-2 for the mode of operation selected. R_{B1} and/or R_{B2} can be replaced with current sources, or by a nonlinear network to compensate for the input current–voltage distortion of the amplifiers.

For the complementary common-emitter amplifier shown in Fig. 8.9(b), the input voltage separation is approximate $[V(+) - V(-) - 2V_{BE}]$ and can be achieved by other means. The method suggested in Fig. 8.9(b) utilizes a second complementary pair to provide, simultaneously, input signals at *both* A and B and the dc bias current for Q-1 and Q-2.

For class A operation of Q-1 and Q-2, each biased in the center of the desired dynamic range, the positive and negative distortions are equalized since alternately the current change in one transistor under signal conditions is equal and opposite to that in the other, as indicated in Fig. 8.10(a). The actual current directions in Q-1 and Q-2 are as indicated in Fig. 8.9(b). Hence the zero signal current in R_L is $I_{C1} - I_{C2} = 0$. Under signal conditions, the differential currents add for $i_{c2} \cong -i_{c1}$, as indicated.

For class B operation, two types of distortion dominate as indicated in Fig. 8.10(b). One is crossover distortion because of the very highly nonlinear input voltage–current relationship at very low collector currents, and the other is peak flattening as each transistor is driven either toward saturation or into the high injection region. For smaller signals, the peak distortion is eliminated as shown by the dashed curves.

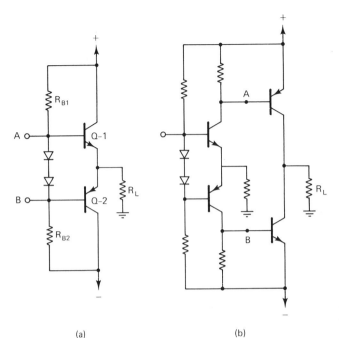

Figure 8.9 Power-output stages using complementary transistors. (a) Output from emitters. (b) Output from collectors.

(a)   (b)

For class AB operation, each transistor is biased with a small current, and for small signals the output current is the differential current, which gradually changes to the current of only one of the transistors, as indicated in Fig. 8.10(b), reducing or almost eliminating the crossover distortion.

The efficiency of class B or class AB operation is much higher than class A, since for zero signal current the power dissipation is negligible.

Comparisons of Emitter-Follower and Common-Emitter Complementary Output Stages

As discussed for the single power output stage, the complementary emitter follower has greatly reduced distortion at the expense of requiring a much larger signed voltage, whose source may also have some distortion. Also, the maximum signal voltage amplitude is limited to a positive peak value of $V(+) - V_{BE1}$ and to a corresponding negative peak value, whereas for the complementary common-emitter pair, the collector–base voltage can be driven into the slightly forward biased region. This may sometimes be important for amplifiers required to operate at relatively low supply voltages. Where a very low output impedance is required, the complementary emitter follower is the choice configuration.

Most amplifiers incorporating power output stages incorporate negative feedback, which reduces output distortion and output impedance in accordance with the relationships derived in Chapter 6.

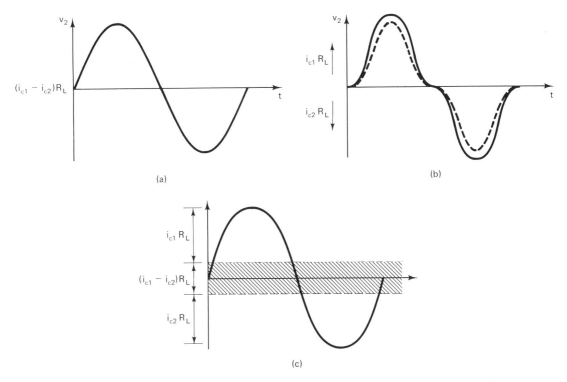

Figure 8.10 Output wave forms for complementary output stages, (a) Class A operation. (b) Class B operation showing crossover distortion. (c) Minimizing distortion using class AB operation.

8.5 TRANSFORMER COUPLING OF POWER OUTPUT STAGES

Transformers are generally avoided in modern integrated-circuit amplifiers in favor of direct coupling to loads because of the size and weight of transformers. However, where large amounts of signal power must be supplied to a very low value of load resistance, neither the efficiency or distortion of the output stages can be optimized. Although output transformers exhibit limitations of both low and high frequencies, the ideal relationships are described as follows: In Fig. 8.11, for the transformer having N_1 turns on the primary side and N_2 turns on the secondary, with corresponding primary and secondary inductances L_1 and L_2, respectively,

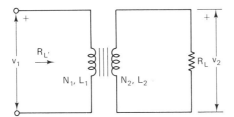

Figure 8.11 Transformer-coupled output.

$$\frac{v_2}{v_1} \cong \frac{N_2}{N_1} \cong \sqrt{\frac{L_2}{L_1}} \tag{8.40}$$

The input impedance R'_L is given by

$$R'_L \cong \frac{L_1}{L_2}, \qquad R_L \cong \left(\frac{N_1}{N_2}\right)^2 R_L \tag{8.41}$$

Thus the transformer can be designed to optimize the load seen by the power output stage for whatever R_L power has to be supplied.

8.6 UTILIZATION OF NONLINEAR AMPLIFIER PROPERTIES

The nonlinearity of amplifying devices such as bipolar and field-effect transistors so far discussed in terms of amplifier distortion may be utilized for various purposes. For example, the output term related to v_i^2 can be exploited for such things as frequency translation of a signal or group of signals in the neighborhood of some designated reference frequency.

In general, for a voltage amplifier with input nonlinearity not a factor, the nonlinearities that have been discussed are described by an equation of the form

$$i_O = I_O + B_1 v_i + B_2 v_i^2 + \text{higher-order terms} \tag{8.42}$$

where I_O is the total output current, v_i is the input signal voltage and the higher-order terms produce combinations of v_i^n and v_O^m.

The input voltage to such an amplifier can be designated as the sum of a sinusoidal reference signal of a frequency ω_R and a band of frequencies expressed as a Fourier series in the neighborhood of a center frequency ω_C. In one easily interpreted form, this can be written as

$$v_i = V_R \cos \omega_R t + \sum_{m=0}^{M} V_m \cos\left[(\omega_c + \omega_m)t \pm \phi_m\right] \tag{8.43}$$

When used in Eq. 8.42 with all higher-order terms neglected, the resultant current is

$$
\begin{aligned}
i_o = I_o &+ B_1\{V_R \cos \omega_R t + \sum_{m=0}^{M} V_m \cos[(\omega_c \pm \omega_m)t \pm \phi_m]\} \\
&+ B_2\left[\frac{V_R^2}{2} + \frac{V_R^2}{2}\cos 2\omega_R t + \sum_{m=0}^{M}\left\{\frac{V_m^2}{2} + \frac{V_m^2}{2}\cos[2(\omega_c + \omega_R)t + 2\theta_m]\right\}\right] \\
&+ B_2\left[V_R \sum_{m=0}^{M} \{V_m \cos[\{(\omega_R + \omega_c) \pm \omega_m\}t \pm \theta_m]\}\right. \\
&\left. + V_R \sum_{m=0}^{M} \{V_m \cos[\{(\omega_R - \omega_c) \pm \omega_m\}t \pm \theta_m]\}\right]
\end{aligned}
\tag{8.44}
$$

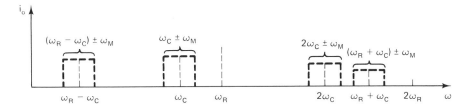

Figure 8.12 Harmonic components from nonlinear amplifiers.

Assuming $\omega_R > (\omega_C \pm \omega_m)$, the resultant spectrum of signals is indicated in Fig. 8.12. With ω_R and ω_C properly spaced, the various spectrum groups can be separated as shown by passing the signal through a bandpass filter of appropriate bandwidth. For example, as shown in Fig. 8.13(a), the reference frequency ω_R can be used to translate the entire spectrum of ω_m centered about ω_c to a lower center frequency $\omega_R - \omega_C$. A similar translation can be achieved for $\omega_R < (\omega_C \pm \omega_m)$.

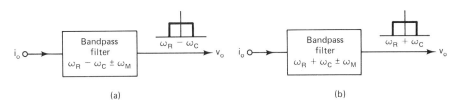

(a)   (b)

Figure 8.13 Selection of frequency spectra using bandpass filters.

Translation of a spectrum of signals to a lower frequency spectrum is referred to as down conversion and is the principle upon which the superheterodyne FM or AM radio receivers and the normal television receiver are based. Translation of a spectrum of signals to a higher frequency is referred to as up conversion and is often used when a signal spectrum is to be transmitted by a UHF, microwave radio link, or by satellite.

A more detailed analysis of conversion processes, including frequencies produced by the higher-order terms, would show signals related to harmonics of both ω_R and ω_C, but they all fall outside the bandpass ranges indicated in Fig. 8.13, provided ω_R and ω_C are not too widely separated. However, it would be desirable that such spurious components not be generated at all because of interference with other signals by radiation or other means of coupling.

If a nonlinear amplifier is replaced by an ideal squaring circuit, the components ω_R and $(\omega_C \pm \omega_m)$ are eliminated, and if an ideal mutiplier can be devised, the frequencies generated are only the desired spectra associated with $(\omega_R - \omega_c)$ and $(\omega_R + \omega_C)$; that is, only the desired frequencies in the up or down conversion processes are generated, with no extraneous or spurious signals produced. Techniques for realizing true multipliers and further applications are discussed in the next section.

A Simple Squaring Circuit

Ideally, a long-channel MOSFET can be represented in low-voltage saturation by an equation of the form

$$i_D = \frac{K}{2}(v_{GS} - V_T)^2 \qquad (8.45)$$

which can be written as

$$i_D = \frac{K}{2}(V_{GS} - V_T)^2 + K(V_{GS} - V_T)v_{gs} + \frac{K}{2}v_{gs}^2 \qquad (8.46)$$

where

$$i_D = I_D + i_d \quad \text{and} \quad v_{GS} = V_{GS} + v_{gs}$$

which is of the form

$$i_D = I_D + B_1 v_{gs} + B_2 v_{gs}^2$$

where

$$B_1 = g_{m0} \text{ and } B_2 = g_m/(V_{GS} - V_T). \qquad (8.47)$$

In the circuit of Fig. 8.14, with balanced input signal voltages $\pm v_g$ and perfectly balanced FETs, the linear gain terms cancel in the common drains and the output voltage is

$$v_o = V_{SS} + h_{FE}R_L(2I_D - I) + 2h_{FE}B_2R_Lv_{gs}^2 \qquad (8.48)$$

which for

$$R_L = \frac{|V_{ss}|}{h_{FE}(I_B)}$$

where $I_B = 2I_D - I$ reduces to its incremental value:

$$v_o = 2h_{FE}B_2R_Lv_{gs}^2 \qquad (8.49)$$

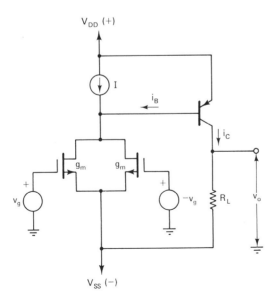

Figure 8.14 Simple FET squaring circuit.

If the FETs are balanced but nonideal, with higher-order terms appearing in Eq. 8.47, all such odd-harmonic terms cancel, leaving only very small residual even-order terms, and the circuit is still a reasonably good squaring circuit.

8.7 ANALOG MULTIPLIER REALIZATIONS

There are many ways in which a true analog multiplier may be implemented, one of which makes use of the squaring circuit as the basic building block to form what is known as the *quarter-square* multiplier based on the equation

$$v_x v_y = \frac{(v_x + v_y)^2}{4} - \frac{(v_x - v_y)^2}{4} \tag{8.50}$$

Conceptually, such an implication is represented by the block diagram in Fig. 8.15.

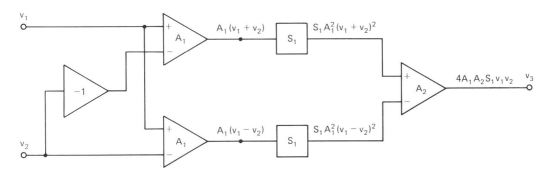

Figure 8.15 Analog multiplier utilizing squaring circuits.

A circuit that is basically an implementation of Fig. 8.15 and that is balanced for all combinations of v_1 and v_2, but that requires very low impedance signal sources because gates and the low input impedance sources of the FETs are driven is shown in Fig. 8.16. In this circuit the dc components are balanced out in the differential amplifier, and the output voltage is

$$v_o = 8B_2 A_2 v_1 v_2 \tag{8.51}$$

Such a circuit is referred to as a four-quadrant multiplier because it is balanced symmetrically for all possible polarity combinations of v_1 and v_2.

An alternative approach to exploiting the somewhat ideal nonlinearity of the FET input–output relationship is to utilize the bias current gain dependence of the BJT to implement the multiplier. This idea was discussed in Chapter 7 in connection with the operational transconductance amplifier, where the bias dependent g_m was used.

A more direct multiplier implementation using the variable transconductance principle may be devised from several component parts as follows: First, two balanced linear voltage-controlled current sources are shown in Fig. 8.17, where the degeneration resistances are used to improve the voltage current linearity of the bipolar transistors. The incremental input resistance of each side (e.g., Q-1) can be written as

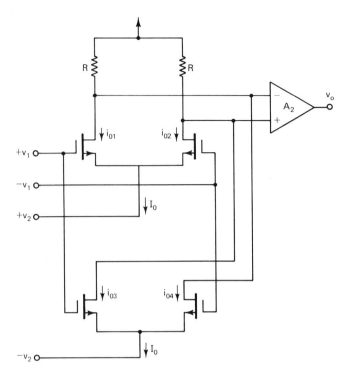

Figure 8.16 Four-quadrant multiplier using FETs.

$$R_{i1} \cong r_{bes1} + \frac{1 + \beta_1}{1 + \beta_2} r_{bes2} + (1 + \beta_1)2R_E \qquad (8.52)$$

With balanced inputs, even without R_E, the input resistance tends to be constant because the current dependence of one r_{bes} is balanced by the opposite current dependence of the other, or

$$R_{i2} \cong r_{bes} + (1 + \beta)2R_E \qquad (8.53)$$

with R_E reducing the input resistance nonlinearity even more, which further linearizes the input voltage–current relationship.

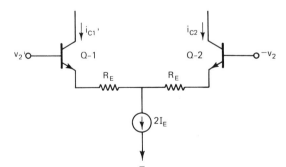

Figure 8.17 Variation of input resistance of differential BJT amplifiers with emitter current.

With the input signal voltages balanced, the voltage at the R_E junctions is constant, and

$$i_{C1} = I_E \frac{h_{FE}}{1 + h_{FE}} + \frac{\beta v_2}{r_{bes1} + (1 + \beta)R_E} \tag{8.54}$$

$$i_{C2} = I_E \frac{h_{FE}}{1 + h_{FE}} - \frac{\beta v_2}{r_{bes2} + (1 + \beta)R_E} \tag{8.55}$$

For simplicity, it will be assumed for the remainder of this discussion that $R_E(1 + \beta) >> r_{bes}$ and that h_{FE} and β are sufficiently large that $i_C \cong I_E$, which allows the preceding equations to be expressed approximately as

$$i_{C1} \cong I_E + \frac{v_2}{R_E} \tag{8.56}$$

$$i_{C2} \cong I_E - \frac{v_2}{R_E} \tag{8.57}$$

Such voltage-controlled current sources can be used to control the operating current in a balanced transconductance amplifier, as indicated in Fig. 8.18(a), with the g_m's of Q-3 and Q-4 being proportional to bias currents and the output current being given by $g_m v_{b'e}$.

Over a large input voltage range, the nonlinearity of the diffusion component of the input resistance must be compensated for. This can be done by replacing the R_L's with the nonlinear resistance of diode-connected transistors, as shown in Fig. 8.18(b).

A four-quadrant multiplier may be devised from the current sources of Fig. 8.17,

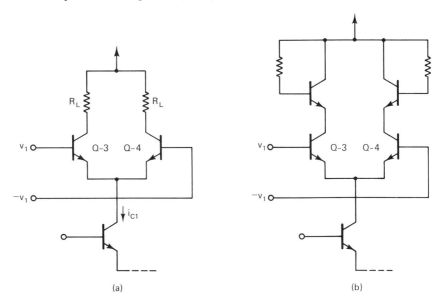

Figure 8.18 Collector signal voltage control in differential amplifier. (a) Resistive load. (b) Diode load.

Sec. 8.7   Analog Multiplier Realizations

363

supplying two balanced circuits like Fig. 8.18 driving a differential amplifier, as shown in Fig. 8.19. The internal transconductance of the Q-3, Q-4 pair is given by

$$g'_m = \frac{i_C}{kT/q} = \frac{i_C}{V_T}$$

as defined by

$$g'_m = \frac{\partial i_C}{\partial v_{B'E}}$$

It will be assumed in the analysis of this circuit that the nonlinear load compensation is such that

$$g_m = \frac{\partial i_C}{\partial v_1}$$

Also, as in the case of the current sources, it will be assumed that h_{FE} is sufficiently high that $i_C \cong i_E$ in the respective transistors.

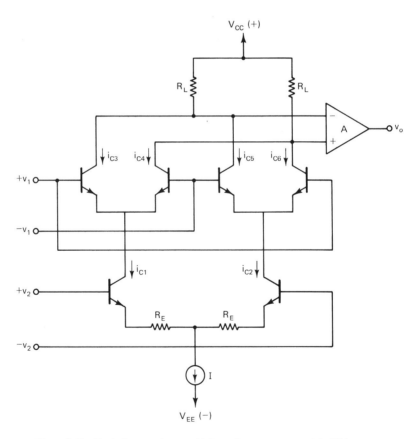

Figure 8.19 Basic four-quadrant multiplier using current control in BJTs.

Under these assumptions, the respective collector currents are of the form

$$i_C = I_C + g_m v_i$$

Then, using Eqs. 8.52 and 8.53,

$$i_{C3} = \frac{I_E}{2} + \frac{1}{2V_T} I_E v_1 + \frac{1}{2V_T R_E} v_1 v_2 \tag{8.58}$$

$$i_{C4} = \frac{I_E}{2} - \frac{1}{2V_T} I_E v_1 - \frac{1}{2V_T R_E} v_1 v_2 \tag{8.59}$$

$$i_{C5} = \frac{I_E}{2} - \frac{1}{2V_T} I_E v_1 + \frac{1}{2V_T R_E} v_1 v_2 \tag{8.60}$$

$$i_{C6} = \frac{I_E}{2} + \frac{1}{2V_T} I_E v_1 - \frac{1}{2V_T R_E} v_1 v_2 \tag{8.61}$$

Then, in the common-collector circuits indicated,

$$v_{C3} = -R_L(i_{C3} + i_{C5}) = -I_E R_L - \frac{R_L}{V_T R_E} v_1 v_2 \tag{8.62}$$

$$v_{C4} = -R_L(i_{C4} + i_{C6}) = I_E + \frac{R_L}{V_T R_E} v_1 v_2 \tag{8.63}$$

and the output of the differential amplifier is

$$v_o = A(v_{C4} - v_{C3})$$

or

$$v_o = \frac{2AR_L}{V_T R_E} v_1 v_2 \tag{8.64}$$

There are in existence a number of analog multipliers with various embellishments but based on the variable g_m principle. Most of them use some sort of variable R_L to compensate for the nonlinear input voltage–current relationship as input currents become sufficiently large that r_b' becomes a substantial fraction of r_{bes}.

8.8 ANALOG MULTIPLIER APPLICATIONS

In addition to the superiority of true multipliers in frequency conversion processes due to the nongeneration of spurious frequency components, which was pointed out in Sec. 8.6, a variety of linear and nonlinear functions can be generated using multipliers in various closed-loop configurations. For example, the circuit of Fig. 8.20(a) is an elementary divider circuit. In this circuit the input voltage is

$$v_o = \frac{Av_1}{1 + AMv_2} \tag{8.65}$$

and for $AMv_2 \gg 1$,

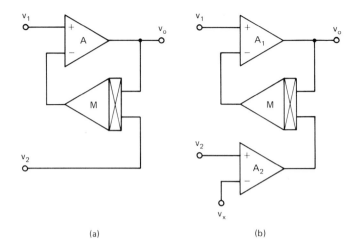

(a)   (b)

Figure 8.20 Divider circuits using analog multipliers. (a) Basic circuit. (b) Improved circuit.

$$v_o \cong \frac{1}{M} \frac{v_1}{v_2}$$ (8.66)

An improved divider circuit that does not depend on a large Amv_2 product is shown in Fig. 8.20(b). In this circuit

$$v_o = \frac{A_1 v_1}{- A_1 A_2 M v_x + A_1 A_2 M v_2}$$ (8.67)

The extra voltage v_x can be set to

$$v_x = \frac{1}{A_1 A_2 M}$$

and when this is done

$$v_o = \frac{A_1}{A_1 A_2 M} \frac{v_1}{v_2}$$ (8.68)

One form of multiplier devised in Sec. 8.7 was derived from squaring circuits. On the other hand, any multiplier becomes a squaring circuit by simply connecting the two inputs. In turn, squaring circuits can be used in a closed-loop configuration to extract the square root of a function. One such circuit is suggested in Fig. 8.21, with v_1 as the input and v_2 the output.

$$v_2 = A_1 A_2 (v_1 - M v_2)$$

or

$$v_2^2 + \frac{v_2}{A_1 A_2 M} - \frac{v_1}{M} = 0$$ (8.69)

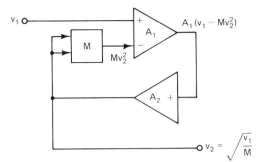

Figure 8.21 Square-root circuit using analog multiplier.

Then for $v_1 \gg v_2/(A_1 A_2)$, the output voltage is given by

$$v_2 \cong \sqrt{\frac{v_1}{M}} \tag{8.70}$$

It is readily apparent that combinations of differential amplifiers, squaring, and square rooting circuits can be used to implement various trigonometric identities as well as many other applications.

8.9 LOGARITHMIC AMPLIFIERS AND APPLICATIONS

The exponential current–voltage characteristics of junction diodes and bipolar transistors may be used to obtain responses of the form $y = M \log x$ or its inverse. A simple embodiment of a logarithmic amplifier is shown in Fig. 8.22. In this circuit the lower transistors Q-1 and Q-2 constitute a current-mirror-type voltage-controlled current source,

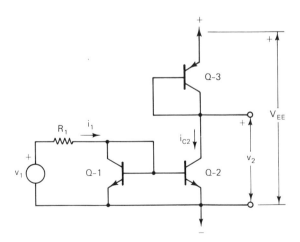

Figure 8.22 Logarithmic amplifier using nonlinear current source.

with the load being the diode-connected transistor, Q-3. In this circuit

$$i_{C2} = K i_1 \tag{8.71}$$

where K is the multiplying factor controlled by the structure of Q-1 and Q-2, and

$$i_1 = \frac{v_1 - V_{BE1}}{R_1} \tag{8.72}$$

for $v_1 > V_{BE1}$. Also,

$$i_{C2} \cong I_{ES3} e^{V_{EB3}/V_T} \tag{8.73}$$

From the three preceding equations,

$$v_{EB3} = V_T \ln(V_1 - V_{BE1}) + V_T \ln \frac{K}{I_{ES3} R_1} \tag{8.74}$$

and
$$v_2 = V_{EE} - V_T \ln(v_1 - V_{BE1}) - V_T \ln \frac{K}{I_{ES3} R_1} \tag{8.75}$$

The variable component of v_2 is

$$\tilde{v}_2 = -V_T \ln(v_1 - V_{BE1})$$

which is proportional to $\ln v_1$ for $v_1 \gg V_{EE}$. This limitation can be minimized by neutralizing the V_{BE} voltage by an approximately equal voltage of the opposite polarity (see Prob. 8.8).

An alternative approach to logarithmic amplifier realization is the use of a diode or diode-connected transistor in the feedback loop of a negative feedback amplifier. An example is shown in Fig. 8.23. For the input signal voltage, $v_1 = 0$, the reference voltage, V_{R1} is chosen to select the input–output offset voltage to put the operating value of $i_D = I_D$ in the range for I_D given most nearly by

$$I_D = I_O \left(e^{v_D/V_T} - 1 \right)$$

This negative offset voltage for v_2 is nullified at the output of the second amplifier using an offset voltage V_{R2} of polarity opposite to V_{R1}, such that $v_o = 0$ for $v_1 = 0$, while at

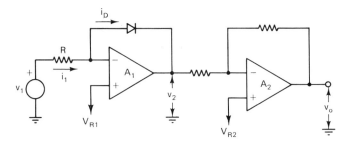

Figure 8.23 Logarithmic nonlinear feedback amplifier.

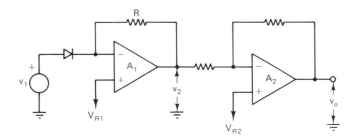

Figure 8.24 Exponential amplifier (inverse log) using nonlinear feedback.

the same time any desired voltage gain is provided by A_2. With operating or bias levels appropriately set, the incremental values of voltages and current relationships are

$$i_1 = \frac{v_1 - v_i}{R} = I_O \, [e^{(v_i - v_2)/V_T}] \qquad \text{and} \qquad v_i = -\frac{v_2}{A_1}$$

from which

$$v_2 = \frac{V_T}{1 + (1/A)} \ln \frac{v_1 + (v_2/A)}{I_O R} \tag{8.76}$$

Then, for $A \ggg 1$,

$$v_2 \cong -V_T \ln \frac{v_1}{I_O R} \tag{8.77}$$

An interchange of the resistor and diode of Fig. 8.23, as indicated in Fig. 8.24, makes the circuit an inverse-log or exponential amplifier. For this circuit

$$v_2 = -\frac{I_O R}{1 + (1/A_1)} \exp \left[\frac{v_1 + (v_2/A)}{V_T} \right] \tag{8.78}$$

which for large A is

$$v_2 \cong I_O R \, e^{v_i/V_T} \tag{8.79}$$

Some of the functions discussed in Sec. 8.9 can be realized by combinations of log and inverse-log amplifiers. For example, a multiplier may be implemented from the relation

$$\ln v_1 v_2 = \ln v_1 + \ln v_2$$

by applying v_1 and v_2 to separate logarithmic amplifiers, adding the outputs, and applying the result to an exponential amplifier. Division may be implemented in such a manner using the relations

$$\ln \frac{v_1}{v_2} = \ln v_1 - \ln v_2$$

PROBLEMS

8.1 Repeat the calculations of the example in Sec. 8.1 for $R_L = 3.5$ kΩ and $V_{gs} = 1.0$ V max. Why is V_{gs} limited to 1.0 V? Could the range be extended slightly by shifting the operating point?

8.2 Measurements are made on a MOSFET at an operating point in the saturation region of $V_{DS} = 14$ V, $V_{GS} = 3.5$ V, and $I_D = 4.3$ mA. At this point, $g_m = 3.3$ mS, $r_d = 30.3$ kΩ, and $\mu = 100$. It is to be assumed that the nonlinearity can be modeled reasonably well by an equation of the form of Eq. 3.67, with $m = 1$, $a = 1$, and $n = 2$, rather than the constant μ model.

    (a) Determine values for V_T, K, and λ in the nonlinear $i_D = f(v_{DS}, v_{GS})$ relationship.

    (b) Plot a set of characteristics for part (a) and compare it in the form with the constant μ model. (This can be done by using the constant μ model shown in Fig. 8.1 with parameters shifted appropriately.)

8.3 Starting with

$$i_D = \frac{K}{2}(v_{GS} - V_T)^2 (1 + \lambda v_{DS})$$

repeat the type of development carried out in Sec. 8.1, arriving at an equation for $v_{DS} = f(v_{GS}, R_L)$. Then recast your result in terms of g_m, r_d, and μ at the operating point.

8.4 A transistor whose collector characteristics are modeled by Fig. 8.3 and expressed by Eq. 2.47 has a nonlinear term of the form of Eq. 2.48. A set of parameter measurements at points 1, 2, 3, and 4 yield values of $k_B = 122$, $x = 1.5$, and $y = 1.2$, where i_B is in amperes and v_{CE} in volts.

    (a) From the graph, estimate h_{fe} and h_{oe} at $I_C = 14$ mA with $v_{CE} = 24$ V. Then calculate values for β_o, g_o, and I_{os} for Eq. 2.47.

    (b) At the operating point specified, determine the coefficients of the nonlinear terms in Eq. 8.17.

8.5 Consider a simple squaring circuit based on that of Fig. 8.14 using nonideal FETs, with the model of Eq. 8.45 modified by the $(1 + \lambda v_{DS})$ multiplier.

    (a) Verify that the drain current can be expressed as

$$I_D + i_d = \frac{K}{2}[(V_{GS} - V_T)^2(1 + \lambda V_{DS})] + K(V_{GS} - V_T)(1 + \lambda V_{DS})v_{gs}$$

$$+ K(V_{GS} - V_T)v_{ds}v_{gs} + \frac{K}{2}(V_{GS} - V_T)^2\lambda v_{ds}$$

$$+ \frac{K}{2}(1 + \lambda V_{DS})v_{gs}^2 + \frac{K}{2}\lambda v_{gs}^2 v_{ds}$$

    (b) Suppose that the stage being driven by the FET's output were an emitter follower. Would the result be a good squaring circuit? Explain. Repeat for a common-base stage.

    (c) Rewrite the equation of part (a) for $R_i \to 0$ looking into the stage being driven for the FETs.

8.6 Write the equation for the output v_o of the circuit shown for $A_1 \to \infty$.

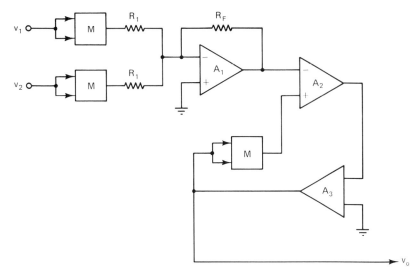

Prob. 8.6

8.7 (a) For the circuit shown, write an equation for v_2 assuming that $R_F = R$.

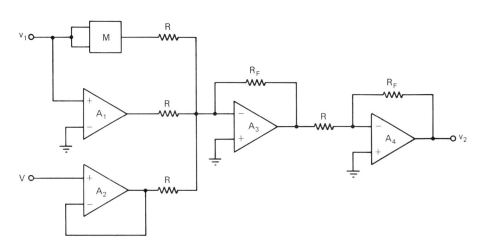

Prob. 8.7

(b) For M, A, and V fixed at predetermined value, write an equation for v_2 in terms of v_1.

(c) If v_1 is varied until $v_2 = 0$, what equation is solved for v_1?

(d) If v_2 cannot be set to zero, what does it tell us about the solution of part (b)?

8.8 For the circuit shown, assume the following circuit parameters and operating conditions.

$$h_{FE} \cong \beta \cong 100, \quad \text{for all transistors}$$

$$I_{E3} = 0.5 \text{ pA}$$

$$R_1 = 2.0 \text{ k}\Omega$$

$$V_{BE1} = V_{BE2} = V_{BE4} = 0.8 \text{ V}$$

$$K = 1$$

$$V_T = 0.026 \text{ V at } T = 300 \text{ K}$$

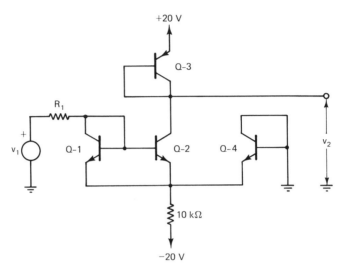

Prob. 8.8

(a) Plot v_2 as a function of v_1 on semilog paper for v_1 ranging from 0 to $+20$ V. What is the range of current i_{C2}?

(b) What, approximately, is the impedance looking into the emitter of Q-4?

REFERENCES

GLASFORD, G. M. "Utilization of Nonlinear Elements for Realization of Linear Functions," *Proc. Fourth Asilomar Conference on Circuits and Systems,* Pacific Grove, Calif., 1970, Western Periodicals, North Hollywood, Calif.

GRAY, P. R., AND R. C. MEYER. *Analog Integrated Circuits,* Wiley, New York, 1984.

MILLMAN, J. *Microelectronics,* McGraw-Hill, New York, 1979.

NARAYANAN, S. "Transistor Distortion Analysis Using Volterra Series Representation," *Bell Syst. Tech. J.,* May–June 1967.

NARAYANAN, S., AND H. C. POON. "An Analysis of Distortion in Bipolar Transistors Using Integral Charge Control Model and Volterra Series," *IEEE Trans. Circuit Theory,* vol. CT-20, no. 4, Jan. 1972.

Poon, H. C. "Modeling of Bipolar Transistors Using Integral Charge Control Model with Application to Third-Order Distortion Studies," *IEEE Trans. Electron Devices,* vol. ED-19, no. 6, June 1972.

Weiner, D. D. and J. F. Spina. *Sinusoidal Analysis and Modeling of Weakly Nonlinear Circuits,* Van Nostrand Reinhold, New York, 1980.

Chapter 9

Analog Switching Circuits, Transmission Channel Time Sharing, Function Generators, and Phase-Locked Loops

INTRODUCTION

There are many applications for which linear and nonlinear analog circuits function in a noncontinuous or switching mode. In some such applications a transmission channel is time shared among groups of analog signals. In other cases, various functions or mathematical operations may be performed by individual circuits at specifically defined time intervals.

A basic building block involved in noncontinuous analog circuits is the *analog switch,* which may be defined as an array of circuit components that in its entirety alternately prevents or allows an analog signal to be passed by a transmission channel or that allows time sharing in a transmission channel among groups of such signals or controls the timing of specific waveforms to be generated within prescribed time intervals. A *transmission gate* is sometimes defined in the same manner as the analog switch, but is more often considered to be the specific circuit component of the analog switch that actually performs the switching or "gating" operations. The term *analog gate* is sometimes used interchangeably with the term analog switch or transmission gate.

This chapter deals with specific elementary circuits as basic building blocks used in analog switches or gating operations and with the various operations performed by and applications of analog switches and related circuits, such as waveform generators, voltage-controlled oscillators, and phase-locked loops and some of their important applications.

9.1 LIMITING, CLIPPING, AND CLAMPING CIRCUITS

A *limiter* or *limiting circuit* may be defined as a circuit that restricts the linear range of an analog circuit to specific voltage ranges by a process of saturation referred to as *hard limiting*. An example of a limiter is the shunt diode in the circuit of Fig. 9.1(a), where the very low forward resistance of the diode becomes the output shunt resistance when the input voltage $v_i \gg V_{REF}$. For an approximate representation, the diode may be modeled by its piecewise linear model with offset voltage, V_0, forward resistance, r_F, and reverse resistance r_B. With $r_B \gg R_1$, $v_2 = v_i$ for $v_i < (V_{REF} + V_0)$ and

$$v_2 = \frac{r_F}{R_1 + r_F} v_1 \tag{9.1}$$

for $v_1 > (V_{REF} + V_0)$. Then, for $r_d \ll R_s$, hard limiting occurs as indicated. Using two diodes, both positive and negative limiting takes place as indicated in Fig. 9.1(b).

A *clipper* or *clipping circuit* may be defined as a circuit that opens a signal transmission path when an input signal exceeds a specific level. A simple example is the circuit of Fig. 9.2.

Using the piecewise diode approximation, the output voltage is

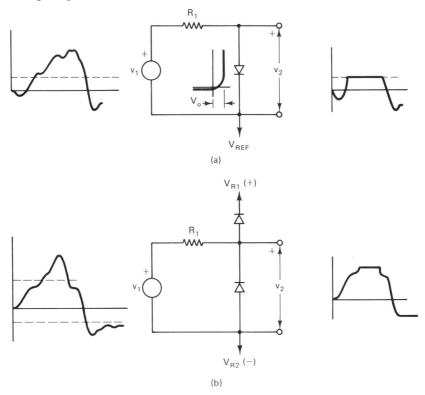

(a)

(b)

Figure 9.1 Diode limiting circuits. (a) Single polarity limiting. (b) Bipolarity limiting.

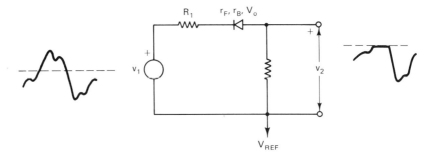

Figure 9.2 Diode clipping circuit.

$$v_2 = \frac{R_2 v_1 + (R_1 + r_F) V_{\text{REF}}}{R_1 + r_F + R_2} \tag{9.2}$$

for $v_1 < (V_{\text{REF}} - V_0)$, and

$$v_2 = \frac{R_2 v_1 + (R_1 + r_B) V_{\text{REF}}}{R_1 + r_B + R_2} \tag{9.3}$$

for $v_1 > (V_{\text{REF}} - V_0)$. The diode is essentially an open circuit for $v_i < (V_{\text{REF}} - V_0)$, since normally $r_B >>> (R_1 + R_2)$ and the output level is limited as indicated.

In this particular application, the output level is limited whether a clipper or limiter is used, and sometimes the two terms are used interchangeably. In many cases, however, there are different side effects depending on whether the amplitude restriction takes place by a saturation process or by an open circuit, and it is preferable to maintain the distinction between a limiter and a clipper.

A single-device amplifier can be simultaneously used as a clipper and limiter as indicated in Fig. 9.3(a), using a single JFET that functions as an amplifier for $\phi_0 < v_G < -V_p$, which *limits* at the gate terminal for $v_G > \phi_0$ and clips due to amplifier cutoff for v_G more negative than $-V_p$.

If an enhancement-mode MOSFET is used as shown in Fig. 9.3(b), limiting takes place as the output current increases along the load line as indicated and moves through the triode region, rather than being limited at the gate, as is the case for the JFET.

The circuits of Fig. 9.3 are not particularly desirable if linearity is to be preserved over the signal transmission range because of the large-signal nonlinearity of the devices. As an alternative, an operational amplifier with feedback may be used to limit the amplitude excursions, as indicated in Fig. 9.4. Over most of the operating range, negative feedback preserves linearity, but limiting occurs at voltages V_{OH} and V_{OL}, usually close to the respective supply voltage levels.

A *clamp* is a simple electronic circuit or component that functions as a switch to connect or disconnect two points in a circuit in some prescribed manner, as indicated in Fig. 9.5(a). One simple embodiment of such a clamp is the single diode limiter of Fig. 9.1, as indicated in Fig. 9.5(b). For $v_{AB} < V_{\text{REF}}$, the switch or clamp is open, and for $v_{AB} > (V_{\text{REF}} + V_D)$, the diode resistance is low and the clamp is closed. The effectiveness of the clamp is determined by the r_F/r_B ratio of the diode and the series resistance of the

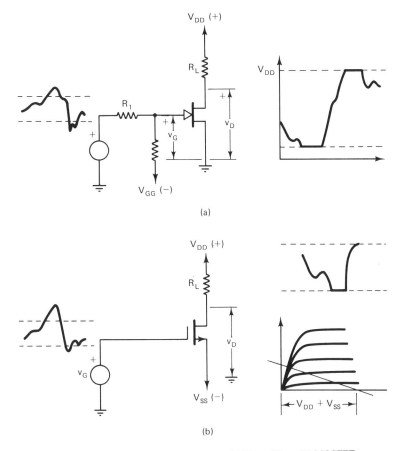

Figure 9.3 Limiting and clipping using FETs. (a) JFET amplifier. (b) MOSFET amplifier.

source to which it is connected. Such a clamp is referred to as a *voltage-controlled clamp* because its opening and closing is controlled by the fixed reference voltage, V_{REF}.

A clamp may be opened or closed at a specific time interval controlled by an external pulse. Such a clamp is referred to as a *keyed clamp* with the pulses referred to as *keying pulses*. In the example shown in Fig. 9.5(c), the clamp is open until the transistor

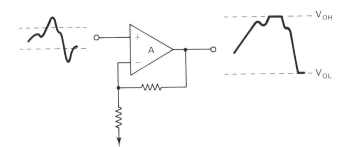

Figure 9.4 Limiting and clipping using overdriven op-amp.

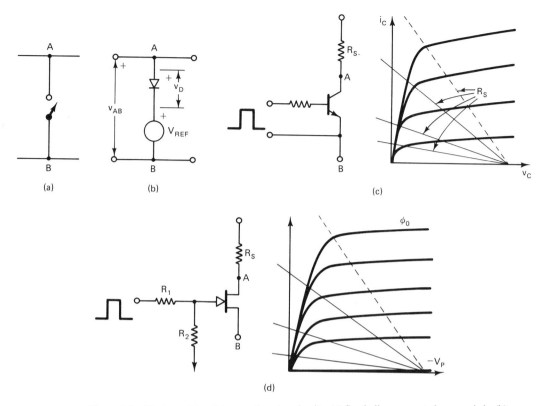

Figure 9.5 Diodes and transistors as clamping circuits. (a) Symbolic representation as switch. (b) Voltage-controlled diode clamp. (c) Pulse-controlled BJT clamp. (d) Pulse-controlled JFET clamp.

is driven rapidly into saturation by an amount dependent on the series resistance of the source to which it is connected. Similarly, an FET may be used as a keyed clamp, as indicated in Fig. 9.5(d). If the amplitude of the pulse at the gate exceeds ϕ_0, the clamp resistance is low, approaching the open-channel conductance of the FET, if the resistance of the voltage source to be clamped is sufficiently high. If the series resistance is too low, the FET will not clamp, as indicated by the dashed load line.

Bidirectional Keyed Clamps

The voltage-controlled diode clamp of Fig. 9.5(b) and the BJT keyed clamp of Fig. 9.5(c) will only clamp voltages of one polarity and are referred to as *unidirectional* clamps. However, if the FET of Fig. 9.5(d) is a symmetrical structure, it can be used to clamp voltages of either polarity. For example, for v_{AB} negative prior to the clamping pulse the FET i_D–v_{DS} characteristic is the same, with source and drain considered as being interchanged. Such a clamp, which clamps points A and B together irrespective of the initial polarity of v_{AB}, is referred to as a *bidirectional* clamp. Bidirectional keyed clamps are often implemented by a more obviously symmetrical structure, such as the diode arrangement shown in Fig. 9.6(a) with symmetrical keying pulses applied at C and D

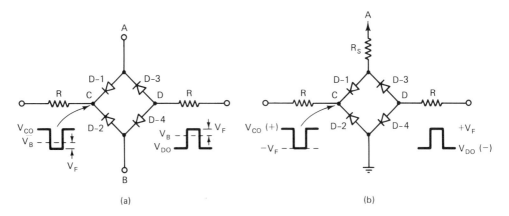

Figure 9.6 Four-diode bidirectional keyed clamps.

through the resistances, R. It is assumed that V_{CO} and V_{DO} are of sufficient magnitude to maintain all diodes nonconducting for the largest value of v_{AB} anticipated prior to clamping. However, if A and B are not connected to external sources and balanced clamping pulses are applied, all diodes conduct equally.

If B is connected to a reference voltage V_B (assumed for convenience to be $V_B = 0$), and the voltage to be clamped is connected through R_S as shown in Fig. 9.6(b), the initial currents when clamping pulses are applied depend on the polarity of v_A prior to clamping.

If $0 > v_A > V_{DO}$ initially, when clamping pulses are applied, current will flow upward through D-3 and R_S, raising the potential at A. When it reaches $v_A = 0$, all diodes will be conducting equally. If $0 < v_A < V_{CO}$ initially, D-1 will conduct when clamping pulses are applied, and v_A will decrease until $v_A = 0$. During the clamping pulses, point A will remain clamped to zero (or whatever v_B is) if the diodes and the clamping pulses are balanced but not depending specifically on how low the forward resistances are. However, if there is shunt capacitance at point A, the speed at which steady state is reached does depend on the forward resistances being low.

Slightly improved versions of the balanced clamp are shown in Fig. 9.7. In Fig.

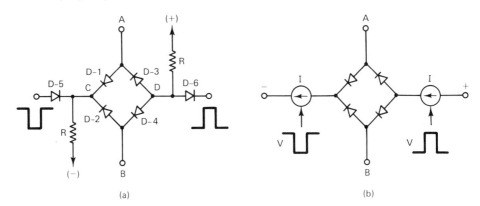

Figure 9.7 Modified four-diode keyed clamps. (a) Diode–source isolation. (b) Current source keying pulses.

9.7(a), during the open interval, D-5 and D-6 are conducting, putting C and D at levels that will maintain all other diodes nonconducting. When clamping pulses are applied, D-5 and D-6 become nonconducting, and all other diodes become forward biased, conducting heavily through the resistance R. The resistance can be replaced by voltage-controlled current sources, as indicated in Fig. 9.7(b). The balance in these improved circuits does not depend on the balance of the clamping pulses themselves, inasmuch as during the time of clamping the clamping pulses are removed.

9.2 TRANSMISSION GATES, ANALOG SWITCHES, AND ELEMENTARY APPLICATIONS

A keyed clamp can be used as a simple transmission gate as indicated in Fig. 9.8. The circuit of Fig. 9.8(a) employs a single FET clamp. The signal source is the output of A_1, which preferably has a low output impedance connected through the clamp Q_1 to a high input impedance load, which is the input impedance of A_2.

The dc level of the clamping or gating waveform relative to the signal is such that when it is most positive D-1 is nonconducting, and gate current flows from R_s through R to hold $V_{GS} \cong \phi_0$. With R_L high, the load line for Q-1 is as shown, and the source and drain are clamped together through the very low drain resistance.

When the negative clamping pulse is applied, D-1 conducts, which lowers the voltage at the FET gate sufficiently to cut off current in Q-1, and R_L is disconnected from the signal source.

In Fig. 9.8(b), the balanced diode clamp shown in Fig. 9.7 is used, with point A being clamped to point B during the clamping pulses, which activate the current sources.

The transmission gate function can be thought of as the section between A and B where the switching actually takes place, whereas the entire circuit, including the optimized source and load impedance using the op-amps, may be thought of as an analog switch in accordance with the preferred terminology introduced at the beginning of the chapter.

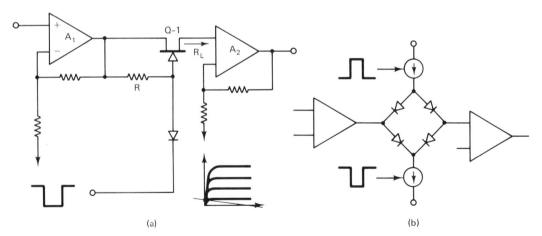

(a) (b)

Figure 9.8 Transmission gates. (a) Using JFET clamps. (b) Using four-diode clamps.

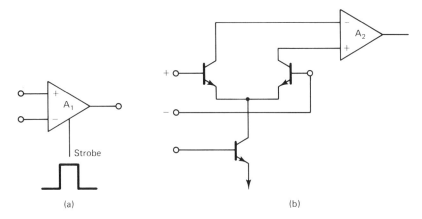

(a)   (b)

Figure 9.9   Operational amplifier with strobe. (a) Symbolic representation. (b) Strobe signal applied to current source of input differential amplifier.

An operational amplifier that can be gated internally, like the operational transconductance amplifier discussed in Chapter 7, can function as a self-contained analog switch, as indicated in Fig. 9.9(a). This is basically a current-controlled differential amplifier with balanced outputs driving a second amplifier differentially to make the output dc voltage level under zero signal conditions independent of input bias current, as indicated in Fig. 9.9(b).

Thus the gating signal can be used to completely disengage the signal from the load, with the amplifier itself functioning as the transmission gate. The gating signal input is sometimes referred to as a *strobe*.

Signal Sampling and Sampled Data Transmission

One specific application of a transmission gate or analog switch is that of converting a continuous analog signal to a discrete signal consisting of periodic samples of the input signal, as indicated in Fig. 9.10. According to the *sampling theorem*, the basic information in the continuous signal is preserved if it is sampled at a rate at least twice that of the higher frequency component of importance in the original signal. However, to create the samples, the bandwidth of the circuits in the transmission gate itself must be considerably higher than the rate. Furthermore, if the distinct nature of the sampled signal is to be preserved in subsequent transmission, the transmission channel must have much greater bandwidth than that of the original signal. Sometimes it is desired to sample a signal

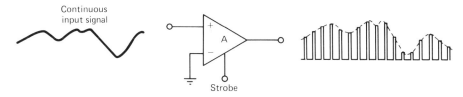

Figure 9.10   Strobed op-amp for data sampling.

when and only when the signal is at a prescribed level, in other words, to detect the presence of a specific signal level.

Sample-and-Hold Circuits

It is sometimes desirable not only to sample a continuous signal but to hold the level of the sample until the time the succeeding sample is taken, as indicated for the signal segment in Fig. 9.11(a). A transmission gate can be converted to a sample-and-hold circuit by the use of a shunt capacitance at the output of the transmission gate, as illustrated by the example of Fig. 9.1, which is basically the circuit of Fig. 9.8(a). The holding capacitor, C_H, is a storage or "memory" element that maintains a constant output between samples. In order that the required signal level be acquired during the period T_1 of the sampling pulse, it is necessary that the charging time constant

$$(R_S + r_d)C_H << T_1$$

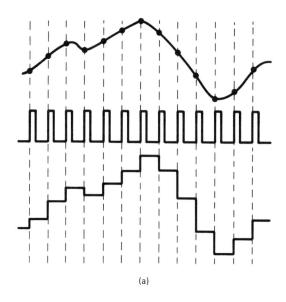

(a)

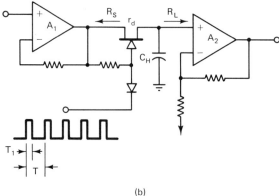

(b)

Figure 9.11 Sample-and-hold function and circuit. (a) Waveform. (b) Simple FET sample-and-hold circuit.

which requires R_S and r_d to be as small as possible. Also, for a true sample to be obtained, it is required that $T_1 \ll T$. In order that the level be maintained during the period between samples, the condition should be met that r_d (OFF) $\rightarrow \infty$ and that $R_L C_H \gg T$. Thus the input resistance of A_2 should be as large as possible, which would be the case if an FET input amplifier were used.

Alternatively, a variable-gain BJT input amplifier can be used for A_2, as suggested in Fig. 9.12. The bias current of the input differential amplifier stage of A_2 is controlled by the input to Q-2 and R_1 and R_2. During the time of the clamping pulses, Q-2 is in saturation, I_{bias} is at its largest value, and A_2 is at maximum bandwidth to preserve the rise time of the samples. During the holding time, $T - T_1$, Q-2 is off, and bias current is supplied through $R_2 \gg R_1$ and is reduced to a very low value, which greatly increases the input resistance. Because of the large holding time relative to acquisition time, the reduced bandwidth of A_2 of the low values of bias current is not so important.

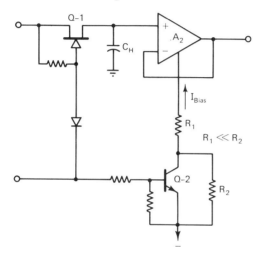

Figure 9.12 Improved sample-and-hold circuit.

The data acquisition time of a sample-and-hold circuit can be decreased by a feedback system as indicated in Fig. 9.13. The amplifier, A_3, is an operational transconductance amplifier as described in Chapter 7. If the output of A_2 lags the input voltage during acquisition, the difference is amplified by A_3 to increase the charging rate. During the holding time, the high-output-impedance OTA is switched off and does not affect the holding time.

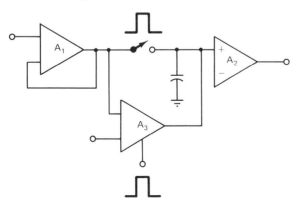

Figure 9.13 Sample-and-hold circuit and fast data acquisition.

An alternative version of the basic sampling component of the sample-and-hold circuit makes use of the balanced diode clamp of Fig. 9.7 as implemented in Fig. 9.14 with a specific form of voltage-controlled current source using enhancement-mode MOS transistors. The clamping pulse levels are selected such that during the holding interval $T - T_1$ the FETs are all nonconducting, and during the sampling period T_1 the R's are chosen to provide the diode current I. This clamp is symmetrically balanced. For example, if just before a clamping pulse is applied, $v_1(t) < V_C$, then, when the pulse is applied, the current I flows through D-1 and into terminal A, with D-2 and D-3 initially being nonconducting. At the same time, the holding capacitor C_H begins to discharge through D-4 at a rate $dv_c/dt = -(I/C)$, which continues until $v_c(t) = v_1(t)$. At the end of T_1, all diodes again become nonconducting, and v_c is held at the value reached at the end of T_1. If, conversely, $v_1(t) > V_C$ when a clamping pulse is applied, the current I flows through D-2 to charge C_H. At the same time, current flows from terminal A into D-3, with D-1 and D-4 remaining nonconducting until the signal level is fully acquired by C_H.

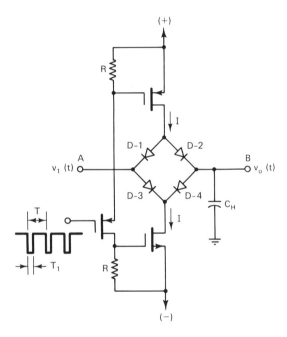

Figure 9.14 Sample-and-hold circuit with current source four-diode clamp.

9.3 SIMPLE BISTABLE, MONOSTABLE, AND ASTABLE CIRCUITS

Any circuit comprising a single device or combination of devices that has an equal probability of having either one of two possible output levels for a fixed input level is said to be bistable.

The high-gain positive feedback regenerative comparator known as the Schmitt trigger described in Chapter 7 is shown again in Fig. 9.15 as a bistable circuit. The

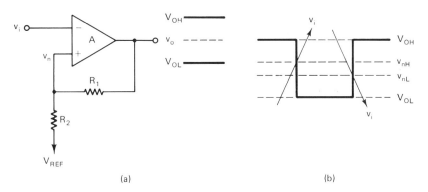

Figure 9.15 Bistable circuit with hysteresis (Schmitt trigger). (a) Op-amp comparator. (b) Waveforms.

comparator is a very high gain operational amplifier which limits at high and low values, V_{OH} and V_{OL}. The design of the amplifier is normally such that when

$$v_i = v_n, \qquad v_o \cong \frac{V_{OH} - V_{OL}}{2}$$

However, any internal perturbation is amplified such that the output is immediately driven either to $v_o = V_{OH}$ or $v_o = V_{OL}$. Thus the circuit is bistable. Initially, if $v_O = V_{OH}$, $V_n = V_{nH}$, given by

$$v_{nH} = \frac{R_2 V_{OH} + R_1 V_{REF}}{R_1 + R_2} \tag{9.4}$$

Then if v_i is $v_i < v_n$ and increases and the amplifier gain is very high, when $v_i \geq v_{nH}$, the amplifier output will rapidly change from V_{OH} to V_{OL}.

If, however, $v_O = V_{OL}$, $V_n = V_{nL}$, given by

$$v_{nL} = \frac{R_2 V_{OL} + R_1 V_{REF}}{R_1 + R_2} \tag{9.5}$$

the amplifier output will rapidly change from V_{OL} to V_{OH}.

The difference in input levels at which amplifier switching occurs is given by

$$v_{nH} - v_{nL} = \frac{R_2}{R_1 + R_L} (V_{OH} - V_{OL}) \tag{9.6}$$

and is called the *hysteresis* of the circuit, as indicated in Fig. 9.15(b).

A Schmitt trigger bistable circuit can be implemented by a simple two-device regenerative amplifier as indicated by the BJT circuits of Fig. 9.16. The positive feedback nature of the amplifier can be verified by observing that Q-1 is a gain-producing inverting amplifier, with Q-2 an emitter follower whose output is applied back to the emitter of Q-1 through the common-emitter resistor R_E in such a direction as to reinforce the change in output voltage.

A set of dc operating conditions can be established such that Q-1 and Q-2 are conducting approximately equally for a prescribed value of V_1. However, if $AB > 1$ at

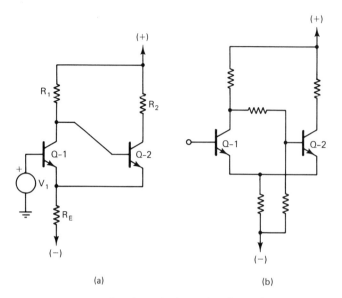

Figure 9.16 Schmitt trigger using discrete devices. (a) Direct-coupled BJTs. (b) Resistance coupled BJTs.

every point through the entire dynamic range, as defined by the given feedback equation of Chapter 6, and the output is driven to one of two possible stable states without any change in V_1, the circuit is bistable. A very exact analysis of Schmitt trigger operation can be carried out using the graphical characteristics of the devices or a quite good approximate analysis using nominal values for $h_{fe} \cong h_{FE}$ and assumed values of V_{BE} for V_{BES}, V_{BEA}, and V_{BEC}. A test for sufficient gain to ensure bistability is to assume Q-1 on and Q-2 off and determine if the actual value of V_{BE2} is sufficiently low to meet the condition; then assume Q-2 on and Q-1 off and determine if this condition is likewise sufficient.

Example

In the circuit of Fig. 9.16(a), assume Q-1 and Q-2 are identical approximately linear transistors with $h_{fe} \cong h_{FE} = 100$ and $V_{BE} \cong 0.75$, and specify a design center current of $I_{C1} = I_{C2} = 2$ mA with $V_{B1} = 0$ (observe that also $V_{B2} = 0$):

$$R_1 = \frac{V_{CC} - V_{B2}}{I_{C1} + I_{B2}} = \frac{5 - 0.0}{2 + (2/100)} = \frac{5.00}{2.02} = 2.48 \text{ k}\Omega$$

$$R_E = \frac{V_E - V_{EE}}{I_{E1} + I_{E2}} = \frac{-0.75 + 5}{2\left(\dfrac{101}{100}\right)} = \frac{4.25}{4.04} = 1.05 \text{ k}\Omega$$

(1) Assume Q-2 nonconducting and Q-1 in saturation with $V_{BES} = 0.8$ V.

$$I_E = \frac{V_1 - V_{BES} - V_{EE}}{R_E} = \frac{0 - 0.8 + 5}{1.05} = \frac{4.2}{1.05} - 4.0 \text{ mA}$$

$$I_C = \frac{V_{CC} - V_C}{R} = \frac{5.0 + 0.6}{2.48} = 2.26 \text{ mA}$$

$$I_B = I_E - I_C = 4.00 - 2.26 = 1.74 \text{ mA}$$

$I_B = 1.74 >> I_C/h_{FE}$; hence *Q-1 in saturation is a stable state with Q-2 cutoff.*

(2) Assume Q-1 nonconducting and Q-2 on, but not in saturation. Write the following equations for Q-2, noting that $I_{E2} = (1 + h_{FE}) I_{B2}$.

$$V_E = V_{CC} - \frac{I_E}{1 + h_{FE}} R_1 - V_{BE}$$

$$V_E = I_E R_E + V_{EE}$$

Solving these equations for V_E,

$$V_E = \frac{V_{CC} - V_{BE} + (R_1/R_E)\,[1/(1 + h_{FE})]\,V_{EE}}{1 + (R_1/R_E)\,[1/(1 + h_{FE})]}$$

$$= \frac{5 - 0.75 + (2.48/1.05)(1/101)(-5)}{1 + (2.48/1.05)(1/101)}$$

$$V_E = 4.0 \text{ V}$$

This is far in excess of the required voltage to maintain Q-1 cut off; hence Q-2 *on and Q-1 off is a stable state*.

(3) Suppose we allow Q-2 to saturate. Determine the maximum possible value for R_2 to let Q-1 be cut off just as much as Q-2 is cut off when it is cut off (i.e., $v_{BE1} = 0.2$ V or $V_E = -0.2$ V).

$$I_E = \frac{V_E - V_{EE}}{R_E} = \frac{-0.2 + 5}{1.05} = 4.57 \text{ mA}$$

$$I_B = \frac{V_{CC} - (V_E + V_{BE})}{R_1} = \frac{5 - (0.2 + 0.8)}{2.48} = 1.85 \text{ mA}$$

$$I_C = I_B + I_E = 4.57 + 1.85 = 6.42$$

$$R_C = \frac{V_{CC} - (V_E + V_{CES})}{I_C} = \frac{5 - (0.2 + 0.2)}{6.42} = 0.78 \text{ k}\Omega$$

(4) With both transistors allowed to be driven into saturation with each transistor cut off equally, as established by part (3), determine approximately the hysteresis of the circuit (assume that an off transistor will begin to conduct substantially when $V_{BE} \cong 0.7$ V).

First, assume v_1 sufficiently negative to maintain Q-1 off and that it is increased until it starts to conduct. For Q-1 on, $V_E = -0.2$; Q-1 will start to conduct when $V_B \cong -0.2 + 0.7 \cong +0.5$ V.

Second, assume v_1 sufficiently positive to hold Q-1 in saturation and that it is decreased until it comes out of saturation (until $V_{CE1} \cong 0.7$ V), which is sufficient to begin to turn Q-2 on. Using the following equation for Q-1,

$$I_E = \frac{V_{CC} - V_{CE} - V_{EE}}{R_E + [h_{FE}/(1 + h_{FE})]R_1} = \frac{5 - 0.7 + 5}{1.05 + (100/101)2.48} = 2.65 \text{ mA}$$

and

$$V_1 = V_{EE1} + I_E R_E + V_{EE}$$

$$= 0.75 + 2.65(1.05) - 5$$

$$V_1 = 1.47 \text{ V}$$

$$\text{Hysteresis} = V_{IH} - V_{IL} = 0.5 + 1.47 \cong 2.0 \text{ V}$$

It is observed that $V_{OH2} = 5.0$ V and $V_{OL2} = 0.0$ V. If the output is taken from Q-2 and the input transistor,

$$V_{iH} = 0.5 \text{ V} \quad \text{and} \quad V_{IL} = -1.47 \text{ V}$$

For output taken from V_1,

$$V_{OH1} = 5.0 \text{ V} \quad \text{and} \quad V_{OL1} = -0.6 \text{ V}$$

In a practical BJT Schmitt trigger, the input loading at the base of Q-1 would have to be taken into account in translating the input voltage levels to the actual voltage sources with their own source resistances. This would not be a problem using MOSFETs, as indicated in Fig. 9.17(a), with typical characteristics given in Fig. 9.17(b). A very wide range of input–output voltage levels can be designed for using direct coupling between stages as indicated. Similar circuits with reversed output V_{OH} and V_{OL} levels can be used employing p-channel MOSFETs. The design should be such that, when Q-1 is on, V_{GS} of Q-2 is below V_T, the threshold voltage, thus driving it well into the triode region, and when Q-2 is on, its source voltage is sufficiently high to maintain V_{GS} below its threshold voltage.

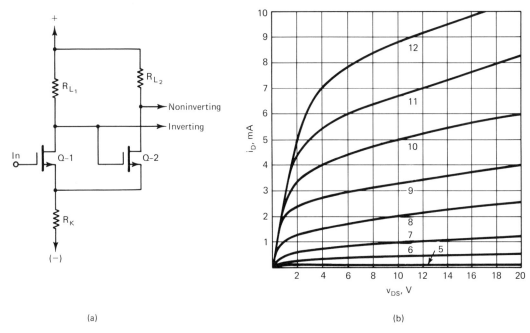

(a)   (b)

Figure 9.17   Schmitt trigger using direct-coupled high threshold MOSFETs. (a) Circuit. (b) MOSFET characteristic.

Symmetrical Bistable Circuits

Circuits may be symmetrically bistable, as indicated in Fig. 9.18(a) for BJTs and in Fig. 9.18(b) for enhancement-mode MOSFETs. In the case of Fig. 9.18(a), coupling may be direct ($R_1 = 0$, $R_2 = \infty$) or through the divider network as shown. In either case, the

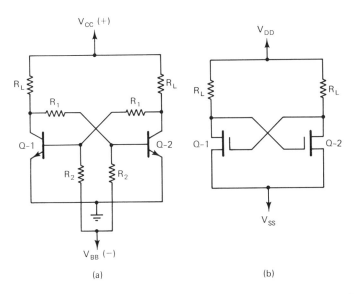

Figure 9.18 Symmetrical bistable circuits (flip-flops). (a) BJT circuit. (b) Direct-coupled FET circuit.

resistors are chosen to force one transistor into saturation, which will maintain the other at cut off. The divider network permits a much greater design flexibility than the direct-coupled version. For the MOSFET version of the bistable circuit, considerable design variation of ON–OFF conditions can be achieved with direct coupling. For any practical use to be made of a bistable circuit, means must be provided to force it to switch from one state to the other, a process referred to as *triggering*. Such a triggered bistable circuit is commonly referred to as a *flip-flop*. Triggering can be accomplished in various ways, one of which is to use auxiliary paralleling amplifiers, as indicated in Fig. 9.19.

If positive pulses are applied simultaneously to the two auxiliary inputs, the two outputs will switch alternatively as shown and the circuit is said to *toggle*. The complete period of the output waveform is twice that of the input pulses, and for this reason the circuit is referred to as a *counter,* which counts by 2 (one output pulse for each two input pulses), or as a *frequency-divider circuit* because the fundamental rate is half that of the input pulses. For the type of triggering shown, the action is relatively independent of pulse amplitude, but critically dependent on pulse width. For example, if it is assumed that Q-1 is the OFF transistor, a positive pulse applied at Q-3 must be sufficient to initiate conduction in Q-3, which functions as an inverting amplifier and begins to turn Q-2 off. The amplified and inverted signal begins to turn Q-1 on. As soon as this process starts, the change of state is completed by the internal regenerative action. The positive pulse at Q-4 has no effect on the output of Q-2 because it is already on. This action takes time, and the input pulse must be ended before the regenerative action is complete. Otherwise, the continuing pulses will force both Q-1 and Q-2 to be off until their termination, at which time a change of state of unpredictable relative polarity will occur. If the input pulses are too wide, they may be differentiated by the short-time-constant RC circuit, as shown in Fig. 9.19(b).

In a circuit such as that of Fig. 9.19, it is important for many applications to control

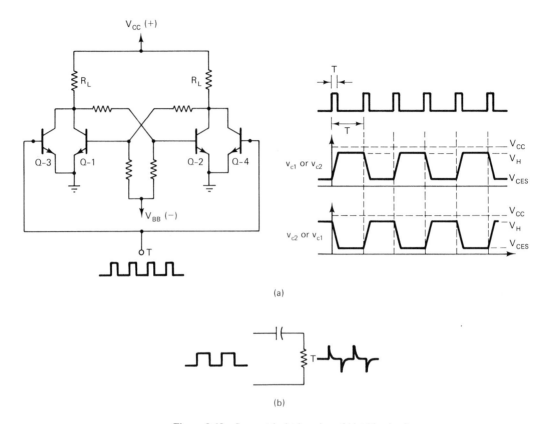

(a)

(b)

Figure 9.19 Symmetrical triggering of bistable circuit.

the relative polarity of the two outputs, which can be done by initially presetting one side or the other by using auxiliary inputs as shown in Fig. 9.20, the additional inputs being referred to as set (S) and reset (R). Also, there are many applications where toggling is not desired in which triggering pulses may be applied separately to the two sides, Q-1 and Q-2.

There are many ways to make the change of state of a flip-flop independent of triggering pulse width other than by capacitive coupling. Feedback can be applied from the output itself back to the inputs to force the trigger to terminate prematurely. More elaborate methods incorporate logic gates into the control circuits or use one flip-flop to control another in a feedback configuration. (Such flip-flops are referred to as *J-K* and master–slave flip-flops and are analyzed in detail in a companion volume, *Digital Electronic Circuits;* see Chapter 10 references.) In some such circuits, the flip-flop transitions actually occur at the *end* or trailing edges of the applied trigger pulses.

Monostable Circuits

A particular class of monostable circuit may be defined as a circuit having positive feedback, which when turned on is driven initially to only one possible stable state, as indicated by the specific example shown in Fig. 9.21. The supply voltage V_{CC} and the

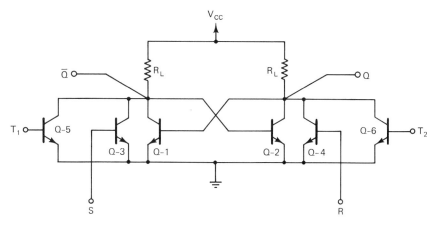

Figure 9.20 Separate triggering of inputs to bistable circuit.

base resistance R_{B2} are chosen to hold Q-2 at the edge of saturation, which in turn maintains Q-1 cut off until a positive pulse is applied to the base of Q-3. This trigger is inverted and amplified at the collector, which is applied through C_C to the base of Q-2, and immediately turns Q-2 off, which in turn drives Q-1 to its edge of saturation. As indicated by the accompanying waveform, the voltage at the base of Q-2 changes by an amount equal to that at the collector of Q-1 as it cuts off. When the pulse ends, the voltage at the base of Q-2 rises toward V_{CC} according to the effective time constant $(R_{L1} + R_{B2})C_C$ until it reaches a point where Q-2 begins to conduct as indicated. At this

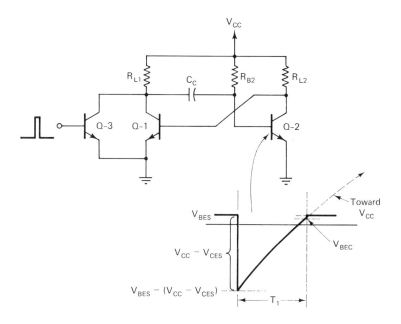

Figure 9.21 Monostable circuit with voltage-controlled recovery time.

time, as Q-2 turns on, regenerative action turns Q-1 off and the circuit returns immediately to its initial state.

The controlling equation of the recovery is given approximately by inspection as

$$v_{G2}(t) = V_{CC} - (2V_{CC} - V_{CES} - V_{BES}) \exp \left[-\frac{t}{(R_{L1} + R_{B2})}C_c \right] \quad (9.7)$$

Then, assuming that the change of state is complete when $v_{G2}(t) = V_{BES2}$, the time T_1 is approximately

$$T_1 = (R_{L1} + R_{G2})C_C \ln \frac{2V_{CC} - V_{CES} - V_{BES}}{V_{CC} - V_{BES}} \quad (9.8)$$

If the inputs to Q-3 of Fig. 9.21 are recurrent pulses at successive periods, $T > T_1$, recurrent waveforms are generated at a rate given by the period T, with the ratio T_1/T controlled by the RC time constant. Such a monostable circuit is useful in generating recurrent waveforms where it is required that specific T_1/T ratios be established.

If the RC time constant is such that $T_1 >> T$ as shown in Fig. 9.22, and if the natural period of recovery T is slightly less than the period of nT pulses, the recurrent rate will be nT and the circuit *counts* by n or is said to be a frequency divider by a factor of n.

There are many possible applications for monostable circuits as specific waveform generators, time-delay circuits, and counters or frequency dividers.

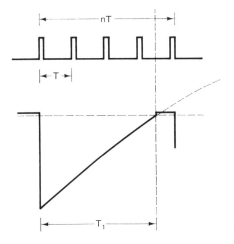

Figure 9.22 Monostable circuit as frequency divider.

Astable Circuits

An *astable circuit* is one that has no stable state but alternately switches between two discrete levels as a result of internal feedback without external intervention. A simple example of such a circuit is shown in Fig. 9.23(a).

The approximate period T_1 for Q-2 off and Q-1 on is given by an equation like Eq. 9.8 for the monostable circuit, while a similar period T_2 for Q-1 off is given by a similar

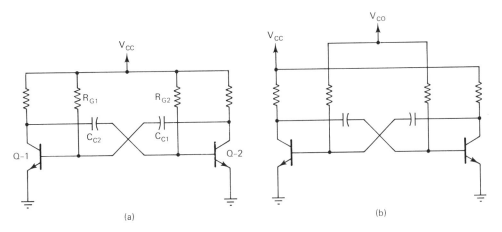

Figure 9.23 Symmetrical astable circuit (multivibrator) with voltage controlled rate.

equation using the time constant $(R_{L2} + R_{G1})C_{c1}$, with the total period given by $T = T_1 + T_2$.

It is sometimes useful to establish some control over the period T independent of the collector supply voltage, as suggested by Fig. 9.23(b). In this case, the period T_1 (like Eq. 9.8) is given approximately by

$$T_1 = R_{GL}C_{c2} \ln \frac{(V_{CO} + V_{CC}) - V_{CES} - V_{BES}}{V_{CO} - V_{CES}} \tag{9.9}$$

The alternative period T_2 is given by a similar expression.

For $V_{CO} = V_{CC}$, Eq. 9.9 reduces to Eq. 9.8, and as V_{CO} increases relative to V_{CC}, the period decreases. In the limit, to $(V_{CC}$ and $V_{CO}) >> V_{CES}$ and V_{BES}

$$T_1 \cong R_{G2}C_{C2} \ln\left(1 - \frac{V_{CC}}{V_{CO}}\right) \tag{9.10}$$

Circuits of this type have historically been referred to as *multivibrators*.

Defining any recurrent waveform-generating circuit as an oscillator, the astable circuit whose rate is determined by the value of the dc voltage V_{CO} represents a simple form of what is commonly referred to as a *voltage-controlled oscillator* or VCO. As a family, VCOs will be discussed at greater length in the following section.

Another astable circuit or VCO that historically has found extensive use is the *blocking oscillator* shown in one of several possible forms in Fig. 9.24. This circuit employs a single device, which may be the BJT shown or a JFET in an identical configuration. Positive feedback is applied through the high-frequency transformer polarized as indicated.

The action of this circuit may be explained starting with the assumption that C is charged negatively to a voltage $-V$ and then left to relax *toward* its equilibrium value, V_{CO}, as suggested by Fig. 9.24(b) with the voltage rise given.

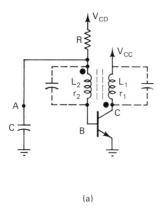

(a)

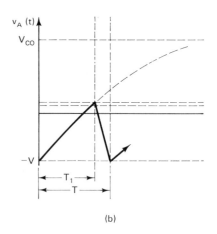

(b)

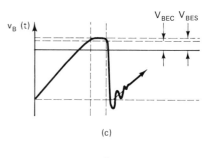

(c)

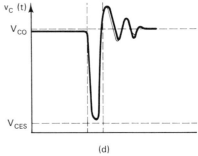

(d)

Figure 9.24 Blocking oscillator. (a) Circuit. (b), (c), (d) Waveforms.

$$v_A(t) = V_{CO} - (V_{CO} + V)e^{-(t/RC)} \qquad (9.11)$$

Since negligible base current is flowing, this rise is duplicated at B until the base voltage reaches its nominal conduction value, V_{BEC}. As soon as base current begins to flow, an amplified current i_C flows toward the collector through L_1, with the collector voltage being given by

$$v_c(t) = V_{CC} = -i_C r_1 - L\frac{di_C(t)}{dt} \qquad (9.12)$$

as indicated by the rapid drop shown in Fig. 9.24(d). This voltage change is coupled back to the base as positive feedback, which in turn rapidly moves the base to the edge of saturation where the current gain is no longer sufficient to sustain positive feedback.

Quantitatively, the various voltage–current relationships during the time that base saturation current flows is complicated by the nonlinearity at the base-current, base-voltage relationship, the resistances and distributed capacitances, and the variability of the coupling coefficient of the transformer if it saturates at high currents. However, qualitatively the action may be explained as follows: Assuming that R is very large compared with the effective input impedance of the transformer as loaded by the low collector resistance of the transistor in saturation in series with the low base resistance, the voltage change across the capacitor is given by

$$v_A(t) \cong V_{BES} - \frac{1}{C} \int_{T_1}^{T} i_B(t) \, dt \qquad (9.13)$$

as indicated in Fig. 9.24b.

As the voltage at A starts to drop, saturation current no longer can be maintained, and as it moves out of saturation through the active region, the transistor again functions as a current amplifier; the $L(di_c/dt)$ component of the voltage of the collector is positive with the feedback resulting in the base and collector waveform at the end of period T indicated in Fig. 9.24(c) and (d). When the voltage $-V$ is reached, the entire process repeats. Because of the distributed capacitance, there are oscillatory overshoots of the voltage across the transformer as the transistor cuts off, which are shown in the waveforms during the beginning of successive periods.

9.4 VOLTAGE-CONTROLLED OSCILLATORS

A voltage-controlled oscillator, or VCO, is a periodic waveform generator, sinusoidal or nonsinusoidal, whose rate of fundamental frequency is controlled by a separately applied voltage or current. In the latter case, where the nature of current control is emphasized, such an oscillator may be referred to as a *current-controlled oscillator*, or CCO. The astable circuits described in Sec. 9.3 are simple specific forms of nonsinusoidal voltage-controlled oscillators.

A sinusoidal VCO may be of the variable reactance type wherein, for example, the controlling capacitance in a sinusoidal oscillator of either the *RLC* or *RC*-feedback type is varied by an external voltage, as suggested in Fig. 9.25(a), where A is the voltage gain of a variable transconductance amplifier whose bias current, which varies the effective transconductance, is controlled by the voltage V. The effective shunt input capacitance is given approximately by

$$C_i \cong C_1 + C_F \left[1 + |A|\right] \qquad (9.14)$$

where A can be varied over a very wide range.

A single FET can function as a variable reactance (capacitance) circuit as indicated in Fig. 9.25(b). Some of the multiplier-controlled variable elements described in Chapter 8 can also be used as variable reactances.

The more usual form of VCO is the relaxation or positive feedback type, normally generating periodic waveforms, usually rectangular and triangular, in which a variable

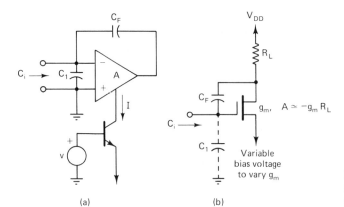

Figure 9.25 Voltage-controlled capacitor. (a) Using current-controlled gain of op-amp. (b) Using variable g_m FET.

reactance may or may not be directly involved. Often a regenerative comparator or Schmitt trigger with controlled hysteresis is a key component in such a VCO. A very simple VCO makes use of two high-gain comparator-type operational amplifiers, as suggested in Fig. 9.26.

The internal gain, frequency response, and slew rate given by $(V_{OH} - V_{OL}/T_1$ in the amplifier of Fig. 9.26(a) are controlled by the bias voltage, V_B. The slew rate can be

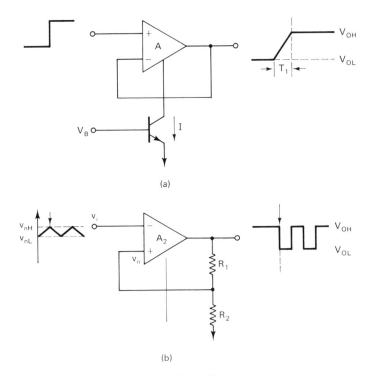

Figure 9.26 Elements of a voltage-controlled oscillator. (a) Variable slew rate op-amp. (b) Schmitt trigger.

varied by several orders of magnitude by variations of V_B. The circuit of Fig. 9.26(b) is a Schmitt trigger of high internal gain, as shown in Fig. 9.15, where the output limits at V_{OH} and V_{OL} as the input v_i exceeds the limits v_{nH} and v_{nL} given by Eqs. 9.4 and 9.5, with hysteresis as defined by Eq. 9.6. The two circuits of Fig. 9.26 can be combined to form the voltage-controlled oscillator shown in Fig. 9.27, with the input voltage v_1 for A_1 being supplied by the output voltage of A_2. The period T of the recurrent waveform is determined by the slew rate $\Delta V_i/\Delta T$ of the amplifier A_1 and the voltage limits determined by the hysteresis of A_2 controlled by the $R_1/(R_1 + R_2)$ ratio. The rectangular and triangular waveforms that are generated by this circuit are symmetrical as indicated.

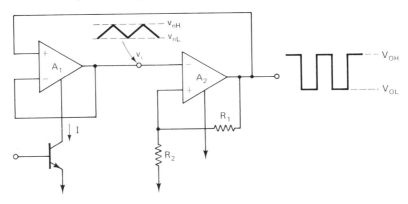

Figure 9.27 Voltage-controlled oscillator combining elements of Fig. 9.26.

Another form of VCO suggested by Fig. 9.28, which also utilizes a Schmitt trigger, may be described as follows: As a starting point, assume that $V_O = V_{OH}$ as a result of $v_c(t)$ having reached a value $v_c(t) \leq v_n$. The output voltage V_{OH} is sufficiently high to maintain Q-3 cut off; hence no current flows through Q-1 or Q-2. The capacitor charges from its low value by the constant current from the voltage-controlled current source flowing through D-2 at a rate

$$v_c(t) = \frac{1}{C} \int i \, dt = \frac{I}{C} t \tag{9.15}$$

When it reaches a value V_{nH} at $t = T_1$, the output switches to V_{OL}, which can be established to be at a sufficiently low level (e.g., $V_{OL} \to 0$) to force Q-3 into saturation. This drops the voltage at all three emitters to a voltage of magnitude $V_{CES} \cong 0.2$ V, which allows the current I to flow through D-1 and Q-1. The diode D-2 will be cut off since the collector voltage of Q-2 will have risen to a value $v_c(t) = v_{nH} \gg V_{BES} + V_{CES}$, noting that V_{BES} is the voltage of the collector of the diode-connected transistor Q-1. However, the current source I and the transistors Q-1 and Q-2 function as a current mirror, where $i_{C2} \cong I$. This current is supplied by the discharge of capacitor C, which discharges at a rate the same as the charging rate until the value v_{nL} is again reached. The result is a periodic waveform at a period $T = T_1 + T_2 \cong 2T_1$, as indicated, triangular at the capacitor and rectangular at the ouptut of A.

An example of a VCO using the voltage-variable capacitor of Fig. 9.25 is shown in Fig. 9.29. The magnified input capacitance, C_i resulting from the inverting gain of A_1

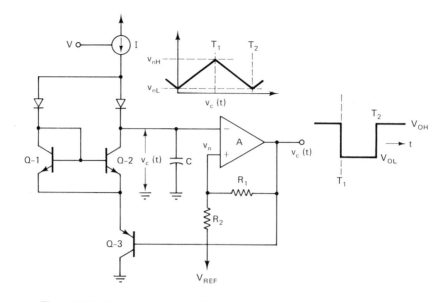

Figure 9.28 Voltage-controlled oscillator using switched current mirror and Schmitt trigger.

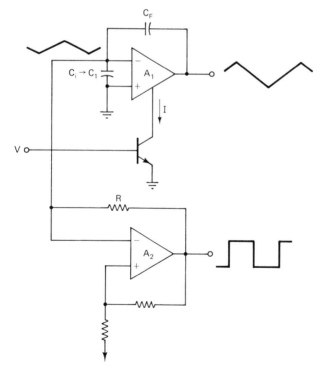

Figure 9.29 VCO using variable capacitance and Schmitt trigger.

is alternately charged to the $\pm$ output voltages V_{OH} and V_{OL} of the Schmitt trigger A_2. In this circuit, the output triangular wave is not precisely linear because the charging rule of C is exponential, charging toward V_{OH} and V_{OL}, respectively.

VCOs as Simple Function Generators

Most of the voltage-controlled oscillators that have been described generate both rectangular and triangular waveforms. From these, other waveforms can be derived by wave-shaping circuits as indicated in Fig. 9.30. For example, if the wave-shaping circuit consists of a sharply turned bandpass RLC or RCC filter tuned to the fundamental frequency, the output is the sine wave as indicated. However, for variable-frequency inputs, it would be difficult to design variable-frequency-tuned filters to track the input. Hence wave-shaping circuits usually consist of resistance-diode networks, where diodes become conducting at different voltage levels in the network. Very arbitrary waveforms can be generated by this method.

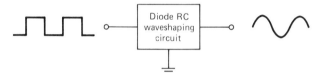

Figure 9.30 Diode RC waveshaping.

9.5 PERIODIC VOLTAGE WAVEFORM GENERATORS

The simple function generators consisting of symmetrical VCOs and wave-shaping networks described in the previous section generate symmetrical waveforms. A more general class of waveform generators is those in which a required time function (or waveform) is generated at controllable time intervals as suggested in Fig. 9.31, where a keyed clamp is often used to alternately activate or deactivate the waveform generator, with T_1 being the active time and T_2 being the clamped or deactivated interval.

The linearly rising waveform or *ramp function* has so many applications that its precise generation deserves special attention. There are many ways in which such a function can be generated, which are variations of capacitor-charging circuits in which the charging current is contrived to be linear. In the simple circuit of Fig. 9.32(a) when the clamp is opened,

$$v_C(t) = V[1 - e^{-(t/)RC}] \qquad (9.16)$$

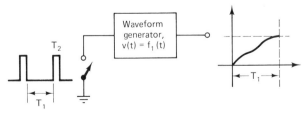

Figure 9.31 Switched waveform generator block.

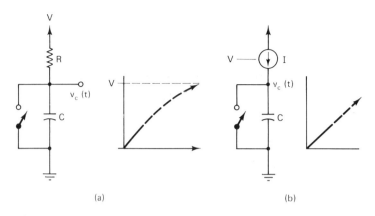

Figure 9.32 Switched capacitor charging. (a) Exponential charging rate. (b) Constant-current charging.

which, for $v_c(t) \ll V$, yields

$$v_C(t) \cong V\,\frac{t}{RC}$$

If R is replaced by a voltage-controlled current generator, as shown in Fig. 9.32(b), the charging is linear over the total range permitted by the particular circuit used.

A specific implementation of the constant-current source based on a circuit of some historical interest is suggested in Fig. 9.33(a). In its original form, the amplifier consisted of a single-device voltage follower with a gain $A \xrightarrow{\leq} 1$, *with the capacitance C_F being so large that no charge is lost during the waveform generating period.*

Initially, the capacitor voltage v_C is held at zero through the low resistance of the clamp. With $A = 1$, v_2 is also zero, and the feedback capacitor is charged in the direction indicated to a voltage $V = V_1 - V_D$, where V_D is the forward voltage of D-1. When the

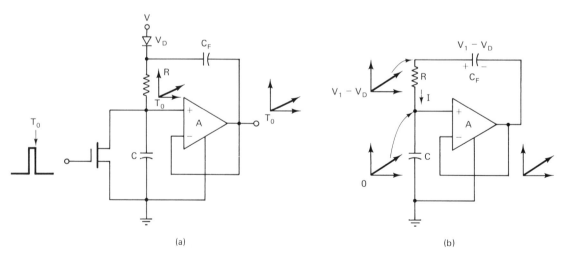

Figure 9.33 Bootstrap sawtooth generator. (a) Circuit. (b) Current paths.

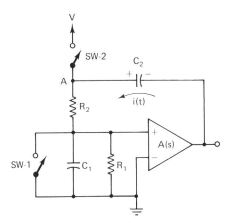

Figure 9.34 Circuit for analysis of bootstrap pulsed waveform generator.

clamping voltage goes negative, Q-1 becomes nonconducting and v_C beings to rise at an initial rate.

$$v_C(t) = \frac{I}{C} dt \qquad (9.17)$$

where $I = (V_1 - V_D)/R$. This rise appears at the output of A, and the same change occurs at the junction of C and R, which immediately causes D-1 to become nonconducting, with the equivalent circuit being that of Fig. 9.33(b). With $A \rightarrow 1$ and C_F sufficiently large, the initial current I is maintained and the output is linear, as indicated until the clamp is again closed at the end of the period.

The original version of this curcuit was referred to as the "bootstrap sawtooth generator," which may be considered as a prototype concept from which a more general class of waveform generators can be derived. Starting with the circuit of Fig. 9.34 (assuming the initial conditions shown), where the amplifier may be frequency dependent, the input resistance is finite, C_2 may not be ideally large, and D-1 is replaced by a keyed clamp that opens and closes in synchronism with the main clamp, a general equation for the ouput starting at the time of the opening of the clamps may be written as

$$V_2(s) = \frac{A(s)V}{R_2C_1\left[s^2 + \left\{ \dfrac{1}{R_1C_1} + \dfrac{1}{R_2C_2} + \dfrac{1}{R_2C_1}[1 - A(s)] \right\}s + \dfrac{1}{R_1C_1R_2C_2} \right]} \qquad (9.18)$$

Several possible solutions will be considered as follows:

(1) $A(s) = A$, nonfrequency dependent and noninverting, and $R_1 \rightarrow \infty$.

$$V_2(s) = \frac{AV}{R_2C_1s[s + \{(1/R_2C_2) + (1/R_2C_1)(1 - A)\}]} \qquad (9.19)$$

and

$$v_2(t) = \frac{AVC_2}{C_1 + C_2(1 - A)}\left\{ 1 - \exp\left[-\frac{C_1 + C_2(1 - A)}{R_2C_2C_1}t \right] \right\} \qquad (9.20)$$

Then, if A represents a unity-gain voltage follower,

$$v_2(t) = \frac{C_2}{C_1} V [1 - e^{-(t/R_2C_2)}] \qquad (9.21)$$

This equation is a rising exponential, as shown in Fig. 9.35(a), with an effective supply voltage of (C_2/C_1) V and a charging time constant R_2C_2.

The actual voltage rise is limited by power-supply considerations to a value $v(t) < V$, is very linear up to its maximum value for $C_2 \gg C_1$, and approaches a true ramp function for $C \to \infty$.

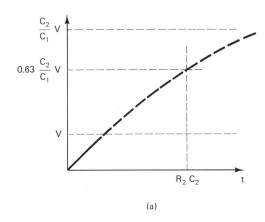

(a)

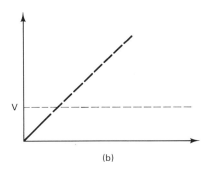

(b)

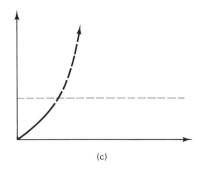

(c)

Figure 9.35 Waveform amplitude limitations and characteristics.

As a practical matter, for integrated-circuit design it is undesirable to require a large value of C_2. Therefore, as a slightly different solution of Eq. 9.20,

$$v_2(t) = \frac{1 + (C_1/C_2)}{R_2 C_1} V(t) \tag{9.22}$$

for

$$A = 1 + \frac{C_1}{C_2}$$

which is a perfect ramp function, as shown in Fig. 9.35(b).

For $A > [1 + (C_1/C_2)]$, Eq. 9.17 shows a pole in the right half-plane and Eq. 9.20 is that of a positive exponential rise, as shown in Fig. 9.35(c). Basically, this represents an unstable condition, and the internal frequency characteristics of the amplifier would have to be considered to show to what extent the response can be controlled for large gain without resulting in uncontrolled instability.

(2) $A(s) = A$ but $R_1 < \infty$, and in Eq. 9.18, eliminating the s coefficient by setting

$$\frac{1}{R_1 C_1} + \frac{1}{R_2 C_2} + \frac{1}{R_2 C_1}(1 - A) = 0$$

the solution is

$$v_2(t) = \frac{AV}{\omega} \sin \omega t \tag{9.23}$$

where

$$\omega = \frac{1}{\sqrt{R_1 C_1 R_2 C_2}}$$

and

$$A = 1 + \frac{R_2}{R_1} + \frac{C_1}{C_2}$$

This represents a pulsed sinusoidal oscillator, which is simply a pulsed version of the basic RC oscillator described in Chapter 6.

(3) $A(s)$ frequency dependent: If the frequency response of the amplifier has to be taken into account, the preceding solution must be modified. For example, in Eq. 9.18, if the amplifier has a single-pole response of the form $A(s) = A\omega_2/(s + \omega_2)$, the solution is that of a third power in s, and the several cases discussed in parts (1) and (2) are modified.

In any case, the basic bootstrap sawtooth generator can be made to generate a variety of pulsed waveforms, including an exponential whose asymptotic value can be controlled, a linear sawtooth, a positive exponential, a pulsed sinusoidal oscillator, and many other waveforms if the frequency response of the amplifier is controlled.

Simplified "Bootstrap" Sawtooth Generator

As a basic concept, the large feedback capacitance C_F in the prototype "bootstrap" sawtooth generator of Fig. 9.33 ideally provides a dc offset voltage and functions ideally as a constant-current source for $C_F \to \infty$. Recognizing this function, it can be reasoned that C_F might be eliminated by using an amplifier that itself has a built-in offset voltage between output and input, together with an initial charge on the sawtooth generating capacitance, as indicated in Fig. 9.36.

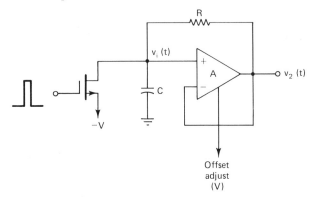

Figure 9.36 Simplified bootstrap sawtooth generator.

The enhancement-mode FET keyed clamp holds $v_c(t) = -V$, while the clamping pulses are positive. At the same time, if the amplifier has a built-in offset such that

$$v_2(t) = Av_i(t) + V \tag{9.24}$$

and $A = 1$, when the clamp is opened,

$$i(t)\frac{V}{R} = I$$

and the voltage across C will be

$$v_c(t) = -V\left(1 - \frac{1}{RC}t\right) \tag{9.25}$$

and

$$v_0(t) = \frac{1}{RC}(t) \tag{9.26}$$

Thus, with this simple circuit, a linear sawtooth can be generated with a unity-gain voltage follower.

Feedback Integrator Sawtooth Generator

An integrator using a very high gain operational amplifier, as discussed in Chapter 7, can be used as a sawtooth generator, as indicated in Fig. 9.37. In Fig. 9.37(a), $v_i(t)$ is clamped to ground through the very low drain resistance of the FET until the beginning

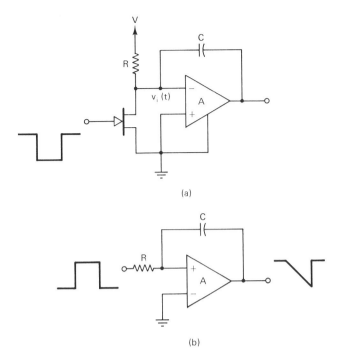

(a)

(b)

Figure 9.37 Feedback integrator ramp function generator.

of the negative pulse at $t = 0$. Opening the clamp is equivalent to applying a step of amplitude V to the circuit of Fig. 9.37(b). As $A \to \infty$, the integrated step is a negative-going sawtooth at the output, as indicated, which ends at the end of the period T when the clamp is again closed.

It is more difficult to generate ideal ramp functions at a high frequency with the integrator type than it is with the "bootstrap" type because of the large gain requirements, which limit the bandwidth of the integrating amplifier as compared with the unity-gain bandwidth of the same amplifier structure.

9.6 SAWTOOTH CURRENT GENERATORS

There are occasions when an electronic system requires the generation of a specific waveform of current through an inductive load, an important example being the linear deflection of a cathode-ray-tube beam, which requires an approximate sawtooth waveform of current applied to a deflecting coil or *yoke*, which includes both inductance and inherent resistance. This is approximated by the lumped circuit of Fig. 9.38(a) with the current–voltage relationship

$$v(t) = Ri(t) + L\frac{di}{dt} \tag{9.27}$$

If $i(t)$ is specified as a ramp function, $i(t) = Kt$, the voltage across the input terminals is

$$v(t) = R_s i(t) + LK \tag{9.28}$$

consisting of a step and sawtooth as shown.

For high-frequency deflection, the distributed capacitance must be considered as well as the resistance of the input current source, as approximated in lumped form in Fig. 9.38(b).

For linear CRT deflection, which is proportional to the electromagnetic field produced by the current, the required waveform of current is only that which flows through the inductive branch of the equivalent circuit of Fig. 9.38(b).

If it is specified that

$$i_L(t) = Kt \tag{9.29}$$

the input current is

$$i(t) = \underbrace{K[1 + (R_s/R_p)]t}_{\text{Ramp}} + \underbrace{K[R_O C_O + (L/R_p)]}_{\text{Step}} + \underbrace{LC_O \delta(t)}_{\text{Impulse}} \tag{9.30}$$

where

$$\delta(t) \equiv \text{Lim} \frac{e^{-(t/\tau)}}{\tau}, \qquad \text{as } \tau \to 0$$

which is a pulse defined as an impulse function that mathematically simultaneously approaches zero duration and infinite amplitude. Such an impulse term cannot be generated, but it can be approximated by a "spike" of very short duration. However, it is instructional to consider the absence of both the step and impulse with a sawtooth current applied to the complete circuit.

For

$$i(t) = K[1 + (R_s/R_p)] t \tag{9.31}$$

the transform of current in the inductive branch can be expressed as

$$I_L(s) = K[1 + (1/Q_s Q_p)]\omega_o^2 \frac{1}{s^2[s^2 + (\omega_o/Q_o) s + \omega_o^2]} \tag{9.32}$$

where

$$\omega_o^2 = \frac{1}{LC}, \qquad \frac{1}{Q_o} = \frac{1}{Q_s} + \frac{1}{Q_p}$$

with

$$Q_p = \frac{R_p}{\omega_o L}, \qquad Q_s = \frac{\omega_o L}{R_s}, \qquad \frac{1}{Q_p Q_s} <<< 1$$

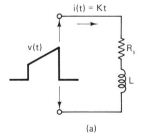

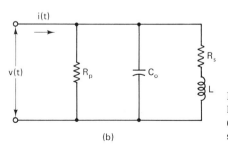

Figure 9.38 Sawtooth current in inductive load. (a) Neglecting distributed capacitance. (b) Distributed capacitance approximated by single capacitor.

The pair of poles resulting from the quadratic are

$$s = -\frac{\omega_o}{2Q_o} \pm \frac{\omega_o}{2Q_o}\sqrt{1 - 4Q_o^2}$$

If the poles are separated, but in the left half-plane, the solution has a decaying oscillatory component and is given by

$$i_L(t) = K'\left[t - \frac{1}{Q_o\omega_o} + \frac{2Q_o/\omega_o}{\sqrt{4Q_o^2 - 1}} e^{-(\omega_o/2Q_o)t} \right.$$
$$\left. \sin\frac{\omega_o}{2Q_o}\left(\sqrt{4Q_o^2 - 1}\, t - 2\tan^{-1}\frac{4Q_o^2 - 1}{-1} \right) \right] \tag{9.33}$$

where $K' = K[1 + (1/Q_sQ_p)]$. The form of this current is sketched in Fig. 9.39(a).

After the slight time delay and the damped oscillation at the beginning, the slope of the curent sawtooth reaches the desired value. By restoring the step, but not the impulse term, the time delay, but not the initial oscillation, is eliminated, as indicated in Fig. 9.39(b).

As critical damping is approached (i.e., $4Q_o^2 \rightarrow 1$), the oscillations are eliminated, and, after an initial delay, the sawtooth approaches its desired value asymptotically, as shown in Fig. 9.39(c). For overdamping (poles on the negative real axis), there is an initial time delay as indicated in Fig. 9.39(d).

The nature of the solution can be controlled by varying R_p. The starting delay time and the initial oscillation can be reduced by adding the "spike" at the input to simulate the impulse function to establish an initial charge in C_O.

There are many applications, such as radar displays that require variations of sweep speeds, where it is not practical to include the step function and/or a simulated impulse but where initial time delay and initial oscillation must be minimized. The time delay T_d is minimized by increasing ω_o, which requires decreasing the inductance and minimizing the distribution capacitance as much as possible. The initial oscillations are damped faster by decreasing Q_o, which is accomplished by decreasing the shunt resistance R_p. For magnetic deflection, which is proportional to the magnetic flux density produced by the coil, which in turn is proportional to Li^2, lowering the inductance requires an increased current for a given deflection and increases the power required (i.e., it lowers the efficiency of the system).

Another problem is the recovery or retrace interval. If the source of current is suddenly removed, as indicated in Fig. 9.40(a), current in the inductance decays in different fashions depending on whether the circuit is oscillatory as shown in Fig. 9.40(b), critically damped as shown in Fig. 9.40(c), or underdamped as shown in Fig. 9.40(d). Critical damping provides the fastest complete recovery time without oscillations extending into the next sweep period. The damping can be controlled by R_p.

There are applications for CRT sweeps where the sweep starts with a negative value of current (deflection to one side of the CRT), progresses through zero value (center of the display), and proceeds to positive values (deflection to the other side of the CRT), and in which sweep speeds are never varied (e.g., the horizontal and vertical deflections in the television system). For such a system, the recovery period can be shortened substantially. In the circuit of Fig. 9.41(a), if the current is removed at the end of the

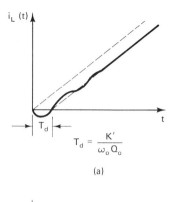

$$T_d = \frac{K'}{\omega_o Q_o}$$

(a)

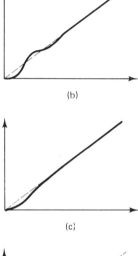

(b)

(c)

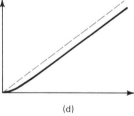

(d)

Figure 9.39 Actual waveforms in inductive branch. (a) With applied sawtooth of current. (b) Applied step and sawtooth (underdamped). (c) Critically damped.

desired sweep period by isolating the shunt resistance R_p, which controlled the initial Q_o to meet active sweep requirements, the decay of the current in the R_s, L, C_o, which is high Q, is oscillatory at the resonant frequency ω_O, as indicated in Fig. 9.41(b). The recovery time is $T_r \cong 1/\omega_o$. If both switches are now closed at the end of T_r, the active sweep progresses as indicated in Fig. 9.41(c). However, the efficiency can be further increased by switching in R_p first, which changes the recovery from oscillatory to critically damped, and gradually supplying an external nonlinear current at a later time to make the sweep linear, as indicated in Fig. 9.41(d). Most high-efficiency high-frequency horizontal deflection in television systems makes use of this principle in one way or another.

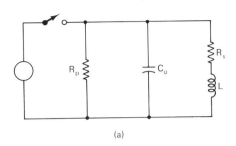

(a)

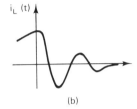

(b)

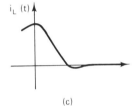

(c)

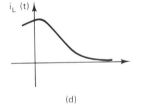

(d)

Figure 9.40 Decay of current in *RLC* circuit with source switched off. (a) Circuit. (b) Oscillatory decay. (c) Critical damping. (d) Overdamped.

9.7 PHASE-LOCKED LOOPS

Some of the basic electronic building blocks that have been described and analyzed in this and preceding chapters, such as the linear amplifier, the four-quadrant multipler, the low-pass filter, and the voltage-controlled oscillator, can be combined in a specific feedback arrangement to create what has come to be known as the *phase-locked loop* (PLL), which forces the frequency of a VCO to be synchronized with that of an input control signal.

The basic function of the PLL can be described with the aid of the block diagram shown in Fig. 9.42. In this circuit, the phase detector generates a signal that is a measure

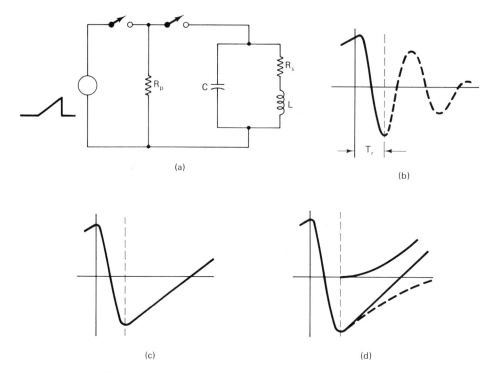

Figure 9.41 Controlled retrace in cathode-ray-tube sweep. (a) Circuit. (b) One half-cycle of oscillatory recovery. (c) Phasing in of new sweep. (d) Delayed switching of new sweep.

of the relative timing (phase) of the input signal $v_i(t)$ and the output of the VCO, $v_o(t)$. Depending on the input signal and the VCO waveforms, the output of the phase detector may contain high-frequency components other than the basic difference or error signal, which is the lowest-frequency component. The extraneous high-frequency components are removed by the low-pass filter, whose output is amplified and used as the control voltage for the VCO, which is designed to be of the correct polarity to change the frequency of the VCO in the right direction to reduce the difference between it and the input signal. In practice, depending on the input and VCO output waveform, the phase detector may be a multiplier or a linear differential amplifier, and the VCO may be a sinusoidal or a nonsinusoidal oscillator.

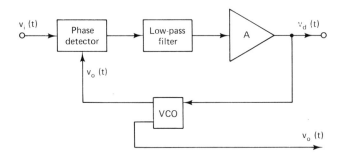

Figure 9.42 Simple phase-locked-loop block diagram.

Simple Phase-Locked-Loop Theory

Regardless of the specific frequency content of the input signal, the waveform generated by the VCO, and the precise configuration of the LPF, a general sinusoidal analysis can be used to construct a simple theory upon which actual designs can be based. In the circuit of Fig. 9.43, it will be initially assumed that the control voltage V_c for the VCO is supplied from an independent source through the switch SW-1, and that the output of the amplifier has a dc level controlled by its offset voltage setting. For a sinusoidal input

$$v_t(t) = V_i \cos \omega_i t \qquad (9.34)$$

and for a VCO output

$$v_o(t) = V_o \cos(\omega_o t - \phi) \qquad (9.35)$$

the output of the multiplier can be written as

$$v_p(t) = \frac{PV_iV_o}{2}[\cos\{(\omega_i - \omega_o)t + \phi\} + \cos\{(\omega_i + \omega_o)t - \phi\}] \qquad (9.36)$$

The low-pass filter removes the upper frequency sideband, so

$$v_D(t) = \frac{PV_iV_o}{2}[\cos\{(\omega_i - \omega_o)\}t + \phi] \qquad (9.37)$$

If the VCO is designed for and its control voltage adjusted such that $\omega_o = \omega_i$, the output of the amplifier will be

$$v_D(t) + V_{DO} = \frac{PV_iV_oA}{2}\cos\phi + V_{DO} \qquad (9.38)$$

If the offset voltage of the amplifier is adjusted to $V_{DO} = V_C$, and SW-1 is switched to the amplifier output when $v_D(t) = 0$ (i.e., $\cos\phi = 0$ or $\phi = 90°$),

$$v_o(t) = V_o \cos(\omega_i t - 90°) \qquad (9.39)$$

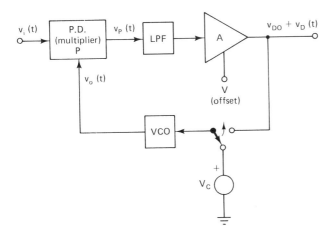

Figure 9.43 Block diagram for phase-locked-loop analysis.

Thus the condition for v_o and v_i to be locked together with zero voltage at the output of the multiplier is that the output of the VCO *lag* the input signal by 90°.

If *SW*-1 is permanently closed with amplifier *A* supplying the control voltage to the VCO, and it differs from the required value for lock conditions, the difference signal from the multiplier will be such as to change $v_D(t)$ to, in turn, change the output frequency of the VCO in the correct direction to decrease the difference and bring the two signals into synchronism.

The *capture range* of the PLL is the maximum range of frequency difference between ω_i and ω_o at the VCO design center value that will permit the VCO to become synchronized with the input signal. The *pull-in time* is the time required for the loop to acquire the signal and achieve final lock conditions. The *hold-in* range is that range of input frequencies for which synchronism is maintained once lock-in conditions have been achieved.

The ability of a PLL to *pull in, lock,* or *capture* a signal, to maintain the output of the VCO to lock if the frequency of the input signal is varying, and to maintain lock in the presence of extraneous components in the signal noise depends in a rather complex way on the overall loop gain and the frequency characteristics of the low-pass filter. If the initial frequency difference, $\omega_i - \omega_o$, exceeds the filter bandwidth, the signal will not be acquired (i.e., the capture range is exceeded). Within the capture range, an increased loop gain decreases the required pull-in time. On the other hand, once lock conditions have been achieved, their maintenance does not depend on bandwidth. Rather, the loop gain as determined primarily by *A* must be sufficient to prevent the relative phase of $v_i(t)$ and $v_o(t)$ from reversing. Otherwise, the correction signal will be such as to increase rather than decrease the frequency difference.

Use of Nonsinusoidal Voltage-Controlled Oscillator

The action of the PLL is relatively independent of its output waveform, which can be demonstrated using a rectangular output waveform as shown in Fig. 9.44, whose fundamental component lags the input by the 90° required for lock. The output of the multiplier is the product signal shown, which is a symmetrical waveform with no dc component having a fundamental frequency twice that of v_i; hence an LPF that cuts off at a substantially lower frequency removes the component, leaving the control voltage to the VCO at its design value for lock maintenance.

If the phase of the input signal begins to lag compared to the position indicated in Fig. 9.44 relative to that of the VCO, as indicated in Fig. 9.45, the product will have a positive dc component as shown, which results in a positive value or error term at the output of the low-pass filter. The amplifier–VCO combination is designed to have the correct polarity to bring the relative phases to the correct lock-in value.

If the phase of the input signal advances relative to that of the VCO, as indicated in Fig. 9.46, an error signal of opposite polarity will be generated, which is again in the appropriate direction to bring the two into lock position.

The net result of the feedback action within the phase-locked loop is to bring the frequencies (fundamental components) of $v_i(t)$ and $v_o(t)$ into synchronization, and when this is done, the 90° phase separation is maintained. The process of bringing the two into

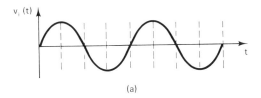

(a)

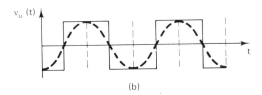

(b)

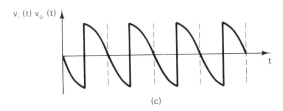

(c)

Figure 9.44 Waveforms for locked PLL with square-wave output of VCO. (a) Input waveform. (b) Current phase relationship with VCO output. (c) Phase detector output for locked condition.

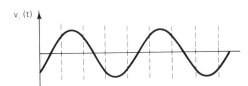

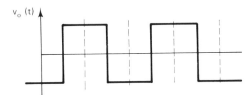

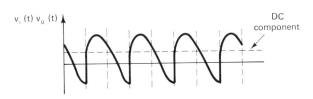

Figure 9.45 PLL waveform for lagging phase of input signal.

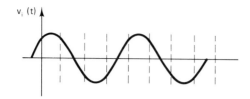

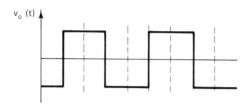

DC
component

Figure 9.46 PLL waveforms for leading phase of input signal.

such a lock condition has been explained as a change in phase that is a shift in instantaneous frequency. Their relationships are

$$\cos(\omega_o t + \phi) = \cos \theta \qquad (9.40)$$

with instantaneous frequency defined as

$$\omega_{\text{inst.}} = \frac{d\theta}{dt} \qquad (9.41)$$

$$\text{or } \omega_{\text{inst.}} = \omega_o + \frac{d\phi}{dt} \qquad (9.42)$$

Thus a change in phase is a shift in instantaneous frequency, and the feedback loop actually shifts the instantaneous frequency of the VCO to force it to track the input frequency with the required 90° phase lag.

It may also be observed that the input signal waveform, like the VCO waveform, also need not be sinusoidal, but need only have an identifiable fundamental component that can be tracked. If such a fundamental component remains fixed in frequency and lock is achieved, the output of A, which is $v_D = V_{DO} + v_D(t)$, is fixed at V_{DO}. However, if $v_i(t)$ is a continuously changing frequency, there will be a "tracking" value of $v_D(t)$ proportional to this change in order to maintain the output of the VCO at its value to maintain lock. Various applications of PLLs make use of one of these two conditions.

Analog Switching Circuits   Chap. 9

9.8 SPECIFIC APPLICATIONS OF PHASE-LOCKED LOOPS

In this section, several applications of PLLs will be considered. These are not intended to be all-inclusive but are selected primarily to illustrate various ways in which the phase-locked loop may be used, sometimes as a modern and preferable alternative to ways in which the same functions historically may have been performed.

Demodulation of Amplitude-Modulated Wave

The conventional amplitude-modulated wave is synthesized in a number of ways, which conceptually can be thought of as first starting with a relatively low frequency continuous signal (audio or video), referred to as the modulation signal, added to a dc level greater than the maximum excursion of the signal, as indicated in Fig. 9.47, represented as

$$v = V_C + \sum_{m=1}^{n} V_m \cos (\omega_m t + \phi_m) \tag{9.43}$$

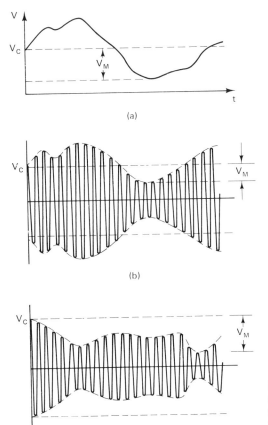

(a)

(b)

(c)

Figure 9.47 Waveforms in amplitude modulation. (a) Signal waveform. (b) Modulated wave referenced to average value of carrier wave. (c) Modulated wave referenced to peak value of carrier.

Sec. 9.8   Specific Applications of Phase-Locked Loops   **415** or

$$v = V_C\left[1 + \frac{V_M}{V_C} \sum \frac{V_m}{V_M} \cos(\omega_m t + \phi_m)\right] \qquad (9.44)$$

where V_M is the maximum amplitude of the signal voltage with the condition imposed that $V_M < V_C$. Then the amplitude-modulated signal is created by varying the amplitude sinusoidally at a very high frequency, with the result given by

$$v = V_C \cos \omega_c t\left[1 + M \sum_{m=1}^{n} \frac{V_m}{V_M} \cos(\omega_m + \phi_m)\right] \qquad (9.45)$$

where $M = V_M/V_C$ is called the *modulation index* or sometimes the *modulation factor.*

The dashed line that follows the form of the modulation signal is called the *modulation envelope,* as indicated in Fig. 9.47(b).

There are applications where it is important to designate the modulation signal as being of a single polarity with respect to some maximum value, as shown by the example of Fig. 9.47(c). In this case, the maximum value of the modulated signal is unmodulated carrier, and the modulation index $M = V_m/V_C$ is referred to as the *modulation* depth.

The modulating information may be recovered in a very simple fashion, as illustrated in Fig. 9.48 for single-tone modulation for simplicity. At carrier peaks, the capacitor charges through the low resistance of the series diode toward peak values. Then, between carrier peaks, C discharges through R until the time of the next carrier peak. If the RC time constant is selected such that the rate of decay approximates the maximum slope of the modulation envelope, the modulating signal is recovered as indicated, except for the slight "sawtooth" irregularities that are of a very high frequency, which can be removed by additional filtering. Such a recovery process is referred to as *peak detection* or *envelope detection.* It is the very simplicity of such a demodulation method that led to the standardization of amplitude modulation in standard AM broadcasting systems.

There are, however, alternative methods of detection or demodulation, that prior

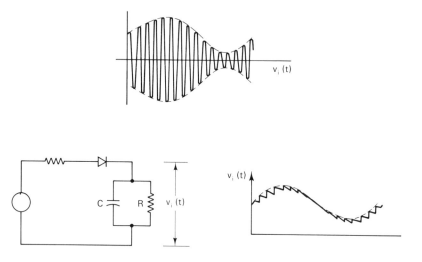

Figure 9.48 Peak detection (demodulation) of AM wave.

to advances in integrated-circuit technology, would have seemed to be too complicated and cumbersome to warrant serious attention. One such system applies a modulated signal and a reference sinusoidal signal "tuned" to the input carrier frequency to a nonlinear two-input amplifier, preferably a true multiplier as shown in Fig. 9.49. For convenience, single-tone modulation is indicated as representing all components in the modulating signal.

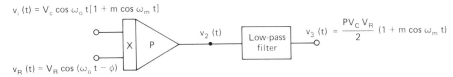

Figure 9.49 Synchronous detection of AM wave.

The low-pass filter removes the higher-frequency components resulting from the multiplication process, and its output is

$$v_3(t) = \frac{PV_CV_R}{4}[m\cos(\omega_m t + \phi) + m\cos(\omega_m t - \phi)] + \frac{PV_CV_R}{2}\cos\phi \quad (9.46)$$

If $\phi = 0$, this equation reduces to

$$v_3(t) = \frac{PV_CV_R}{2}(1 + m\cos\omega_m t) \quad (9.47)$$

which is the form of the original modulation if the dc pedestal is removed.

This process, which is referred to as *synchronous detection*, requires a reference signal precisely equal in frequency and phase to that of the carrier. Even if the frequency is equal but the phase is incorrect, there will be a phase error. It is a simple process to manually "find" a signal in the spectrum by tuning an oscillator that generates the reference signal. However, controlling the relative phase represents a more difficult problem.

A phase-locked loop is suggested as a method of generating the required reference signal, as indicated in Fig. 9.50(a). The output of the PLL can be designed to lock to the frequency of the incoming carrier frequency, but with the 90° phase lag inherent in the VCO output. If such an output is used as the reference voltage in the synchronous detector of Fig. 9.49 with $\phi = 90°$, the modulation information will be lost, and a compensating 90° phase shift must be included in the loop, as indicated in Fig. 9.50(b).

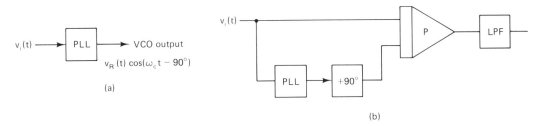

Figure 9.50 Use of phase-locked loop as reference signal for synchronous detection. (a) Locked PLL. (b) Incorporation into synchronous detector.

As a practical matter, obtaining a precise 90° phase shift over a wide frequency range is not simple. However, if the output voltage of the PLL is a rectangular waveform, as is commonly the case, there are a number of possible techniques. One suggested method is shown in Fig. 9.51 in which the PLL is implemented by a VCO with a design frequency twice that of the input frequency, with a counter (frequency divider) that might be a simple bistable circuit, as described in Sec. 9.3, with the waveform at A and B as shown. These outputs are applied to the multiplier, which produces the 90° phase-shifted output as C, as indicated. The multiplier need not be a precision analog multiplier as described in Chapter 8, but can be a relatively crude nonlinear differential amplifier overdriven to the extent that the output levels are constant at the ± saturation levels.

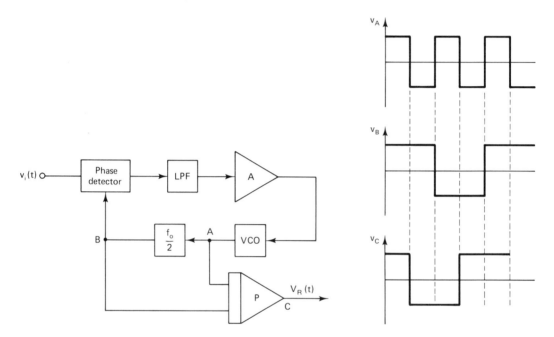

Figure 9.51 Method of obtaining 90° phase shift in PLL for correct synchronous detection.

Frequency Modulation and Demodulation

Frequency modulation is normally defined as the variation of the frequency of a carrier signal, rather than its amplitude, in accordance with modulation information in a prescribed manner, which may be described using single-tone modulation as an example by the equation

$$v_{fm} = V_c \cos \left(\omega_c t + k_f \frac{V_m}{\omega_m} \sin \omega_m t \right) \tag{9.48}$$

In this equation a generalized angle may be defined as

$$\phi = \left(\omega t + k_f \frac{V_m}{\omega_m} \sin \omega_m t \right) \tag{9.49}$$

with an instantaneous frequency defined as

$$\omega_i = \frac{d\phi}{dt} = \omega_c + k_f V_m \cos \omega_m t \tag{9.50}$$

Thus the frequency varies about the center frequency ω_c in accordance with the modulation, with the frequency limits given by

$$f_c \pm \Delta f = f_c \pm k_f \frac{V_M}{2\pi} \tag{9.51}$$

where V_M is the maximum permissible value of the modulation voltage, with Δf being defined as the frequency deviation.

The ratio of the frequency deviation to the modulating frequency may be defined as the *modulation* index, m_f, given by

$$m_f = \frac{k_f V_M}{\omega_m} \tag{9.52}$$

Hence Eq. 9.48 may be written as

$$v_{fm} = V_c \cos(\omega_c t + m_f \sin \omega_m t) \tag{9.53}$$

There are many ways of generating a frequency-modulated signal, and the use of a phase-locked loop for this purpose is illustrated in Fig. 9.52. In this circuit, the center frequency of the output of the VCO is designed to be at the carrier oscillator frequency. The modulation signal is integrated as indicated and added to the input control voltage of the VCO. The bandwidth of the LPF must be sufficient to include the highest frequency in the modulation signal.

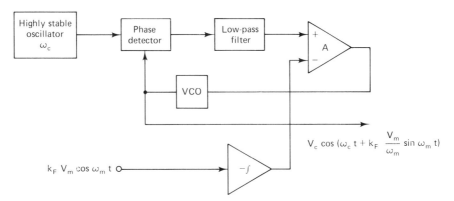

Figure 9.52   Basic method of using PLL to generate a frequency-modulated signal.

In a wide-band FM system, it is necessary to include a "counter" in the loop, as illustrated in Fig. 9.53, in order to maintain the phase within the range of the center value $\pm 180°$ to maintain synchronism. Conceptually, demodulation of an FM signal can be achieved by applying the FM signal to a high-frequency bandpass filter whose slope is linear in the vicinity of the carrier frequency, as indicated in Fig. 9.54. Thus the output of such a filter will track the instantaneous frequency, with the carrier frequency itself

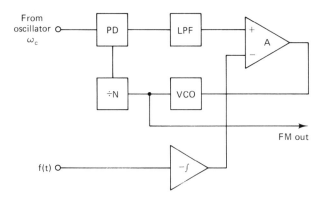

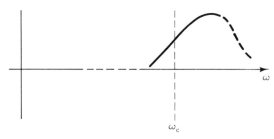

Figure 9.53 Use of a counter to control phase range in PLL output for wide-band FM.

Figure 9.54 Slope demodulation of FM signal.

removed with a subsequent low-pass filter. There are several ways in which the principle is made use of in balanced peak detector circuits, such as the *discriminator* and the *ratio detector*.

Alternatively, a properly designed phase-locked loop offers a simple means of demodulation, as proposed in Fig. 9.55, where the bandwidth of the LPF is sufficiently wide to allow the output of the VCO to track the variation in input frequency, $v_i(f)$. Then the input to the VCO will vary as $Ak_f(V_m/\omega_m) \sin \omega_m t$, and differentiation will then reproduce the original modulation.

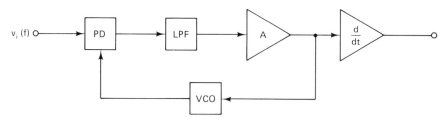

Figure 9.55 Phase-locked loop in FM demodulation.

Frequency Synthesizers Using Phase-Locked Loops

A phase-locked loop can be used to generate a multiplicity of precise frequencies related to a single control or carrier frequency, as illustrated in Fig. 9.56 using a VCO nominally tuned to a multiple N of the frequency, which is to be applied to the phase detector and compared with a frequency that is a fraction M of the precise control or carrier frequency,

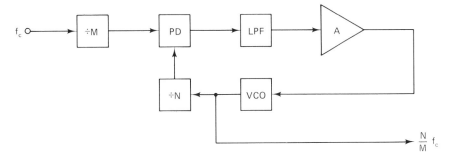

Figure 9.56 Phase-locked loop frequency synthesizer.

f_c. The counts of M and N would normally be integers; however, in the circuit shown, in the output of the VCO, $(N/M)f_c$ can be a fractional value of f_c. Other frequencies $<(N/M)f_c$ can be obtained from various stages in the counter N, as indicated; if simple bistable circuits are cascaded, the count N is 2^n, where n is the number of stages and the available frequencies are related by powers of 2. In the normal VCO, the outputs are rectangular, but sinusoidal frequencies can be obtained using either tuned circuits for each of the desired frequencies or diode resistor wave-shaping circuits.

Some Television System Applications

One notable application of the phase-locked loop, which even predates the name phase-locked loop, is to the control of the horizontal deflection frequency in a television receiver. In a conventional TV system, synchronizing signals are added to the video signal during the blanking (retrace) interval as indicated in Fig. 9.57(a) for one horizontal period T_h. Limiting or clipping circuits are used to separate the video and sync levels, leaving the horizontal sync pulses as shown in Fig. 9.57(b). In principle, the leading edges of the sync pulses as obtained from the RC "differentiator" can be used as "triggers" to imitate the required horizontal sweep circuits, as indicated in Fig. 9.57(c).

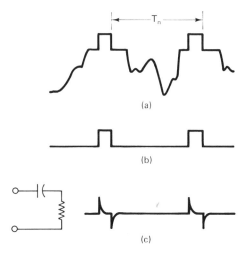

Figure 9.57 Primitive horizontal synchronizing signal recovery method for television receiver. (a) Signal with horizontal sync pulses. (b) Clipping of sync pulses from signal. (c) Simple RC circuit for defining leading edges.

As a matter of fact, when television standards were first established this method was first suggested as a means for control and actually appeared in early television receivers. The result was a disaster, because it was found that "noise" pulses occurring in advance of such leading edges could trigger the horizontal sweep oscillator prematurely, resulting in displacement or "tearing" of the lines in the display.

As an alternative to direct leading edge triggering, it was suggested that the rate and relative phase of the horizontal oscillator might be controlled from a slowly varying dc voltage proportional to the average rate of the incoming horizontal synchronizing pulses. A simple version of such a *horizontal automatic frequency control* circuit is shown in Fig. 9.58. A sawtooth horizontal-sweep waveform and input sync pulses are applied to the phase detector with the desired relative phase as indicated. The phase detector may be represented by a variety of configuration, such as a simple differential or summing amplifier or a multiplier. It may have a balanced output to make it relatively independent of noise or minor variation in sync pulse amplitude. The effective result, however, is an output at *A* whose peak value depends on the relative phasing of the two input signals; for example, if the sawtooth tends to lag the sync pulses, the peak value increases and thus increases the rectified value at the output of the *RC* circuit, which can be thought of as a low-pass filter, as indicated. The variable dc can be amplified if necessary by an amplifier with a variable offset voltage such that its output is at a level to control a VCO that generates the required sawtooth waveform for the horizontal sweep. There are many possible variations of the horizontal AFC, all of which fit the general framework of the phase-locked loop.

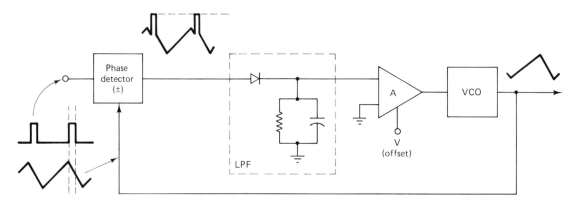

Figure 9.58 Elements of horizontal automatic frequency control in television receiver.

Horizontal AFC is but one of many possible PLL applications in modern television systems. Two obvious examples are FM audio and AM video demodulation, whose principles have been discussed. Another application is the duplication of the transmitted chrominance subcarrier needed to decode the color information. In the NTSC color system, the color subcarrier frequency is an odd multiple of half the line scanning frequency (actually the 455th multiple), which turns out to be approximately 3.58 MHz. Samples of this frequency are transmitted immediately following the horizontal synchronizing

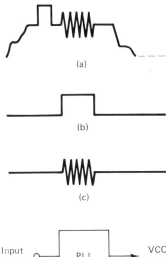

(a)

(b)

(c)

Input from (c) ○ → PLL → VCO output

(d)

Figure 9.59 Regeneration of color subcarrier in television receiver. (a) Waveform with color subcarrier. (b) Gating waveform. (c) Phase-locked loop to regenerate color subcarrier.

pulses, as shown in Fig. 9.59(a), and bear a precise phase relationship to the leading edges of the horizontal sync pulses.

Once horizontal synchronization is achieved, a gating signal, shown in Fig. 9.59(b) may be derived and applied to a comparator to isolate the color bursts, as indicated in Fig. 9.59(c). This signal with a dominant 3.58-MHz component is then applied to a PLL, as shown in Fig. 9.59(d), whose VCO is tuned to the subcarrier frequency. Then the VCO output is locked precisely to the incoming 3.58-MHz signal.

PROBLEMS

9.1 For the circuit shown, assume that $v_1 = 10 \sin \omega t$ and that the diode characteristic can be approximated by $i = I_0 e^{v/V_T}$, where $I_0 = 1 \times 10^{-12}$ A. (This assumes negligible reverse bias current.)

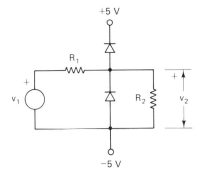

Prob. 9.1

(a) Write three equations relating v_1 and v_2:
   (1) For neither diode conducting.
   (2) For the upper diode conducting.
   (3) For the lower diode conducting.
(b) Plot v_2 over one complete period of the input waveform.

9.2 The enhancement-mode FET shown here has the parameters $V_T = 2.0$ V and $K = 2.0$ mA/V^2.

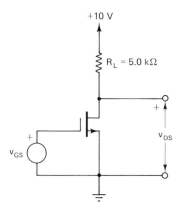

Prob. 9.2

(a) Find an analytical expression for $v_{DS} = f(v_{GS})$ using the basic equation for the FET with $\lambda = 0$.
(b) For $v_{GS} = 10 \sin \omega T$, plot v_{DS} over one complete input period.

9.3 A recurrent waveform is applied to the capacitively coupled circuit as shown:

$$R_1 = 2.0 \text{ k}\Omega \qquad R_2 = 100 \text{ k}\Omega \qquad R_F = 500 \text{ }\Omega$$

$$C = 400 \text{ pF} \qquad T_1 = 4 \text{ }\mu\text{s} \qquad T_2 = 60 \text{ }\mu\text{s}$$

The diode, when forward biased, is approximated by the "piecewise" linear model indicated.

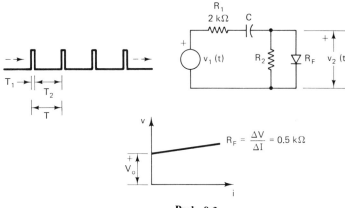

Prob. 9.3

(a) Sketch the overall equivalent circuit for $v_1 > V_0$.
(b) For the steady state (after many periods T of the input waveform), sketch $v_2(t)$ for one period, calculating approximately the voltages at all transition points.
(c) Repeat part (b) for R_2 increased by a factor of 5.

9.4 A waveform is applied to the transistor as shown. The transistor has the very approximate characteristics: $h_{FE} \cong \beta \cong 100$, $V_{CES} \cong 0.1$ V, and $V_{BE} \cong 0.75$ when the transistor is conducting, normally or in saturation.

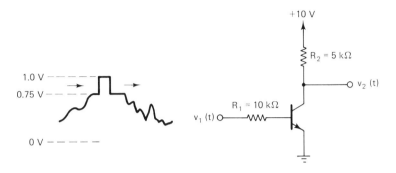

Prob. 9.4

    (a) Sketch the output waveform indicating actual voltage levels.
    (b) Where do limiting and clipping occur according to the definitions in the chapter?

9.5 Waveforms are applied to the circuit as indicated.

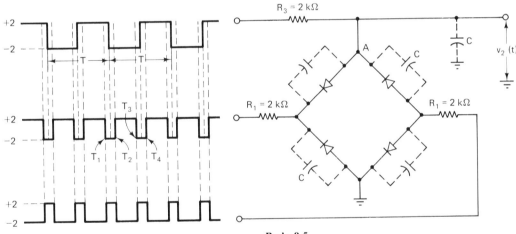

Prob. 9.5

    (a) Ignoring the indicated stray capacitances, which diodes conduct during the interval $T_1 < t < T/2$? during the interval $T/2 < t < T_2$? Sketch the waveform $v_2(t)$. (Assume ideal diodes with forward voltage $V_D = 0.6$ V.) What are the magnitudes of all currents?
    (b) With the signal input open, what is the voltage at point A during the clamping interval?

9.6 Assume that the diodes in the clamping circuit of Prob. 9.5 are identical and can be modeled in the relatively low forward current region by the equation

$$I_D = I_o e^{v_D/mV_T}$$

where $I_O = 0.5$ pA and $m = 1.5$.

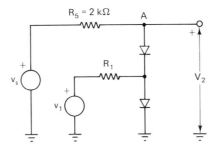

Prob. 9.6

(a) Consider the situation when a positive signal is to be clamped by using the half-clamp shown. For $V_s = 2.0$ V and $V_1 = -2.0$ V, set up the condition for and determine the value of v_D to make $V_2 = 0$ V exactly, and determine the required value for R_1.

(b) Now consider the entire clamping circuit when V_s drops to zero. What is the value for V_2, the voltage across each diode, and the current in the R_1's?

9.7 Consider the diode clamping circuit shown. Assume that the forward voltage across each diode when conducting is $V_D \cong 0.6$ V. Each R is $R = 2.2$ kΩ.

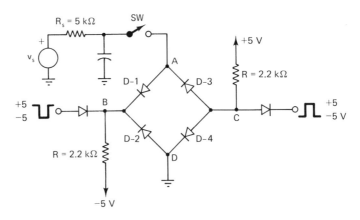

Prob. 9.7

(a) With the signal source disconnected as shown, what is the potential at point A and what is the current through each diode during the time clamping pulses are applied?

(b) With $V_s = +4$ V, $R_s = 5$ kΩ, and SW closed, what are the voltages at points B and C before the clamping pulses are applied?

(c) When clamping pulses are applied, which diodes conduct initially and how much current flows in each conducting diode?

(d) What is the final voltage at point A, and approximately how long does it take to reach it?

9.8 For the circuit shown, the op-amp has a very low output impedance, a very high open-loop gain, and zero offset voltage. The MOSFET is modeled approximately by its basic equations with $V_T = 2.5$ V and $K = 3$.

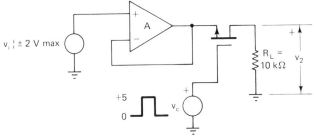

Prob. 9.8

(a) What exactly is the output voltage, v_2, when $v_i = +2$ and $v_c = +5$? when $v_i = +2$ and $v_c = 0$? when $v_i = -2$ and $v_c = +5$? when $v_i = +2$ and $v_c = 0$?

(b) Repeat part (a) for $R_L = 100$ kΩ and for $R_L = 1$ kΩ.

9.9 In the accompanying circuit, Q-1 is a buried layer MOSFET for both enhancement- and depletion-mode operation and $V_T = -1.5$ V; Q-2 is an enhancement-mode MOSFET with $V_T = 1.0$ V and $K = 2$.

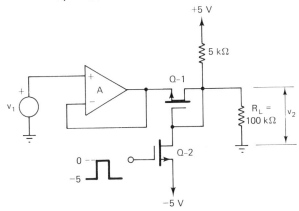

Prob. 9.9

(a) It is desired that the circuit operate as a transmission gate for signal voltage swings of ± 3.0 V. Specify a K factor for Q-1 that will permit Q-1 to clamp properly over the entire range.

(b) For your design, calculate accurately v_2 for $v_1 = +3$ and $v_1 = -3$.

9.10 In the sample-and-hold circuit shown, Q-1 is a buried layer MOSFET that can operate in both enhancement and depletion modes. Assume the basic FET equation for $K = 2$ mA/V^2 and $V_T = -2.0$ V. For Q-2, $K = 2$ and $V_T = +1.0$ V.

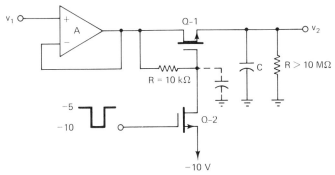

Prob. 9.10

(a) Write the equation for $i_{DS} = f(v_{DS})$ for $v_{GS} = v_{DS}$ and for $r_d = 1/(\partial i_D/\partial v_{DS})$, and determine its value as $v_{DS} \to 0$.

(b) What is approximately the most negative value for v_1 for which the circuit will function properly?

(c) For $C = 100$ pF and an abrupt change in v_1, determine the time required for the voltage across C to reach 99% of the new value, assuming the gate voltage of Q-2 can change instantly.

(d) For an effective capacitance at the drain of Q-2 of 1.5 pF, approximately how long does it take Q-1 to turn on when the negative-going clamping pulse is applied to its gate? What would it be if R were reduced to 1 kΩ? What other effect would this reduction have on circuit operation?

9.11 A current-controlled four-diode clamp is used in the sample-and-hold circuit shown. Assume that Q-2 and Q-3 are exactly complementary transistors with $h_{FE} \cong \beta \cong 50$, and that Q-1 has $\beta = 100$. Also assume that $V_{BE} \cong 0.8$ for all conducting transistors. For exact clamp balance under zero signal condition, $I_{C3} = I_{C2} = 5$ mA.

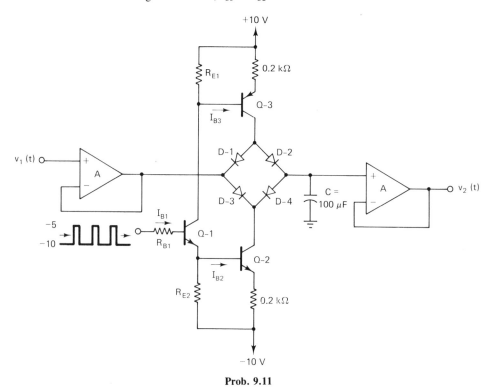

Prob. 9.11

(a) Specify I_{B1} at the clamping pulse peaks, and also I_{B2}, I_{B3}, R_{E1}, and R_{E2}.

(b) For an abrupt change in v_1 of $+2$ V, which diodes conduct immediately and which are immediately made nonconducting?

(c) Until the unbalance is corrected, what is the current that flows through C and in what direction?

(d) Approximately how long does it take the voltage across C to reach a value equal to the new value of $v_1(t)$?

9.12 Two exactly complementary enhancement-mode MOSFETs, Q-2 and Q-3, are used in the current-controlled clamp in the sample-and-hold circuit shown, with Q-1 being identical to Q-2. All FETs are to operate in their saturation region using the basic square-law equations, with $|V_T| = 1$ V and $K = 2$.

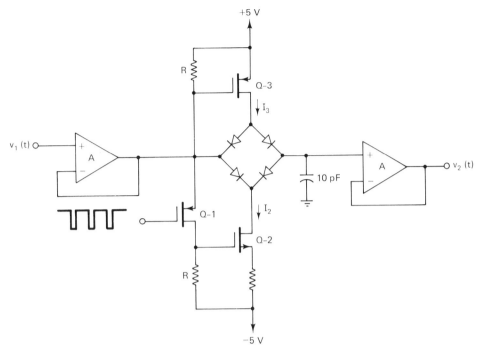

Prob. 9.12

(a) Determine the value of R for $V_{G3} = +2$ V and $V = -2$ V and the value of I_3.

(b) With negative clamping pulses applied as shown, when v_1 changes abruptly, how long does it take the output to equal this change?

(c) For $v_1(t)$ sinusoidal with a frequency of 1 MHz and very narrow clamping pulses, let clamping pulses occur at $0°$, $90°$, $180°$ . . . throughout a period and plot $v_2(t)$ over one input period.

9.13 A Schmitt trigger circuit such as shown in Fig. 9.15 utilizes a relatively low gain linear amplifier.

(a) Write an equation for v_i just as the output limits at V_{OH}, and then as the output limits at V_{OH}.

(b) For $A = 10$, $V_{REF} = 0$, and $R_1 = R_2$, determine v_i when $v_o = V_{OH}$ and when $v_o = V_{OL}$.

(c) Assuming perfect linearity up to the limiting values, what is the minimum possible value of $R_2/(R_1 + R_2)$ for bistability to exist?

9.14 The circuit shown here is to be designed as a Schmitt trigger using the FET of Fig. 9.17(b). It is desired that the output voltage V_o be able to assume the value of either $+10$ V or -10 V with $V_{G1} = 0$, with a design condition established that for either transistor cutoff its $V_{GS} = +2$ V (about halfway between zero and its threshold voltage).

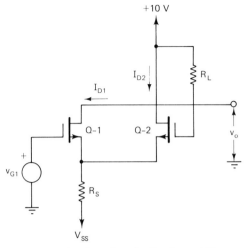

+10 V

I_{D2}

R_L

I_{D1}

Q-1   Q-2

v_o

+

v_{G1}

R_S

V_{SS}

Prob. 9.14

(a) Determine the values for R_L, R_S, and V_{SS} required to meet the output conditions specified, and state the values for I_{D1} and I_{D2} when the respective transistors are conducting.

(b) Explain what happens when v_{G1} starts at a highly negative value and increases linearly as a ramp function. At approximately what level does a transition occur? Repeat for v_{G1} starting at a positive value and decreasing linearly.

9.15 Consider the simple bistable circuit shown in Fig. 9.20 with $V_{CC} = +5.0$ V and $R_L = 4.2$ kΩ, with voltages at T_1 and S derived from outputs of similar circuits as indicated. Assume for each transistor that $V_{CES} \cong 0.2$ V, $v_{BES} \cong 0.8$ V, and $h_{FE} \cong 100$.

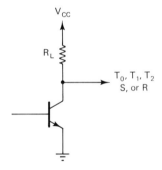

V_{CC}

R_L

T_0, T_1, T_2
S, or R

Prob. 9.15

(a) When the driving transistor is cut off, what is the current I applied to T, S, or R? What will be the output voltage of the transistor that is driven by such a source?

(b) When $V(T_1)$, $V(T_2)$, and $V(R)$ are all at $V = V_{CES}$ 0.2 V and the base of Q-3 is driven with $I_{B3} = 1.0$ mA, what are the voltages at collectors Q-1 and Q-2?

(c) Now suppose the voltage at (S) is reduced to V_{CES}. What will the collector voltages be?

(d) Now suppose voltages are applied simultaneously to (S) and (R) such that $I_{B3} = I_{B4} = 1.0$ mA. What are the voltages at the collectors? If the voltages at (S) and (R) are suddenly reduced to 0.2 V, what will happen at the collectors? Can you predict which will remain in conduction? Why?

9.16 Consider the dual-emitter common-base transistor driven by two sources, each of which switch between $V_H \cong +5$ V and $V_L \cong 0.2$ V, and driving the base of another transistor as shown. Let it be assumed that when either base–emitter junction is forward biased $V_{BE} \approx 0.8$ and that when the base–collector junction is forward biased $V_{BE} \cong 0.7$ V.

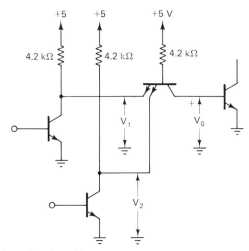

Prob. 9.16

(a) Complete the table:

V_1	V_2	V_0
0.2	0.2	
0.2	5.0	
5.0	5.0	
5.0	5.0	

If you have done this correctly you have identified a two-input AND gate whose output is V_H only when both inputs are high.

(b) For the two inputs as indicated here, sketch the voltage V_O.

9.17 Two cascaded flip-flops of the type shown in Fig. 9.20 connected in the toggle mode, as shown symbolically here, are separated by an AND gate as identified in Prob. 9.16. The Q output is identified as the collector output of Q-2, and the $\overline{Q}$ output is the collector output of Q-1. Before any pulses are applied at T_1, the FFs are initially set by applying pulses directly to S_1 and S_2. This initially sets Q-1 and Q-2 in their high state, at which time the pulses at

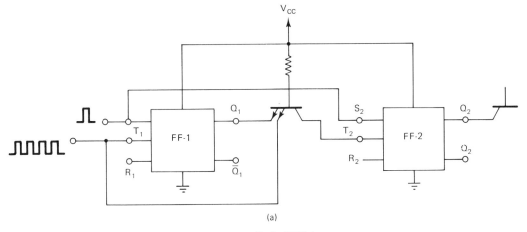

(a)

Prob. 9.17(a)

S_1 and S_2 are removed. After the initial set, a sequence of pulses is applied. These pulses are sufficiently narrow and the transition delays in the FFs are sufficiently long that a substantial portion of each transition occurs at approximately the trailing edge of each clock pulse.

(a) Plot carefully the waveforms of $V(Q_1)$ and $V(T_2)$, and $V(Q_2)$ with respect to the timing of the input $V(T_1)$:

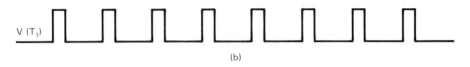

(b)

Prob. 9.17(b)

(b) What is the ratio of the number of pulses at T_3 to the number at T_1? This ratio n is identified as a count of n.

(c) If n such stages are cascaded, what is the count at the end of such n stages if each stage counts by 2?

9.18 Two MOSFETs, each with $K = 2.0$ and $V_T = 4.0$ V, are used in the circuit shown.

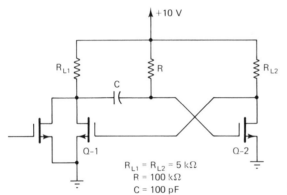

$$R_{L1} = R_{L2} = 5 \text{ k}\Omega$$
$$R = 100 \text{ k}\Omega$$
$$C = 100 \text{ pF}$$

Prob. 9.18

(a) Plot the i_D–v_{DS} characteristic of the transistor in the saturation region for values of v_{DS} up to $+10$ V and i_D up to 10 mA.

(b) With one narrow pulse applied to the input as shown, sketch the waveforms appearing at the drain of Q-1 and the gate and drain of Q-2, including the actual voltage levels.

(c) Determine as accurately as you can the period of the waveform generated.

9.19 MOSFETs like those in Prob. 9.16 are used in the accompanying circuit with the same R_{L1} and R values as in Prob. 9.18.

(a) Without any input signal applied, sketch the output waveforms and determine the total period $2T$ for each complete output waveform.

(b) Approximately what should be the period T between the input pulses of a recurrent pulse train to reduce the period $2T$ in part (a) by a factor of 2?

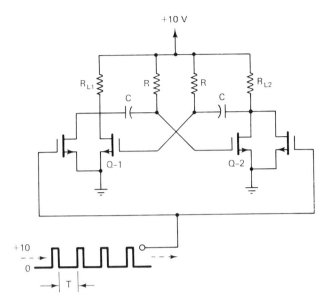

Prob. 9.19

9.20 Two high-gain ($A \to \infty$) op-amps are used in the accompanying circuit. The noninverting op-amp, when connected as shown, has a slew rate dv_1/dt that is directly proportional to the bias current, I_{set}. The limiting values of the output of the inverting op-amp are $V_{2H} = 5.0$ V and $V_{2L} = 0$ V.

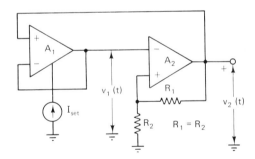

Prob. 9.20

(a) Sketch the waveforms $v_1(t)$ and $v_2(t)$, showing their amplitude limits and relative timing.

(b) For a slew rate of 2.5 V/μs at $I_{set} = 10$ μA, determine the time T at which the waveform $v_1(t)$ repeats. What is T if I_{set} is changed to 2.0 μA?

9.21 For the voltage-controlled oscillator shown in Fig. 9.28, the voltage-controlled current source is the current mirror shown here, making use of identical transistors each with $h_{FE} = 100$ for $V = 0$ V. (Assume all forward-bias conducting junctions have a forward voltage of $v = 0.75$ V.) The output levels of the Schmitt trigger are $V_{OH} = 5.0$ V and $V_{OL} = 0.0$ V with the transitions occurring for $v_c = 1.5$ V and 3.5 V.

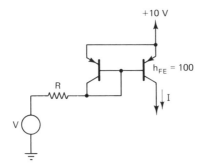

+10 V

$h_{FE} = 100$

R

V

Prob. 9.21

(a) Sketch the voltage waveforms $v_c(t)$ and $v_o(t)$.

(b) For $I = 5$ mA and $C = 100$ pF, determine the fundamental frequency of the output waveform.

(c) Determine R_1, R_2, and V_{REF} for the Schmitt trigger.

(d) Determine R for $I = 5$ mA with $V = 0$.

(e) Determine the range of output frequencies for a range of $V = \pm 5$ V and plot frequency versus V.

9.22 A VCO is to be designed along the principles illustrated in Fig. 9.28, but using MOS devices for the current source. First, let us examine the characteristics of a voltage-controlled current generator using the current mirror in the accompanying circuit with identical MOSFETs.

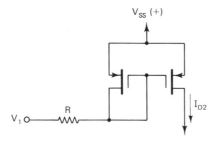

V_{SS} (+)

R

V_1

I_{D2}

Prob. 9.22

(a) Using the usual ideal square-law MOSFET characteristic, prove that the controlled current i_2 can be written as

$$i_{d2} = \frac{(V_s - V_1) + V_T}{R} + \frac{1}{R^2 K} - \sqrt{\frac{2(V_s - V_1 + V_T)}{KR^3} + \frac{1}{K^2 R^4}}$$

(b) Using FETs with $K = 2$ mA/V^2 and $V_T = -1$, $V_S = +10$ V, determine R to establish an operating current of $I_{D2} = 4$ mA.

(c) Plot i_{D2} versus V_1 over a range of $V_1 = \pm 5$ V and compare your result with a linear relationship.

(d) Repeat part (b) for $I_{D2} = 4$ mA for $V_1 = -5$ V and then part (c) for a range of V_1 from -10 to 0 V.

9.23 A pulsed waveform generator of the type shown in Fig. 9.34 is intended as a sinusoidal oscillator using keyed clamps for both switches. Assume that the voltage supplies to the amplifier are $\pm V$ and that oscillations up to these peaks can be generated by appropriate amplifier soft limiting. For the condition $R_1 = R_2$ and $C_1 = C_2$, determine the voltage at point A starting at the time the switches are simultaneously opened (assume zero resistance when switches are closed). Write your result in terms of the initial value and its magnitude and relative phase.

9.24 A pulsed waveform generator like that shown in Fig. 9.34 uses keyed clamps for both switches as shown here. It is suggested that Q-1 be a relatively high threshold enhancement-mode device, while Q-2 be a depletion-mode device in order that a single keying pulse may be used. Determine FET parameters K and V_T such that the waveform generator can be used as a "bootstrap" sawtooth generator for values of A up to $A = 1$ and an RC oscillator with $A \geqslant 3$. Assume that the amplitude of the oscillations at $v_2(t)$ is limited to ± 5 V. (You might want to plot the low-current characteristics of your FETs for clarification in your own mind.)

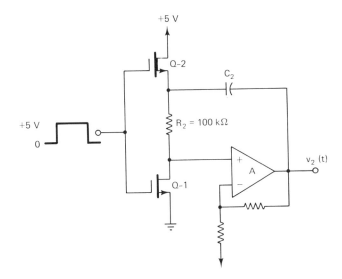

Prob. 9.24

9.25 In the circuit shown, Q-1 is a MOSFET having $V_T \cong 4.0$ V and $K = 2$ (If you have completed Prob. 9.16, you have a plot of this characteristic, which will be convenient to use; alternatively you can just use the parameters given without bothering with a plot.) The amplifier has supply voltages of $+12$ V and has a built-in dc offset voltage of 10 v, making $V_2 = V_1 + 10$ V. The input voltage, $v_s(t)$, is given by $v_s(t) = 0$ for $t < 0$ and $t > T_1$, and $v_s(t) = -10$ V for $0 < t < T_1$.

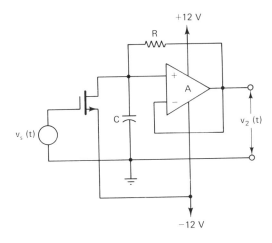

Prob. 9.25

(a) For $T_1 = 100\ \mu s$, $R_F = 100\ k\Omega$, and $C = 1000\ pF$, determine V_1 for $t < 0$ and for $t > T_1$.

(b) Write the equations for $v_1(t)$ and $v_2(t)$ during the interval T_1, and plot $v_s(t)$, $v_1(t)$, and $v_2(t)$.

9.26 The four-quadrant multiplier shown in Fig. 8.19 is suggested as a phase detector in the circuit shown here. The VCO output levels are such that the upper transistors are normally conducting when the respective input (base) voltages are high but are switched completely off when the respective input voltages are low.

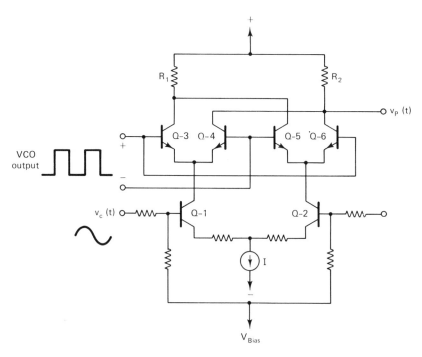

Prob. 9.26

(a) Indicate the conducting period of each transistor when the signal–VCO phase relationships are those of Figs. 9.44, 9.45, and 9.46.

(b) Sketch the output waveform corresponding to part (a).

9.27 A differential amplifier as shown is suggested as a phase detector for the input signal consisting of a series of pulses whose peaks are clamped at 0 V and whose amplitudes are sufficiently high that Q-1 is cut off during the between-pulse interval. The output of the VCO is sinusoidal and is referenced as indicated for the locked condition.

(a) Suppose, over a period of time equal to $12T$, the relative phase of the VCO output changes linearly from $-90°$ to $+90°$ relative to its locked position. Plot the output from Q-1 over this $12T$ interval using a $15°$ change between points.

(b) If the pulse output from Q-2 is now filtered by a low-pass filter, and if the frequency of the VCO is increased with increasing control voltage, is the change in the right direction to pull the VCO output into lock position? If not, how can it be corrected?

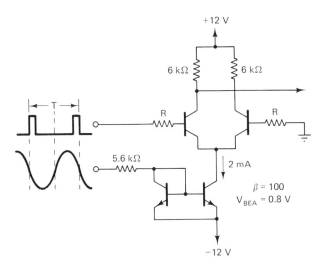

+12 V

6 kΩ   6 kΩ

R   R

T

5.6 kΩ   2 mA

$\beta = 100$
$V_{BEA} = 0.8$ V

−12 V   **Prob. 9.27**

9.28 It was shown in the text that to demodulate a DSB AM signal by synchronous detection a phase-locked loop was necessary to generate a sufficiently stable reference signal to eliminate phase errors in the demodulation process (see Eq. 9.46). Otherwise, a simple variable frequency local oscillator could be used to "find" the desired signal in the spectrum. Now consider a single-sideband AM system whereby only one set of sidebands of the carrier frequency is transmitted (e.g., the upper group), which may be written as

$$v = \frac{V_c V_m}{2} \sum_{m=1}^{N} \cos\left[(\omega_c + \omega_m)t + \phi_m\right]$$

In this signal, the carrier frequency itself is not present. Now let us "find" this signal in a spectrum by tuning a locally generated carrier frequency of the form $v = V_0 \cos(\omega_c t + \phi)$, where ϕ is an arbitrary phase that is not controllable relative to that of the input signal.

(a) Write the equation of the output of a multiplier to which these two signals are applied as actual frequency components.

(b) Discuss the relative importance of the phase error term relative to that of the synchronously detected DSB signal.

REFERENCES[1]

GLASFORD, G. M. *Fundamentals of Television Engineering*, McGraw-Hill, New York, 1955.

MILLMAN, J., AND H. TAUB. *Pulse, Digital, and Switching Waveforms*, McGraw-Hill, New York, 1965.

[1]*Notes on references:* The references are listed in chronological order of publication; all contain material relating to various topics in this chapter. There are no current books that deal comprehensively with major portions of the subject matter. Various periodicals, notably the *IEEE Journal of Solid State Circuits,* contain papers that deal with specific circuit designs and novel implementation of circuits outlined in this chapter, including sample-and-hold circuits for various applications, voltage-controlled oscillators, voltage regulators, and phase-locked loops. These are too numerous and comprehensive to include in a list of references. Special issues of this journal, usually the December issue devoted specifically to analog circuits, contain papers on many related topics.

CONNELLY, J. A. *Analog Integrated Circuits,* Wiley, New York, 1975.

STRAUSS, L. *Wave Generation and Shaping,* 2nd ed., McGraw-Hill, New York, 1976.

TAUB, H., AND D. SCHILLING. *Digital Integrated Electronics,* McGraw-Hill, New York, 1977.

MILLMAN, J. *Microelectronics,* McGraw-Hill, New York, 1979.

ALLEN, P. E., AND E. SANCHEZ-SINENCIO. *Switched Capacity Circuits,* Van Nostrand Reinhold, New York, 1984.

Chapter 10

Analog-to-Digital and Digital-to-Analog Conversion Fundamentals

INTRODUCTION

Most information derived from natural phenomena or man-made processes exists in analog form and is continuous and of varying intensity, at least over specific time intervals. Chapters 4 through 7 dealt with the linear amplification of electrical time-varying signals derived from such processes, while Chapter 8 considered all aspects of nonlinearity in devices and circuits handling such signals, and Chapter 9 considered specific switching operations to permit time sharing of signals, circuits to generate specific functions over discrete time intervals, and feedback control systems defined as phase-locked loops to synchronize the rate of two input signals, along with their various applications.

In this chapter we now consider the representation of analog signals by digital codes or, more specifically, the conversion of analog information to digital form. This involves the quantization of amplitudes of a signal to a number of discrete levels by the sampling of the signal at recurrent time intervals and the generation of a digital code consisting of 1's and 0's assigned to each level, a process referred to as analog-to-digital (A/D) conversion. Conversely, a digital-to-analog (D/A) converter reverses this process and reconstitutes the original analog signal from the coded information. However, more broadly considered, the D/A converter constructs signals at prescribed amplitudes from digital codes regardless of the origin of such codes.

The emphasis in this chapter is on the analog-to-digital conversion process. Similar material in a companion volume, *Digital Electronic Circuits*, has an emphasis on the digital-to-analog conversion processes and treats these processes in more detail (see chapter references). However, it will be seen that the two processes are not completely separable inasmuch as there are a number of A/D conversion techniques whereby D/A converters appear within feedback loops to implement the A/D conversion processes.

10.1 SAMPLING RATE AND BANDWIDTH REQUIREMENTS

As suggested in Sec. 9.2 and shown in Fig. 9.10, a continuous signal can be sampled at a periodic rate, with the result that the signal is represented by a series of discrete amplitude levels. Such a continuous input signal, shown in Fig. 10.1(a), can be represented over a specific time interval T_p by a harmonic series of the form

$$v_i(t) = \overline{V}_i\left[1 + \sum_{n=1}^{N} \frac{V_n}{\overline{V}_i} \cos(n\omega_i t + \phi_n)\right] \tag{10.1}$$

where $\omega_i = 2\pi/T_p$, $\overline{V}_i$ is the average or dc value over the interval and N is the highest harmonic of ω_i that makes any significant contribution to the waveform.

A series of pulses used to sample the signal amplitudes as indicated in Fig. 10.1(b) can likewise be expressed as a harmonic series of the form

$$v_s(t) = \frac{T_1}{T}V_s\left\{1 + 2\sum_{n=1}^{\infty} \frac{\sin[n\pi(T_1/T)]}{n\pi(T_1/T)} \cos n\omega_s t\right\} \tag{10.2}$$

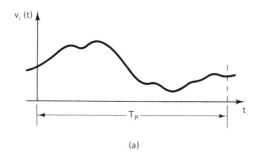

(a)

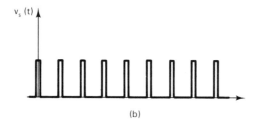

(b)

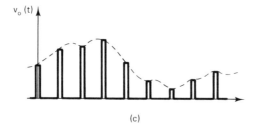

(c)

Figure 10.1 Signal sampling. (a) Signal voltage waveform. (b) Sampling pulses. (c) Sampled signal.

For very narrow sampling pulses, $T_1 << T$,

$$\frac{\sin[n\pi(T_1/T)]}{n\pi\,(T_1/T)} \to 1$$

and

$$v_s(t) \cong \frac{T_1}{T} V_s \left(1 + 2 \sum_{n=1}^{\infty} \cos n\omega_s t\right) \tag{10.3}$$

The sampled waveform shown in Fig. 10.1(c) can be expressed as the product of Eqs. 10.1 and 10.2 as

$$v_o(t) = V_i V_s \frac{T_1}{T} \left\{ \left[1 + 2 \sum_{n=1}^{\infty} \cos n\omega_s t \right] \right.$$

$$\left. \left[1 + \sum_{n=1}^{N} \frac{V_n}{V_i} \cos(n\omega_i t + \phi_n) \right] \right\} \tag{10.4}$$

If the average value of the signal integrated over the period T_p is zero (no d-c component), as given by

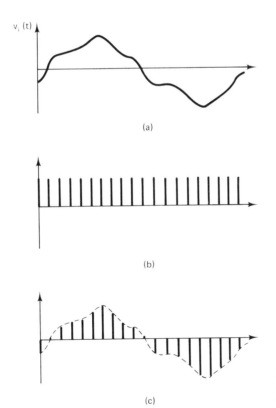

(a)

(b)

(c)

Figure 10.2 Bipolar sampling. (a) Bipolar signal. (b) Sampling pulses. (c) Sampled signal.

$$v_i(t) = \sum_{n=1}^{N} V_n \cos(n\omega_i t + \phi_n) \tag{10.5}$$

and is sampled by the signal as before using

$$v_s(t) = \frac{T_1}{T} V_s \left(1 + 2 \sum_{n=1}^{\infty} \cos n\omega_s t \right)$$

as indicated in Fig. 10.2, the resultant samples are positive when the original signal is positive and negative when the signal is negative.

If the highest frequency component in the signal to be preserved is

$$\cos(N\omega_i t + \phi_N)$$

the sampling rate should be

$$\omega_s \geq 2N\omega_i$$

Otherwise, extraneous components at the lower sidebands of the sampling frequency of the form $\omega_s - n\omega_i$ will be introduced into the channel, which will prevent the highest-frequency components from being preserved properly. The requirement that $\omega_s \geq 2N\omega_i$ is known as the *sampling theorem,* as previously mentioned in Chapter 9.

The sampling theorem represents a statistically limiting condition and does not imply that each and every frequency component $f_s/2$, regardless of its phase position, is properly preserved in amplitude and phase, but only that the *existence* of such components is preserved. For example, in Fig. 10.3(a), samples happen to exist at the positive and negative peaks of such a frequency component, and if the resultant samples are filtered by a perfect filter that cuts off just beyond ω_i, there will be a component of the filter output in phase with the original component.

If, however, the samples happen to be positioned arbitrarily as indicated in Fig. 10.3(b), the amplitude of the signal after filtering will be reduced, and its phase will be shifted. The limiting case is the one in which samples happen to occur at the axis crossings, in which case the existence of the component will not be detected at all.

From a slightly different point of view, the required sampling rate relates to alternate changes in slope. In the case of Fig. 10.3(b), there is one positive and one negative slope in each period, and the samples sense that there are two slopes regardless of their relative position. This corresponds to detecting alternating slopes in the sawtooth waveform shown in Fig. 10.3(c), whose fundamental component is that of the maximum frequency sine wave. The required sampling frequency can thus further be related to sensing the maximum slope in a signal waveform as indicated in Fig. 10.3(d), where the slope shown corresponds to the maximum slope of the highest-frequency component to be preserved.

Transmission Channel Bandwidth Requirements

If the transmission channel after sampling has almost unlimited bandwidth, the sampled signal will be preserved as narrow samples. Otherwise, to preserve the signal information without contamination introduced by the lower sidebands of the sampling signal, the

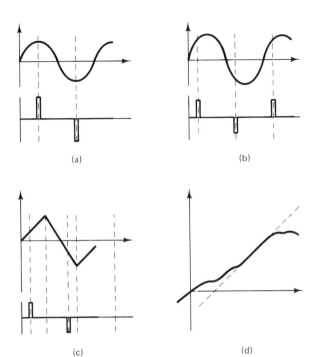

Figure 10.3 Illustration of sampling theorem. (a) Maximum frequency signal component with sampling with no phase error. (b) Arbitrary phase of sampling signal. (c) Slope detection for triangular waveform. (d) Sensing of maximum slope by sampling.

transmission channel must cut off just beyond the frequency $N\omega_i$. Then, the original signal with frequencies up to $N\omega_i$ is retained in the baseband transmission channel.

However, in the sampled signal there are other spectra that contain the original signal information. For example, if the signal is sampled at $2N\omega_i$, where $N\omega_i$ is the highest signal frequency to be preserved, there is a spectrum of bandwidth $2N\omega_i$ centered about the sampling frequency $\omega_s = 2N\omega_i$, as shown in Fig. 10.4, where the signal information exists as upper and lower sidebands symmetrically spaced about the sampling frequency. This exists as a double-sideband amplitude-modulated signal, as discussed in Chapter 8, and carries the original modulating information. Similarly, if the sampling frequency is at a higher harmonic of $N\omega_i$, double-sideband spectra are centered about this sampling frequency. Thus, sampling and filtering are equivalent to other methods of generating amplitude-modulated signals.

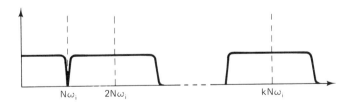

Figure 10.4 Transmission channel bandwidth requirements for sampled signal.

10.2 QUANTIZING AND DIGITIZING ANALOG INFORMATION

An analog signal of continuously varying amplitude may be approximated by a sequence of discrete levels. The number of such levels between zero and the maximum value is determined by the required accuracy in preserving the levels of information as based on the specific use to which it is to be put. Such a process is referred to as *quantization*. In a particular system, the number of levels is usually selected to be some power of 2 because of the ultimate goal of representing each level by a binary code. If, for a particular application, k quantizing levels are selected, where $k = 2^n$, the result is referred to as n-bit quantization. For example, if 16 levels are selected, it is 4-bit quantizing.

We will introduce two examples of quantizing, one in which the signal is at a single polarity and one in which amplitude excursions are positive and negative about an average value, such signals being referred to as *bipolar* signals.

The signal indicated in Fig. 10.5(a) is divided or quantized into eight levels (including zero), so it is referred to as a 3-bit quantizer. The process of quantization may be accomplished by a sample-and-hold circuit with samples occurring at a fixed recurrent

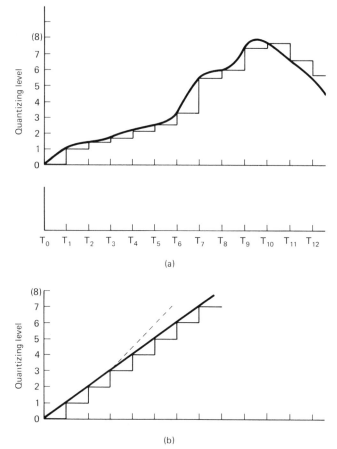

(a)

(b)

Figure 10.5 Principles of quantization.

rate, with the resultant hold levels superimposed on the signal waveform as indicated. If only eight discrete levels are to be assigned, then all samples between T_1 and T_4 will be assigned level 1. The samples taken at T_4 and T_5 will be assigned level 2. For the signal shown, all quantization levels exist except level 4; that is, there is no sample taken at a time that identifies the existence of level 4. In other words, the sampling rate is not sufficient to detect a signal of such rapid variation. This slope relates to the maximum slope of the highest-frequency component in a signal, as discussed in the previous section. For simplicity of illustration, let us consider a ramp function whose slope does compare with the required sampling rate, as indicated in Fig. 10.5(b), where one sample exists for each quantization level. However, if at some point the slope is increased, as indicated by the dashed line, quantization levels will be missed, just as shown in Fig. 10.5(a).

Examples of Digital Codes

Each of the k (8) quantized levels including zero can be assigned a specific n (3) bit code as indicated in the third column of Table 10.1. This is referred to as the natural binary code. The fourth column is the 4-bit natural binary code, while the fifth column is known as the natural binary-coded decimal code (NBCD). The NBCD code requires a 4-bit code to represent 10 levels, whereas in the 4-bit natural binary code, 16 levels can be represented.

If a digitized analog signal is to be transmitted in real time over a single transmission channel, the digitizing process is equivalent to creating an increased number of samples equal to the original sampling rate multiplied by the bit rate (rate at which successive 1's and 0's are generated). Thus, if the signal is sampled at twice the rate corresponding

TABLE 10.1

Decimal Notation	Signal Range	3-Bit Natural Binary Code	4-Bit Natural Binary Code	4-Bit NBCD Code
15	15–16	—	1111	—
14	14–15	—	1110	—
13	13–14	—	1101	—
12	12–13	—	1100	—
11	11–12	—	1011	—
10	10–11	—	1010	—
9	9–10	—	1001	1001
8	8–9	—	1000	1000
7	7–8	111	0111	0111
6	6–7	110	0110	0110
5	5–6	101	0101	0101
4	4–5	100	0100	0100
3	3–4	011	0011	0011
2	2–3	010	0010	0010
1	1–2	001	0001	0001
0	0–1	000	0000	0000

to the highest important frequency component in the signal, with transmission bandwidth half the sampling rate, the required bandwidth is increased by a factor equal to the bit number. The required transmission bandwidth is thus

$$BW = \frac{\text{signal bandwidth} \times 2}{2} \text{ (bit rate)}$$

So to transmit a 3-bit binary-coded signal corresponding to an analog signal with eight quantized levels requires three times the bandwidth required to transmit the sampled signal itself.

Quantizing and Coding of Bipolar Signals

If a time-varying signal has both positive and negative values, as shown in Fig. 10.6, its samples will be either positive or negative as indicated; and if such a signal is digitized, one digit in the selected code can be used to indicate the polarity, while the others represent the magnitude.

Suppose, for example, that the ramp function is identified as having both positive and negative values, as indicated in Fig. 10.7. Here we indicate 3-bit quantizing of both positive and negative values (8 levels). However, the 16 levels indicated require a 4-bit quantizer, with the additional bit used to identify the polarity. This, however, is a matter of terminology. If we are thinking of the peak-to-peak amplitude of the signal in the same sense that we considered the full range of a unipolar signal, we have a 4-bit quantizer, and it is just convenient that one particular bit identifies the polarity.

Many possible codes can be used to represent the 16 peak-to-peak levels of quantizing, two of which are identified in Table 10.2. For the sign + magnitude code, a 3-bit natural binary code identifies the magnitude of the signal relative to zero, while the additional bit identifies the polarity. For these codes there are two zeros separately identified as (0–1) and −(0–1). There is a smooth transition between positive and negative values since only one digit changes at a time in the vicinity of zero.

The second code, referred to as one's complement, retains the same sign bit as the sign + magnitude and the natural binary code for positive values, but for negative values it uses the complement of the corresponding positive values.

Both the sign + magnitude and the one's complement code have two values for

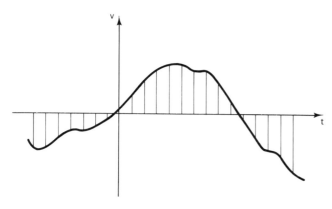

Figure 10.6 Sampling of bipolar signal.

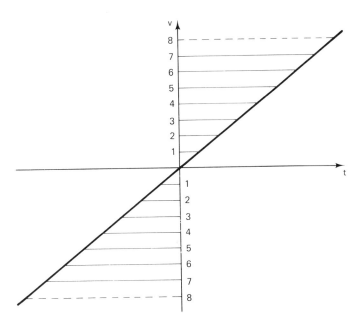

Figure 10.7 Bipolar signal quantizing.

zero, one being the 0 to 1 range and the other the 0 to −1 range, sometimes referred to as ambiguous zeros. Conceptually, this creates no problems because the desired levels of quantization are uniformly spaced and the double designation is simply a matter of terminology. However, it can be shown that it does create some complexities in the digital processing of converted analog signals or groups of such signals.

TABLE 10.2

Signal Level	Sign + Magnitude	One's Complement
7–8	0 111	0111
6–7	0 110	0110
5–6	0 101	0101
4–5	0 100	0100
3–4	0 011	0011
2–3	0 010	0010
1–2	0 001	0001
0–1	0 000	0000
−(0–1)	1 000	1111
−(1–2)	1 001	1110
−(2–3)	1 010	1101
−(3–4)	1 011	1100
−(4–5)	1 100	1011
−(5–6)	1 101	1010
−(6–7)	1 110	1001
−(7–8)	1 111	1000

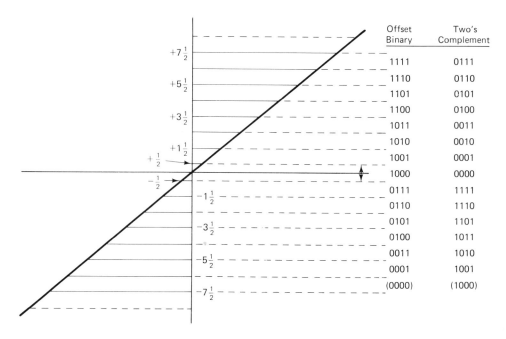

	Offset Binary	Two's Complement
$+7\frac{1}{2}$		
	1111	0111
$+5\frac{1}{2}$	1110	0110
	1101	0101
	1100	0100
$+3\frac{1}{2}$	1011	0011
	1010	0010
$+1\frac{1}{2}$	1001	0001
$+\frac{1}{2}$	1000	0000
$-\frac{1}{2}$	0111	1111
	0110	1110
$-1\frac{1}{2}$	0101	1101
$-3\frac{1}{2}$	0100	1011
	0011	1010
$-5\frac{1}{2}$	0001	1001
$-7\frac{1}{2}$	(0000)	(1000)

Figure 10.8 Examples of code generation.

Two alternative codes that eliminate the two-valued zero are shown in Fig. 10.8, with the signal range divided symmetrically about its zero level as indicated. The offset binary code is simply the natural binary code shown in Table 10.1, offset to start at the most negative value with only one bit range assigned as zero. The two's complement differs from the offset binary only in that the sign bits are interchanged. For either code, if zero is taken as the exact center of the range, as indicated in Fig. 10.8, and complete $\pm$ symmetry is expected to be preserved, one quantization level is lost (15 instead of 16; i.e., one is left over). This can be restored if the zero level is arbitrarily shifted down by $\frac{1}{2}$-bit, which then identifies zero as the 0 to 1 range. This creates no problem as long as track is kept of it in subsequent digital processing. The offset error due to lack of symmetry is simply compensated for in the reverse digital-to-analog processing.

All the codes described, as well as other modifications, pose no unique problems in the process of quantization and code identification, but there may be substantial differences in the subsequent processing and organization of the resultant digital information, particularly where computer interfacing is involved.

10.3 SIMPLE IMPLEMENTATION OF A/D CONVERTERS USING PARALLEL COMPARATORS

In processing analog signals for digital coding, circuits that have been discussed in detail in previous chapters, such as comparators and sample-and-hold circuits, are normally involved. A particularly interesting and useful form of a high-gain comparator is the

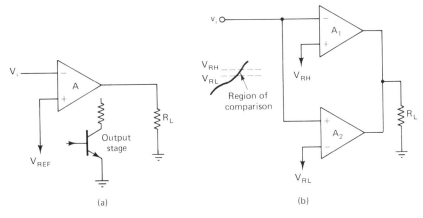

Figure 10.9 Use of comparators as sense amplifiers (window detectors). (a) Required output characteristic. (b) Two-comparator sense amplifier.

circuit referred to variously as a *slicer, sense amplifier,* or *window detector,* which senses the existence of a specific range of signal levels. Such a circuit is described as follows: Shown in Fig. 10.9(a) is a high-gain inverting comparator with an output stage that is on and in saturation at its low level V_{OL} and cut off at its high level V_{OH}. This amplifier is combined with an identical amplifier used as a noninverting amplifier, as shown in Fig. 10.9(b). In this combination, if $v_i < V_{RL} < V_{RH}$, A_1 is off but A_2 is on and $v_o = V_{OL}$. Also, if $v_i > V_{RH} > V_{RL}$, A_2 is off but A_1 is on, and the output is also $v_o = V_{OL}$. But if $V_{RL} < v_i < V_{RH}$, both amplifiers are off and the output is at V_{OH}. Thus the output is at V_{OL} except when the input is between V_{RL} and V_{RH}, in which case the output is at V_{OH}.

The comparators can be designed to be gated as indicated in Fig. 10.10(a), which is represented symbolically as a sense amplifier (SA), as shown in Fig. 10.10(b). The V_{REF} voltage establishes the center of the "window," and the $V_H–V_L$ adjust is a control that sets the size of the range. Thus, if the input signal is within the window during the

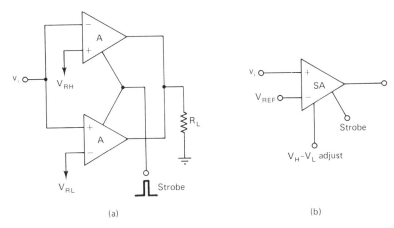

Figure 10.10 Sense amplifier with strobe. (a) Configuration. (b) Symbol.

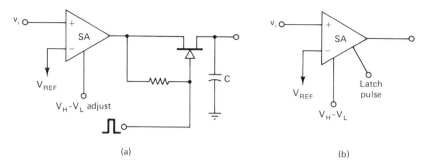

Figure 10.11 Sample-and-hold sense amplifiers. (a) With separate sample-and-hold switch. (b) With internal latch.

time of the strobe pulse, a signal V_{OH} will appear at the output. Otherwise, the output will be at V_{OL}, the levels identifiable in logic terminology as $V_{OH} = V_O(1)$ and $V_{OL} = V_O(0)$. A sense amplifier can be used with a sample-and-hold circuit to hold any level between strobe pulses, as indicated in Fig. 10.11(a), or it may have an internal latch as described in Chapter 9, which achieves the same result.

A/D Converter Utilizing Parallel Sense Amplifiers

The circuit of Fig. 10.12 indicates a means of converting analog information to digital codes utilizing sense amplifiers. The circuit is that of a 4-bit converter requiring 16 sense amplifiers with latches. During a sampling interval T, an output will exist at one and only one SA output, depending on the level of the input signal. Sixteen sequence generators (SGs) triggered with each sampling pulse generate all possible combinations of four 1's and 0's during each sampling interval T, with each sequence being assigned a particular signal level. The digital encoder consists of the SA and SG outputs combined in AND gates, as indicated in Fig. 10.12(b).

Thus a 4-digit code corresponding to one of 16 signal ranges appears at the output during each successive sampling interval. The number sequences can be obtained in a variety of ways. For example, they may be generated by type-*D* flip-flops with feedback, or they may be stored permanently in a read-only memory and accessed with appropriate read circuits.

A/D Converter Using Simple Parallel Comparators

The sense amplifiers of Fig. 10.12 can be replaced by the somewhat simpler high-gain latching comparators with the addition of EXCLUSIVE-OR gates (see Prob. 10.4) as shown in Fig. 10.13, with one such gate for each comparator. Each comparator at a lower level than the signal level will have an output, but an output will exist from the EXCLUSIVE-OR gate of the highest level reached. Thus the comparator plus EX-OR may be thought of as another implementation of the sense amplifier. The digital portion is simply indicated as a logic block to indicate that there are a number of ways that the required digital sequences can be formed.

A/D converters of the forms just described are referred to as *parallel converters*

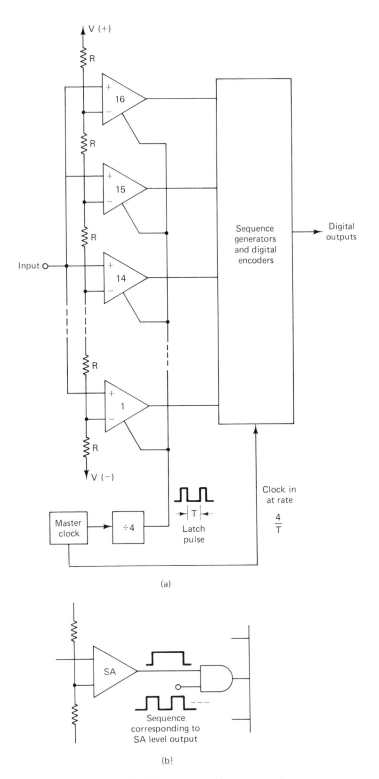

(a)

(b)

Figure 10.12 A/D converter with sense amplifiers.

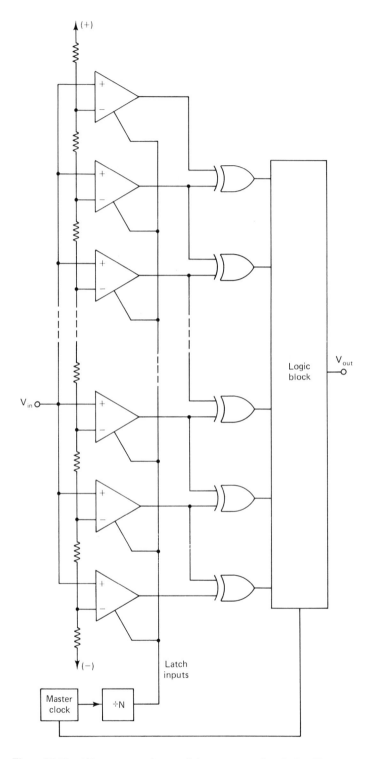

Figure 10.13 A/D converter using parallel converters and exclusive-OR gates.

and require a number of parallel sense amplifiers or comparators equal to the number of quantization levels. Such converters are expensive in terms of circuit hardware but have the greatest speed, because all the existing outputs are capable of generating the required codes simultaneously. Other forms of A/D converters require only one comparator, but more processing time.

10.4 COUNTER-RAMP A/D CONVERTER

One possible implementation of a single comparator converter is shown in Fig. 10.14, which may be described as follows: We first assume that a linear ramp starts to be generated with some particular sample taken at starting time T_0 when the signal has reached a level V_A, with this input level being held for the duration of the between-sample period, and we further assume that the output of the comparator is switched to V_{OH} at this time. As long as the output of A is at V_{OH}, the Q output from the D flip-flop is also high at $V(Q) = V(1)$, while that at $\overline{Q}$ is $V(\overline{Q}) = V(0)$ and the clamp at the amplifier input remains open. The Q output and master clock pulses are applied simultaneously to the AND gate, from which the gated master clock pulses are applied to an n counter. Regardless of how the counter is implemented, there must be a sufficient number of flip-flops in the counter to permit a maximum count of n, where n is the bit number required. This count is to be reached for an input level corresponding to V_A (Max). However, for a particular $V_A < V_A$ (Max) at T_0, the output of the comparator switches to V_{OL} in a time T corresponding to this level, which in turn stops the count at a number $< n$. At the same time, $\overline{Q}$ switches to $V(\overline{Q}) = V(1)$, the clamp closes, and the input ramp returns

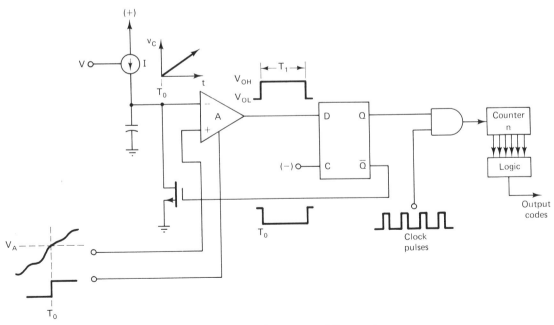

Figure 10.14 Counter ramp A/D converter.

to zero, which switches the output back to V_{OL}, where it remains until the next sample is taken. The output from each counter stage is applied to the logic block, which in turn generates the appropriate code for that particular level. The total time required to code the data is equal to the time of total quantization of k levels. Thus, for a 4-bit converter, $k = 16$, and the time must encompass 16 master clock pulses and require $n = 4$ toggle-type flip-flops to generate the 4-bit code. The logic block notes where the count has stopped and generates the appropriate code for the corresponding level of quantization. Converters of this type are referred to as *single-slope counter-ramp* A/D converters, and their accuracy depends on the precision to which a single ramp function at a specific slope can be generated.

Dual-Slope Counter-Ramp A/D Converter

The circuit shown in Fig. 10.15 is one of many possible forms of counting-type converter that eliminate dependence on the accuracy of generation of the ramp function. In this particular circuit, it will be assumed that at the time T_0 at which a sample V_A of the quantized waveform is held in an appropriate sample-and-hold circuit, a negative-going ramp is generated by the integrator, whose output is given by

$$v_o(t)]_{t > T_0} = -\frac{V_A}{\tau} t \tag{10.6}$$

where $\tau = RC$.

As long as v_0 is negative, the output of the comparator is high at $v_0 = V_{OH}$, and the counter begins to count clock pulses as indicated. If the counter is a ripple or synchronous counter, the time required for the last stage to reset is $2^N T_C$, where T_C is the clock pulse interval, and at this time

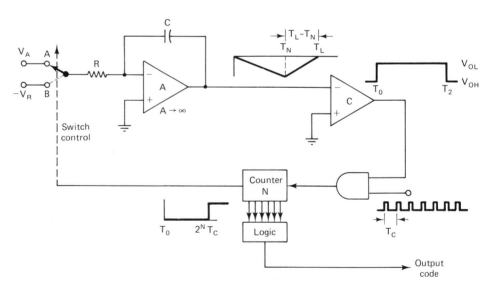

Figure 10.15 Dual-slope counter-ramp A/D converter.

$$v_o(T_N) = -\frac{V_A}{\tau} 2^N T_C \qquad (10.7)$$

At this time, when the last stage of the counter resets, the input resistor R is switched to $-V_R$, which is a fixed reference voltage of magnitude V_R, and at the same time all stages of the counter are reset to their initial values. Then, beginning at this time T_N, the output ramp from the integrator progresses upward as indicated until it again reaches zero at time T_2. This is the interval

$$T_2 - T_N = \frac{V_A}{V_R} 2^N T_C \qquad (10.8)$$

During this interval, the count again progresses, and when the input to the comparator reaches zero, its output reverses to V_{OL} and the count is stopped. Thus the count reached in the interval $T_2 - T_N$ is

$$n = \frac{V_A}{V_R} 2^N \qquad (10.9)$$

which is seen to be independent of the time constant τ. The input and output of the integrator A are reset to zero at the time the next sample is to be taken at a new value of V_A. This particular form of dual-slope A/D converter requires up to twice the conversion time of the single-slope converter because the count progresses for the entire N count *before* the logic processing starts.

10.5 DIGITAL-TO-ANALOG CONVERSION PROCESSES AND TECHNIQUES

When a digital-to-analog converter is used to reverse the process of converting an analog signal to a digital code and reassembling the code into the original quantized levels, the original process and coding system must be known. Sometimes, however, a D/A converter is used to convert a sequence of codes into arbitrary amplitude levels, perhaps for process control or other applications, without reference to any original A/D process, in which case no prior history of conversion is involved.

D/A processes are inherently simpler and less consuming of hardware than A/D converters, consisting primarily of a resistive network, switches, and perhaps a high-gain operational amplifier used as a feedback summing amplifier for improved stability. Such arrangements are referred to as D/A decoders. An example of a resistive network used in a D/A conversion process is shown in Fig. 10.16.

There are $k = 2^N$ possible values of V_O for all possible combinations of switch positions. For example, for $N = 4$, $k = 16$, which are the quantizing levels for a 4-bit converter. If these levels are to represent uniform voltage increments, the resistors must have unequal values (i.e., they must be weighted).

If the weights are chosen according to Table 10.3, the voltage across R_L is

$$V_O = \frac{R_L V_R}{R + (2^N - 1)R_L} (S_{N-1} 2^{N-1} + S_{N-2} 2^{N-2} + \ldots + S_0 2^0) \qquad (10.10)$$

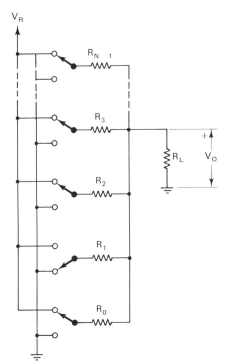

Figure 10.16 Weighted resistive network for D/A converter.

where the S_{N-n} notation means that the output of that particular switch is either grounded (0) or at V_A ($S = 1$), which indicates the availability of 2^{N-1} outputs other than zero.

TABLE 10.3

R	Relative Value
R_0	R
R_1	$R/2$
R_2	$R/2^2$
R_3	$R/2^3$
.	.
.	.
.	.
R_{N-1}	$R/2^{N-1}$

A slight modification of the weighted resistor D/A converter uses the very high gain ($A \to \infty$) op-amp as shown in Fig. 10.17. In this case, using the same weights given in Table 10.3,

$$V_O = -\frac{V_R R_F}{R} (S_{N-1} 2^{N-1)} + S_{N-2} 2^{N-2} + \ldots + S_0) \qquad (10.11)$$

for $R_F \gg R$.

The wide range of resistance values required for a high-bit-number converter creates problems in accuracy and processing for integrated implementation, and there are a number

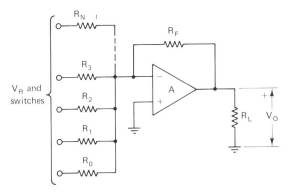

Figure 10.17 Weighted resistor D/A converter with feedback amplifier.

of possible modifications that relieve this problem, but at the expense of using more resistors.

One such modification, of which there are several possible variations, is known as the R–$2R$ ladder network shown in Fig. 10.18(a). With the exception of R_L, which is included to minimize the effect of capacitance loading of the high-resistance series ladder network, only two distinct values of resistance are used. In this particular circuit, a sign bit switch is included, which could have been done for the weighted resistance network as well.

The voltage across R_L is given by

$$V_O = \frac{\pm V_R R_L}{2^N(R + R_L)} (S_{N-1}2^{N-1} + \ldots + S_0 2^0) \qquad (10.12)$$

for 2^N possible equal analog increments.

An inverting feedback amplifier is often used with the R–$2R$ ladder network, as indicated in Fig. 10.18(b). In this case, R_L of Fig. 10.18(a) is the input resistance of the feedback amplifier given by

$$R_L = R_i = \frac{R_F}{1 + |A|} \qquad (10.13)$$

if the input resistance of the amplifier without feedback is very high. Hence, from Eq. 10.11, the output voltage from the amplifier is

$$V_2 = \frac{\pm V_R \dfrac{R_F}{1 + |A|} |A|}{2^N \left[R + \dfrac{R_F}{1 + |A|} \right]} (S_{N-1} 2^{N-1} + S_{N-2} 2^{N-2} + \ldots S_0 2^0) \qquad (10.14)$$

which for very high gain with $R \gg R_F/(1 + |A|)$ is

$$V_2 = \frac{\pm V_R}{2^N} \frac{R_F}{R} (S_{N-1}2^{N-1} + S_{N-2} 2^{N-2} + \ldots + S_0 2^0) \qquad (10.15)$$

There are many variations of R–$2R$ networks and weighted resistance networks and combinations thereof to meet specific design accuracy criteria. In addition, there are arrays of current-driven rather than voltage-driven networks.

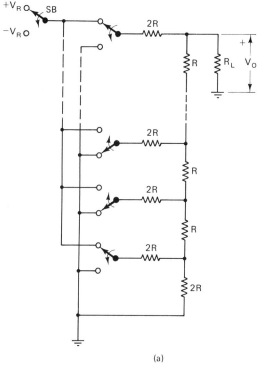

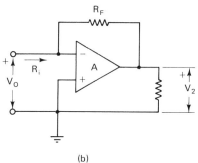

Figure 10.18 *R–2R* ladder network D/A converter. (a) Network with resistive load. (b) Network with op-amp load.

One modification of the *R–2R* converter with summing amplifier is the *inverted ladder* converter shown in Fig. 10.19, where the switches are at the output rather than at the input of the network. In this particular configuration, with A sufficiently high, R_i is very low, and the currents through the various elements of the ladder do not change with switch position changes; hence switching transients are minimized.

Switches for D/A Converters

There are many possible switching arrangements for D/A decoders. For illustration, a 3-bit decoder without sign bit using the natural binary code is shown in Table 10.4 using the weighted resistor or *R–2R* networks. The bit line data may exist either sequentially

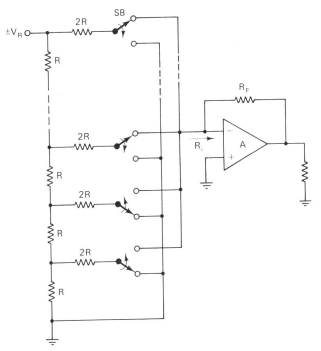

Figure 10.19 Inverted ladder network D/A converter.

or simultaneously. For example, if data are derived from a real-time communication channel, as described in Sec. 10.3, they will exist sequentially. If on the other hand, the digital data are stored in a random-access memory for later access, the output of the bit lines can be accessed simultaneously.

In any case, it is assumed for switching purposes that the bit line data exist simultaneously so that the sums can be switched all at the same time. If they are derived from sequential inputs, they can be delayed in registers or sample-and-hold circuits such that all switches are opened or closed together to produce the required output.

Switching may be implemented in a variety of ways utilizing various forms of clamp circuits. For example, consider the MOS circuit shown in Fig. 10.20. In this circuit, Q-1 is a depletion-mode device and Q-2 is a low-threshold enhancement-mode

TABLE 10.4

Quantization Range	Bit Line			Switch Position		
	1	2	3	S_2	S_1	S_0
7–8	1	1	1	V_R	V_R	V_R
6–7	1	1	0	V_R	V_R	0
5–6	1	0	1	V_R	0	0
4–5	1	0	0	V_R	0	0
3–4	0	1	1	0	V_R	V_R
2–3	0	1	0	0	V_R	0
1–2	0	0	1	0	0	V_R
0–1	0	0	0	0	0	0

$V_0 = K\,(4S_2 + 2S_1 + S_0)$

Sec. 10.5   Digital-to-Analog Conversion Processes and Techniques   **459**

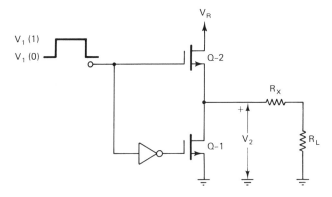

Figure 10.20 MOSFET switch for D/A converter.

device. Therefore, with $V_1 = V_1(0) = 0$, Q-2 is cutoff and Q-1 is on at $V_{GS} = 0$. Then with R_x high, Q-1 is in its very low voltage triode region, with $V_2 = V_{DS} \rightarrow 0$. Then, when $V_1 = V_1(1)$ at sufficiently high voltage level, Q-2 will be turned on in *its* low-voltage triode region with $V_2 = V_R - V_{DS} \cong V_R$.

A switch for an inverted ladder converter is shown in Fig. 10.21. With the switching gate at logic 1, Q-1 saturates, and the particular network element is connected to the amplifier input, and at logic level 0 the amplifier is disconnected from the network. Also, Q-2 is cutoff and Q-3 saturates, which connects the $2R$ resistor to ground through the low saturation resistance of Q-2.

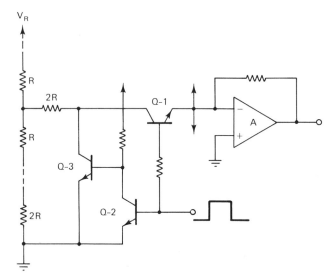

Figure 10.21 Switch for inverted ladder D/A converter.

Switches for Bipolar Conversions

In most digital codes used in bipolar conversions, the first digit is the sign bit, which is used simply to switch the input to the decoder network from $+V_R$ to $-V_R$. A simple CMOS sign-bit switch is shown in Fig. 10.22(a). The $V_1(0)$ and $V_1(1)$ gating signal is centered symmetrically about ground. The transistors are low-threshold enhancement-

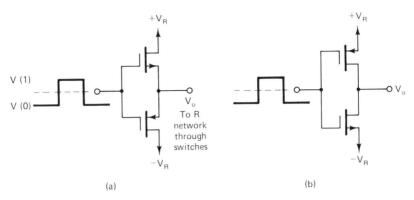

Figure 10.22 Switches for bipolar D/A converters. (a) Low-threshold enhancement-mode MOSFET switch. (b) High-threshold enhancement-mode MOSFET switch.

mode devices. When $V = V(1)$, the output is switched to $V_0 \rightarrow V_R$ if the load resistances are sufficiently high to drive the n-channel transistor into its very low voltage triode region. Similarly, when $V = V(0)$, the output is switched to $V_0 \rightarrow -V_R$ through the p-channel transistor. Alternatively, the complementary transistors shown in Fig. 10.22(b) with high-threshold transistors may be used. In this case, the output polarities are reversed, which must be accounted for in subsequent decoder action.

One of the simplest codes to process using the sign bit as the first digit is the natural binary sign bit + magnitude code shown in Table 10.3, because the negative magnitudes, level by level, are identical to equal positive magnitudes. This suggests two possible sign-bit switching methods. One of these is the direct method, for example, adding the sign bit to the R–$2R$ circuit of Fig. 10.18 using switches as suggested in Fig. 10.22. However, the magnitude switches of Fig. 10.20 or 10.21 must be modified to accommodate bipolar input signals. An alternative method is shown in Fig. 10.23, where the

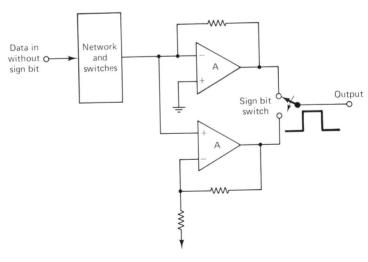

Figure 10.23 Sign-bit switch at D/A converter output.

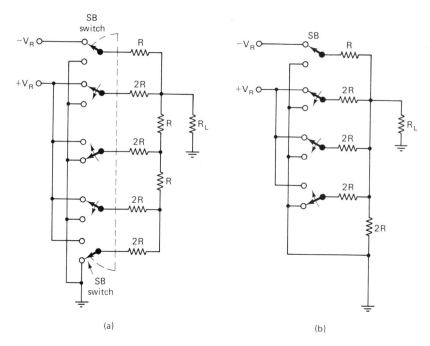

Figure 10.24 Switching for specific digital codes. (a) One's compolement code. (b) Two's complement code.

amplifiers have identical gains, with one inverting and the other noninverting and the sign bit deleted from the input to the R network. The sign-bit switch is used to switch the output polarity.

Negative values for other than the sign bit + magnitude code can be implemented by using the sign-bit switch in a slightly different manner. For example, for the one's complement code shown in Table 10.3, the arrangement for a 3-bit converter + sign bit shown in Fig. 10.24(a) can be used, while the arrangement shown in Fig. 10.24(b) is an implementation of the two's complement code shown in Fig. 10.8.

10.6 ANALOG-TO-DIGITAL CONVERTERS USING DIGITAL-TO-ANALOG CONVERTERS

There are various ways in which D/A converters can be employed in feedback loops to implement A/D converters. One is referred to as the *successive approximation type;* it is used extensively and there are many possible variations. These will not be discussed in detail, but the basic principles are outlined with reference to Fig. 10.25 as follows: At the time of a particular sample, V_A, of the analog signal, a D/A converter begins to generate a succession of analog amplitudes according to the prescribed code, starting with the MSB level and working downward, and apply them to the comparator along with the quantized input signal. If V_A is at or higher than the level corresponding to the MSB, an output is obtained from the comparator. Due to this, a control logic signal is

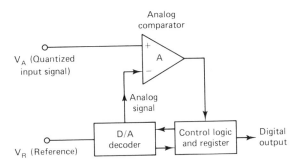

Figure 10.25 Successive-approximation D/A feedback A/D converter.

applied to the D/A decoder, which causes that code to be retained, to be passed through appropriate register circuits, and to appear as the digital output. On the other hand, if the level corresponding to the MSB is not reached by the sample V_A, no output will be obtained from the comparator, and the control logic will prevent the code from reaching the output and at the same time will remove the analog signal from the D/A decoder and replace it with a signal level corresponding to the NH (next highest) MSB. The comparison process is repeated successively until the amplitude of V_A is "found" and the corresponding output code generated. The total conversion time is that required to generate successively the bit groups corresponding to *all* levels of quantization. Shift registers in the logic block are required to align all possible groups of bits within the same time frame.

A slightly different form of feedback A/D converter, known as the *counter-comparator type,* utilizes the D/A converter in a slightly different manner and has some of the elements of the counter-ramp type previously described. There are many possible variations, but the basic principles are described as follows with reference to Fig. 10.26. At the beginning of the sample time at which level V_A is reached, the counters are started and the count continues until the analog signal from the decoder reaches the input level, V_A. In this particular implementation, the analog signal from the decoder starts with the LSB and works upward sequentially. When the decoder output reaches V_A, the count is stopped, and the digital code corresponding to this count is generated. This differs from the counter-ramp type in that the analog reference signal is obtained from the D/A decoder rather than from a linear ramp generator. Even so, such a circuit is sometimes referred to as a counter-ramp comparator since the result is essentially equivalent.

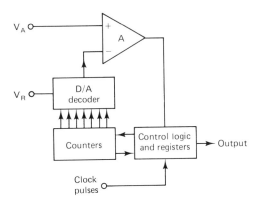

Figure 10.26 Counter-comparator A/D converter using D/A decoder.

10.7 CONCLUDING COMMENTS ON A/D AND D/A CONVERTERS

This chapter has been but a brief introduction to the vast subject of analog-to-digital and digital-to-analog data-conversion processes. Each system involving such processes is unique with respect to the requirements to be met, and the available design strategies to meet such requirements are varied and almost unlimited.

For many applications where analog information of a specific time duration is to be quantized, coded, and processed digitally, the bandwidth requirement is not an issue, because in a closed system quantization levels need not be preserved sufficiently long to generate the corresponding code. The level may simply be detected and its information stored and processed after expiration of the signal duration. However, if data are to be processed continuously in real time, as in a communication system, the bandwidth requirement as a function of the bit rate as specified by Eq. 10.1 must be met. For very high speed processing, some form of parallel converter, as illustrated by Figs. 10.12 and 10.13, is probably necessary, and the transmission channel must have the requisite bandwidth. On the other hand, even for high-speed real-time signal processing, as in a television broadcast plant, when video signals from various sources are combined and routed to different points in the system, although the bit rate requirement must be met, bandwidth per se is not an issue because signal routing is by coaxial cable, which is capable of extended bandwidth. In contrast, after preparation for broadcast, the digital information must be converted back to an analog signal because the extended bandwidth required for digital transmission is not available in over-the-air channels. But, again, when the signal is received and processed for video display in a receiver, real-time digital processing becomes practical after analog-to-digital conversion for all facets of signal-decoding processes, such as horizontal and vertical synchronization, color separation, and sound processing. Digital processing of information in a television receiver offers many opportunities for enhancement of both picture and sound quality not possible using all-analog methods.

In the area of high fidelity sound recording and reproduction, quantizing and coding of speech or music, recording the resulting digital information on a disk, reading the codes by a laser beam, and converting back to analog information for final sound reproduction represents the most important advance since the invention of the phonograph.

As the advantages of digital signal processing as the preferred method for data handling become more apparent for many applications both exotic and mundane, from satellite communication to control of microwave oven functions, the need for A/D and D/A converter designs and innovation likewise increases, and almost everything studied in this book is germane to such processes. The creative design of analog-to-digital and digital-to-analog converters is one of the most challenging tasks for the electronic circuit and system designer.

PROBLEMS

10.1 A multiplier is used as a phase detector in a phase-locked loop as indicated. Suppose the input signal is a periodic waveform expressed as Eq. 10.1, with its dc component removed, and that the VCO output is a symmetrical square wave as expressed by Eq. 10.2, without its dc component, and $T_1 = T/2$.

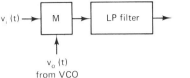

$v_i(t) \rightarrow$ M $\rightarrow$ LP filter $\rightarrow$

$v_o(t)$
from VCO

Prob. 10.1

(a) Rewrite Eqs. 10.1 and 10.2 as described.

(b) Mathematically, determine the angle ϕ_1 in the fundamental component of the input signal in order that there be no dc component at the output of M for $\omega_s = \omega_1$.

(c) Noting that there are product terms at the output of the multiplier of the general form $\cos{(q\omega_i t + \phi_q)} \cos{k\omega_s t}$ that can be expressed in terms of sum and difference frequencies, what is the maximum cutoff frequency of the low-pass filter for no such components to appear at the output?

10.2 The circuit shown is that of a simple comparator designed to have output levels limited at $V_{OH} = 5$ V and $V_{OL} = 0.1$ V, with the design center values of dc currents as indicated with $V_{il} = V_{iN} = 0$.

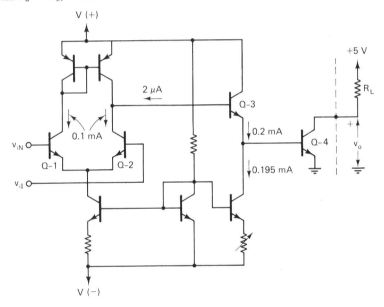

Prob. 10.2

(a) With the base of Q-2 held at zero, write an approximate equation for i_{C3} versus v_{il} in terms of g_{m1}, g_{m2}, and β_3. Repeat for the base of Q-1 held at zero with v_{il} variable.

(b) Assume that, at the operating levels specified, $g_{m1} = g_{m2} = 3$ mS, $h_{FE3} \cong \beta_3 \cong 100$, and $h_{FE4} \cong \beta_4 \cong 100$. For V_{iN} held constant, determine approximately the change in v_{il} required to switch the output current of Q-4 from zero to its saturation value [*Note:* Assume constant g_m's, which is partly justified by the fact that, as g_{m1} decreases, g_{m2} increases, and vice versa. See part (a).]

(c) For $V(+) = -V(-) = 10$ V, estimate the range of input signal levels that can be accommodated by the circuit between output limiting values V_{OL} and V_{OH}.

10.3 Two amplifiers like that of Prob. 10.2 with a common-load resistor are to be used as a sense amplifier like the one shown in Fig. 10.9. As an example, one specific sampling range is to be considered as shown.

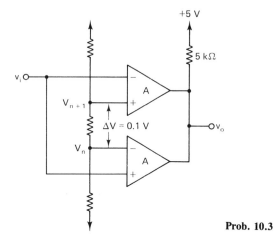

Prob. 10.3

(a) From the specific circuit of Prob. 10.2, verify that the operation is in accordance with the description given in Sec. 10.3, with output $V_{OL} \cong 0$ V and $V_{OH} = 5.0$ V.
(b) Over what range of input levels v_i will the output be V_{OH}, and over what range will it be V_{OL}?
(c) Sketch the total circuit for a 3-bit A/D converter with the required number of amplifiers and the required total supply voltage.

10.4 For the circuit shown, all transistors are identical, each with $h_{FE} = 100$, $V_{BES} = 0.8$ V, and $V_{CES} = 0.1$ V. All R's are $R = 5$ kΩ. The two inputs V_A and V_B each may assume one of two discrete values, $V = 5$ V or $V = 0.1$ V.

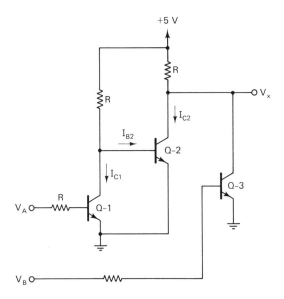

Prob. 10.4

(a) $V_A = 5V$ and $V_B = 0.1$ V. Determine approximately I_{B1}, I_{C1}, V_{CE1} I_{B2}, I_{C2}, and V_{BE2}.

(b) Complete the truth table shown here, showing values of V_x for all combinations of V_A and V_B.

V_A	V_B	V_x
0.1	0.1	
5.0	0.1	
0.1	5.0	
5.0	5.0	

10.5 In the accompanying diagram, the two blocks represent circuits of the type shown in Prob. 10.4. The devices shown have the same characteristics as those used in Prob. 10.4 with the R's having the same resistance, R, value.

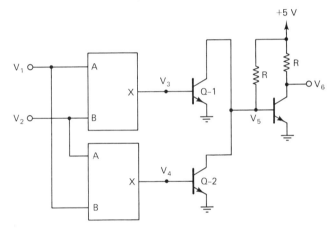

Prob. 10.5

(a) Complete the truth table.

V_1	V_2	V_3	V_4	V_5	V_6
0.1	0.1				
5.0	0.1				
0.1	5.0				
5.0	5.0				

(b) Make your own definition of an exclusive-OR gate, assuming that it is defined by the truth table constructed in part (a).

10.6 The circuit shown is to be considered as one quantizing level of a high-speed parallel A/D converter. The A's are very high gain comparators whose outputs limit at $V_{OL} = 0$ V and $V_{OH} = 5.0$ V. The outputs of the exclusive-OR gate are also $V_{OL} = 0.0$ V and $V_{OH} = 5.0$ V. Let it be further assumed that the input impedance of the EX-OR gate is sufficient that no appreciable charge is lost from the sample-and-hold capacitor during the between-sampling interval, T. The voltage-divider reference network is such that, if the voltage levels V are applied to the amplifier as indicated, the level to the amplifier above it is $V = +0.1$ V and the level to the amplifier below it is $V = -0.1$ V.

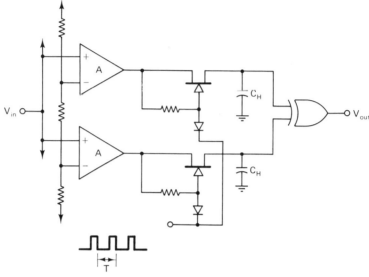

Prob. 10.6

(a) For the following input levels, determine approximately the output of the EX-OR gate during the between sample interval, T: $V_{in} < V$; $V < V_{in} < (V + 0.1)$; $v_{in} > (V + 0.1)$.

(b) What is the required total $\pm$ voltage range to be applied to the voltage-divider reference network to implement a 256-bit converter; also, how many comparators and exclusive-OR gates are needed?

10.7 The circuit shown is suggested as a 4-bit counter-ramp-type A/D converter. A is a very high gain op-amp comparator. The FETs are all low-threshold enhancement-mode FET switches.

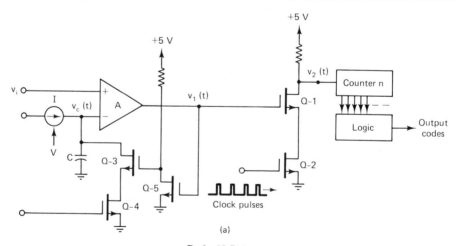

(a)

Prob. 10.7(a)

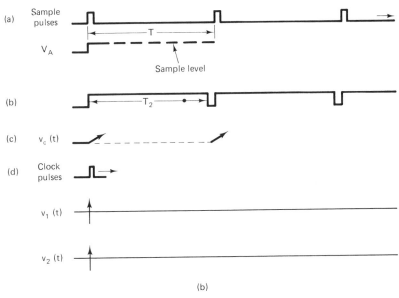

(b)

Prob. 10.7(b)

The input signal is the output of a sample-and-hold circuit derived from samples of a unipolar analog signal, with samples taken at intervals of $T = 0.1$ μs as indicated by waveform (a). At the same time, a waveform of period $T_2 < T$, as shown in (b) is generated and applied to the input of Q-4. It is suggested that a linear ramp, $v_c(t)$, is to be generated starting at the beginning of a particular sampling period until the level V_A is reached in a time $T_A < T_2$.

(a) What is required maximum count rate for a 4-bit converter?

(b) What should the clock pulse rate be?

(c) Complete all waveforms shown, assuming a sampling level for $V_A \cong \frac{3}{4}$, the maximum amplitude of the input signal.

10.8 A variable-width gate is applied to a high-gain strobed comparator as indicated.

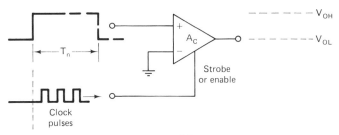

Prob. 10.8

(a) Sketch the output waveform if there are n clock pulses within the gate of time duration T_n.

(b) Can this comparator be defined as an AND gate (see Prob. 9.14 for a very simple AND gate)?

(c) Suppose that the input to the strobed comparator is the waveform applied to the comparator of Fig. 10.15. Can the circuit given here be used to replace the combination comparator and AND gate of Fig. 10.15?

10.9 A variable-length sequence of clock pulses limited by a gate to a maximum of 8 is applied to the three cascaded toggle flip-flops as indicated (see Figs. 9.19 and 9.20) for individual FF circuits. The flip flops are initially set such that the outputs Q_1, Q_2, and Q_3 are all low V_{OM} before any clock pulses are applied.

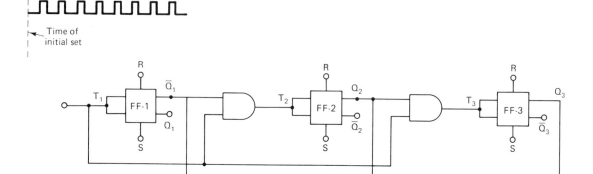

Prob. 10.9

(a) Sketch the waveforms at Q_1, T_2, Q_2, T_3, and Q_4 carefully, assuming that the actual flip-flop transistors occur at the end of each clock pulse. (See Prob. 9.16 and 9.17 for a crude implementation of flip-flops of this nature.)

(b) Repeat the waveform of part (a) if only the first pulse is applied, and then if only the first two are applied, and so on.

(c) Assume that outputs from all three FFs are available simultaneously after the end of the total period, and complete the table with 1's and 0's for highs and lows, showing the output at the end of the period as a function of the number of pulses actually generated.

Output at end of gate period			
Pulse Number	FF-1	FF-2	FF-3
0			
1			
2			
3			
4			
5			
6			
7			
8			

10.10 For the three toggle flip-flops used in Prob. 10.9, assume that either of the outputs Q or $\overline{Q}$ can be connected to the following stage and also that either Q or $\overline{Q}$ can be used as outputs.

(a) Devise a set of connections that will generate the 3-bit natural binary code.

(b) Verify your design by constructing a table like that of Prob. 10.9(c).

REFERENCES[1]

CONNELLY, J. A. *Analog Integrated Circuits,* Wiley, New York, 1975.

GLASFORD, G. M. *Digital Electronic Circuits,* Prentice-Hall, Englewood Cliffs, N.J., 1987.

HOESCHELE, D. F., JR. *Analog-to-Digital and Digital-to-Analog Conversion Techniques,* Wiley, New York, 1960.

SHIENGOLD, D. H. *Analog–Digital Conversion Handbook,* Analog Devices, Norwood, Mass., 1972.

———. *Analog–Digital Conversion Notes,* Analog Devices, Norwood, Mass., 1977.

TAUB, H., AND D. SCHILLING. *Digital Integrated Electronics,* McGraw-Hill, New York, 1977.

[1]*Notes on references:* Many issues of the *IEEE Journal of Solid State Circuits* include papers on analog-to-digital and digital-to-analog converters for specific applications.

Index

A

Abrupt junction, 9
Ambiguous zeros, 467
Amplification factor, 107, 121
Amplifier:
  bandpass, 322–332 (*see also* Filter, active
    RC)
  composite-compound, 180–224
    cascode, 182–187
    class A, 205
    class B, 206
    class AB, 207
    complementary pair, 205, 209
    Darlington, 191–193
    differential, 195–205 (*see also*
      Differential amplifier)
    emitter coupled, 195–200
    source coupled, 201–205
    voltage follower, 188–191
  current sources for biasing, 209–224 (*see*
    also Current mirrors *and* Current
    sources and loads)
  feedback (*see* Feedback amplifiers)
  gain in dB, 132–133
  gain–bandwidth product, 245–250
  micropower, 314–315
  operational, 279–334 (*see also*
    Operational amplifiers)
  power, 350–358
  single device, 130–171
    common base, 140–171
    common collector, 137–139
    common drain, 143–146
    common emitter, 135–137
    common gate, 146–147
    common source, 143–144
    frequency response, 148–169
    transient response, 148–169
    two-input, 142, 148
  transconductance, 309
  voltage gain, 131
Amplitude modulation and demodulation,
    415–418
  peak detection of, 416
  phase-locked loops in, 417–418
  synchronous detection, 417
Analog-to-digital converters, 439–464
  counter-comparator, 463–464
  counter-ramp, 453–455
  digital-to-analog converter in, 462–464
  digital codes, 445–447
  parallel comparator, 448–454
  quantizing, 444, 446

Analog-to-digital converters (*cont.*)
  signal sampling, 440–441
  successive approximation, 462–463
  transmission bandwidth, 442–444
Analog multipliers, 361–369
  applications, 365–367
  four quadrant, 362–365
  logarithmic and exponential amplifier, 369
  squaring circuits in, 361–362
  variable transconductance, 362–365
Analog switch, 374 (*see also* Transmission
    gate)
Astable circuits, 392–395
  blocking oscillator, 393–395
  symmetrical, 392–393
  voltage controlled oscillator, 395–398
Avalanche breakdown, 26

B

Bandgap voltage, 38–39
Bandpass amplifier (*see* Filter, active)
Base resistance, 72, 86, 91
Base width variation with voltage, 84
Biasing using current sources (*see* Current
    sources and loads *and* Current
    mirrors)
Bipolar junction transistor (BJT), 51–96
  amplifier (*see* Amplifiers)
  base recombination current, 83
  breakdown voltage, 86–87
  capacitances, 75, 79
  circuit models, 68–71
    current dependence, 77, 79
    high-frequency, 75–77
    h-parameter, 71–73
    hybrid-π, 72–74
    T-model, 73–74
    temperature dependence, 77–78
  current components, 82–84
  depletion layer recombination, 84
  Ebers–Moll model, 56–58
    common emitter form, 58–62
    parameter measurements, 62–63
  high voltage and current effects, 85–87
    conductivity modulation, 86
    emitter crowding, 86
    Kirk effect, 87
  mathematical models, 52–62
    empirical, 66–67, 87
    normal active region, 64–68
  multipliers (*see* Multipliers)
  noise sources, 90–91

nonlinearities, 80–87
  current gain, 82–84
  empirical models for, 66–67, 87
  operational amplifiers (*see* Operational
    amplifiers)
  parameters, 68–79
    current dependent, 77–79
    h-parameter, 68–70
    hybrid-π, 72
    T-parameter, 74
    temperature dependent, 77–79
  recombination currents, 83–84
  structures, 51–52
  structural parameters, 80–86
    current gain dependence, 82–85
  transit time for minority carriers, 83
  transition frequency, 77
Bistable circuits, 384–390
  operational amplifier comparator, 385
  Schmitt trigger, 385–388
  symmetrical (flip-flop), 388–391
  triggering, 390–391
Blocking oscillator, 393–395
Bode plots, 160, 289
Boltzmann's constant, 6
Bootstrap sawtooth generator, 399–401
Breakdown voltage, 26
Built-in potential, 9

C

Capacitance, 27–32
  channel (FET), 104, 114
  depletion layer, 27–30, 75, 79
  diffusion, 30–32, 75, 79
Cascode amplifier, 183–187
  frequency response, 185–187
  input admittance, 185
Clamping circuits (clamp), 377–380
  bidirectional, 378–379
  keyed, 377–379
  voltage controlled, 377
Clipping circuit (clipper), 375
Common base amplifier, 140–141, 165
Common collector amplifier, 137–139,
    167
Common drain amplifier, 143–146, 166–167
Common emitter amplifier, 135–137, 155
Common gate amplifier, 146–147, 163–164
Common source amplifier, 143–145, 154,
    159
Comparator voltage, 316–317, 384–388
Comparator parallel, 450–452

Complementary pair amplifier, 205–209
  CMOS, 208–209
  npn–pnp transistor, 208–209
  power, 355–357
Conductivity modulation, 86
Continuity equation, 7, 18, 33
Converters, AID (*see* Analog-to-digital
  converters)
Current, 2–6
  base recombination, 82–83
  depletion layer recombination, 83–84
  diffusion, 5, 6
  drift, 2
Current (series) feedback, 242–245
Current gain, bipolar transistor:
  common base,
  common collector,
  common emitter,
Current mirrors, 210–224
  BJT, 210–220
  FET, 220–224
  supply voltage sensitivity, 213–214
  temperature sensitivity, 215–219
Current sawtooth generators, 405–408
Current sources and loads, 209–224

D

Darlington amplifier, 191–193
  biasing, 193
  h-parameter model, 192
Decibel, 133, 284, 288
Depletion layer at *pn* junction:
  nonuniform impurity distribution,
    11–13
  recombination current in, 82–83
  uniform impurity distribution, 9–11
Detector:
  peak, 416
  phase, 410
Differential amplifier, 195–205
  balance conditions, 196–198
  biasing (BJT), 219
  biasing (FET), 224
  current sources for, 198–199
  differential-cascode, 199–203
  emitter coupled, 195–199
  frequency response, 203–205
  source coupled, 201–203
Diffusion constant, 6
Diffusion current, 5, 6
Diffusion length, 16

Digital-to-analog converters, 455–462
  A/D converter, use in, 462–464
  inverted ladder network, 458
  R–2*R* ladder network, 457–458
  sign-bit switches, 461
  weighted resistor type, 455–466
Digital codes, 445–448
  bipolar signals, 446–447
  natural binary, 446
  NBCD, 446
  one's complement, 446
  sign + magnitude, 447
Diode, *pn* junction, 21–46
  capacitance, 27–32
  circuit models, 32–33
  depletion layer generation, 25
  equation, modifications, 25–26
  nonideal, 25–26
  noise in, 41–45
  temperature characteristics, 38–40
  transit time, minority carrier, 31
  storage time-constant, 34
  structural parameters, 23–24
  switching characteristics, 33–37
Distortion:
  bipolar transistor amplifier, 346–350
    input circuit, 353, 354
    power amplifier, 350–351
  crossover, 355–356
  FET amplifier, 341–345
  harmonic, 342–345
Divider circuit (analog), 366
Divider, voltage, IGFET, 223
Doping in semiconductors, 3–5
Drift, input offset, 283
Drift current, 2

E

Ebers–Moll equations, 55–62
  common emitter form, 58–62
  modifications, 80–86
  npn transistor, 55–57
  parameter extraction, 62–63
  pnp transistor, 77–78
  temperature effects, 87–90
Einstein relationship, 6
Emitter-coupled amplifier, 194–200
Emitter coupled bistable circuit, 385–387
Emitter crowding, 86
Emitter follower, 135–137, 167 (*see also*
  Common collector amplifier)

Equivalent circuits, 32, 68, 107
  bipolar transistor, 68–70
  diode, 32
  field effect transistor, 107–112
Exclusive-OR gate, 450
Exponential amplifier, 369

F

Feedback:
  internal transistors (see h-parameter circuit
    models)
  single bipolar transistor amplifier, 134 ff.
Feedback amplifiers, 233–270
  block representation, 233–235
  classification, 244–247
  current (series), 242–247
    input admittance, 245
    output impedance, 243
  differential, 247
  distortion reduction, 267–268
  frequency response, 249–266
  gain bandwidth product, 245–250
  gain equations, 235–276
  general representation, 233–235
  input admittance, 240–242
  multiple feedback paths, 248
  noise in, 268
  open loop gain (definition), 236–237
  oscillations in, 260–266, 269–270
  stability, 260–266
  time-domain response, 259–263
  transient response, 259–263
  voltage (shunt), 237–241
    open loop gain, 237–238
    output impedance, 239
Feedback RC sinusoidal oscillators, 269–270
Field effect transistors, 98–125
  figure of merit, 111–112
  insulated gate (IGFET), 112–125
    capacitances, 114–118
    depletion mode, 115, 116, 118
    empirical models, 120–124
    enhancement mode, 116–118
    E–D mode, 119
    threshold voltage, 115, 117, 118
  junction (JFET), 98–112
    built-in voltage, 100
    channel capacitance, 104
    empirical models, 106–107
    equations for, 100–105
    incremental circuit models, 107–112
    p-channel, 109

    pinchoff voltage, 100, 103
    saturation current, 100, 101, 103
  MOSFET (see Insulated gate FET)
  noise, 123–125
  temperature effects, 123
  transconductance, 120, 121
Filter, active RC, 318–332
  bandpass, 332–335
    biquadratic (biquad), 322–325,
      328–329
    fractional bandwidth, 323
    gain-bandwidth product, 324
    multiple amplifier, 328–332
    multiple pole, 330–331
    Q-factor, 322
    single amplifier, 322–325
  high-pass, 325–327
    maximally flat response, 325
    peaked response, 326–327
  low-pass, 318, 321
    maximally flat n-pole 318–319
    transient response, 320
Filters, single tuned circuit, 322, 323
Fractional frequency deviation, 323
Frequency modulation, 418–420
  phase-locked loops in, 419–420
Frequency response:
  cascode amplifier, 185–187
  feedback amplifier, 249–266
  inverting single device amplifier,
    154–162
  inverting voltage follower, 162–163
  noninverting amplifier, 163–167
  operational amplifier, 285–289
  single-pole network, 148–152
  source-coupled stage, 203–205
Frequency synthesizer, 420–421
Frequency translation, 358–359
Function generator, 399

G

Gain:
  current, bipolar transistor (see Current
    gain)
  power, 132–133
    in decibels (dB), 133
  voltage (see Voltage gain)
Gain–bandwidth product, 245–250, 301,
    324
Graded junction, 11, 12, 28, 30
Gyrator, 332–333

H

Harmonic distortion, 343–345, 353
High-frequency response (*see* Frequency response)
High-level injection, 85–87, 350
High-pass filter, 325–327
High-voltage effects in transistors, 85–87
h-parameters, 68–70
h-parameter circuit model, 71–73, 76
Hybrid-π circuit model, 72–73, 75
Hysteresis, 317, 385

I

Impurities in semiconductors, 3–5
    nonuniform distribution, 11–12, 31, 82–83, 103
    uniform distribution, 9–11, 32, 83, 100
Input admittance:
    cascode FET amplifier, 185
    feedback amplifier (impedance), 240, 241, 246
    inverting amplifier (single device), 157–158
    inverting voltage follower, 162–163
    noninverting amplifier, 164
    noninverting voltage follower, 166–167
Input impedance, 131, 134, 135, 138, 241, 246
    amplifier (single device), 131, 134, 135,138
Input resistance, 137
    cascode amplifier, 183
    Darlington amplifier, 183
    operational amplifier, 292, 295
    single device amplifier, 137, 139, 141, 147, 181
    voltage follower, 189
Instability, 260-266, 285–289
Insulated gate FET (*see* Field effect transistor)
Integrator:
    active filter, 328–329
    feedback amplifier, 252–254, 276
    frequency modulation generation, 419
    sawtooth generator, 404
Intrinsic carrier concentration, 3, 38
Inverted ladder network, 458

J

Junction:
    capacitance, 27, 30, 75, 79, 104
    diode, 21–37

metal–semiconductor, 19–21
temperature characteristics, 38–40

K

K-factor:
    feedback equations in, 235, 247
    FET equations, 104, 114, 221
Kirk effect, 87

L

Large-signal models:
    bipolar transistor, 87, 347
    field-effect transistor, 121, 122
Lifetime, minority carrier, 7, 16, 18, 25, 26, 33, 83, 84, 99
Limiting circuit (limiter), 375
Linearly graded junction, 11, 12, 28, 30
Logarithmic amplifier, 367-369
Loop gain (*see also* Open loop gain):
    feedback amplifier, 236–237
    operational amplifier, 284
    phase-locked loop, 412
Low-pass filter, 318–321
Low-pass *RC* network, 151–152

M

Maximally flat response:
    bandpass filter (amplifier), 330–331
    feedback amplifier, 258
    high-pass filter, 326–327
    low-pass filter, 318–320
Metal–semiconductor junction, 19–21
Micropower operational amplifier, 314–316
Miller (capacitance) effect, 158
Minority carrier lifetime, 7, 16, 18, 25, 26, 33, 83, 84, 89
Mobility, 2, 4
    temperature dependent, 38, 39
Modulation and demodulation:
    amplitude, 415–417
    frequency, 418–420
    phase-locked loop in, 415, 420
Monostable circuits, 390–392
MOSFET (*see* Field effect transistors)
Multiplier, analog, 312, 361–364, 369

N

Negative feedback (*see* Feedback *and* Feedback amplifier)
Negative impedance converter, 333
Noise:
  characteristics, 41–43
  bipolar transistor, 168–169
  burst, 45
  excess low frequency, 45
  excess high frequency, 45
  feedback amplifier, 268
  figure, 169–170
  flicker, 45
  frequency-dependent model, 170, 171
  shot (Schottky), 44
  thermal (Johnson), 43–44
Nonlinear device models:
  BJTs, 66–67, 87, 348–349
  FETs, 120–124
Nonlinearity in amplifiers, 340–369
  bipolar transistor amplifier, 346–350
  field-effect transistor, 341–345
  power series approximation, 341–345
  utilization:
    divider, 366
    exponential amplifier, 369
    frequency translation, 358–359
    logarithmic amplifier, 367–368
    multiplier, 361–367
    squaring circuit, 359–360
Nyquist plot, 264
Nyquist stability criterion, 264

O

Offset current, 282
Offset voltage, 282, 397, 404, 411
Open-loop bandwidth, 285
Open-loop gain:
  feedback amplifier, 236–237
  operational amplifier, 284
Operational amplifiers, 279–334
  active filters use, 318–332
  bipolar transistor, 290–298
  Bode plot, 287–288
  common mode rejection ratio (CMRR), 283–284
  drift, 283
  extended bandwidth, 298–305, 309
  FET input, 300–302
  frequency response, 285–290
  gain-stabilized transconductance amplifier, 313–314

IGFET (MOSFET), 305–308
  input bias current, 281
  input offset current, 282
  input offset voltage, 282
  micropower, 314–315
  slew rate, 290
  stability, 287–288
  transconductance amplifier, 309
  video, 298–305
  voltage comparator, 316–318
Oscillations, in feedback amplifiers, 260–264
Oscillators:
  feedback *RC* sinusoidal, 403
  pulsed sinusoidal, 403
  voltage-controlled oscillator (VCO), 395–399 (*see also* Phase-locked loop)
Output impedance:
  bipolar transistor amplifier, 134–135, 138, 140
  current source, 209
  Darlington amplifier, 191–192
  definition, 131
  feedback amplifier, 239, 243, 246
  field-effect transistor amplifier, 145, 147
  voltage follower, 189, 191

P

Parallel comparator, 448–454
Peak detector, 416
Permittivity, 6
Phase detector, 410
Phase-locked loop, 409–423
  AM demodulator, 415–418
  capture range, 412
  color subcarrier detector, 423
  FM demodulator, 420
  frequency synthesizer, 420–421
  hold-in range, 412
  horizontal AFC, 422
  pull-in time, 412
  simple theory, 411–412
Phase shift (90°), 418
Phase-shift oscillator, 269–270
Pinchoff (channel), 100–104
  current, 101, 105
  voltage, 100, 103
Potential, 6
  metal–semiconductor junction, 20
  pn junction, 9
Power amplifier distortion, 350–357

Q

Q-factor, 322, 406
Quantizing, signal, 440–441
  bipolar signals, 446–447

R

R–2R ladder network, 457–458
Recombination, 16–18
  current, 82–84
Resistance, 2
  sheet, 2
Rise time:
  inverting amplifier, 162
  maximally flat network, 320
  RC network (single pole), 153

S

Sample-and-hold circuits, 382–384
Sampling, 181, 440–442
  rate, 441
  theorem, 442
  transmission bandwidth, 442
  waveform (equation), 440
Saturation region:
  bipolar transistor,
  field-effect transistor, 106–107, 120–121
Sawtooth waveform generator:
  current, 405–409
  voltage, 402–404
Schmitt trigger, 384–388
Semiconductor junction, 7–21
  abrupt, 9
  built-in potential, 9, 10, 12
  depletion layer, 11–12
  diode (see Diode, semiconductor)
  linearity graded, 11–12
  voltage–current relationship, 13–15
Sense amplifier, 449–451
Series (current) feedback, 242–246
Sheet resistance, 2
Shunt (voltage) feedback, 242–246
Sign bit, 460–461
Signal waveform:
  coding, 445–446
  sampling, 440–442
  quantizing, 440–441
Slew rate, 290
Source follower (see Common drain
  amplifier)
Square-root circuit, 367

Squaring circuit, 359–360
Stability, 260–267 (see also Feedback
  amplifiers and Operational amplifiers)
Storage:
  delay time, 35–36
  minority carrier, 33–36
  time constant, 34
Switches:
  bipolar conversion, 460–461
  D/A converters, 458–462
  sign-bit, 461
Synchronous detection, 416–417

T

T-equivalent circuit, 73–74
Television, 421–423
Temperature characteristics:
  bipolar transistor, 77–79, 87, 88–89
  compensation with current sources,
    215–219
  field-effect transistors, 123
  intrinsic carrier concentration, 38
  mobility, 38–39
Thermal noise, 43–44
Threshold voltage, 115, 117, 118
Transconductance:
  bipolar transistor, 72–73, 78
  field-effect transistor, 103, 105, 120–121
Transconductance amplifier, 309
Transformer coupling, 357–358
Transient response:
  feedback amplifier, 259–263
  inverting amplifier (single device),
    161–163
  single-pole network, 148–152
Transistor (see Bipolar junction transistor
  and Field effect transistor)
Transistor noise (see Noise)
Transit time (minority carrier), 31, 83
Transition frequency, 77

U

Uniform impurity distribution:
  bipolar transistor base, 83–84
  channel, JFET, 100–103
  pn junction, 9–11
Unity gain amplifier (see Voltage follower)
Unity gain bandwidth (op amp), 285
Up-converter (frequency), 359

V

Variable capacitance, 395–396
Variable transconductance amplifier, 309
Voltage amplification (*see* Voltage gain)
Voltage comparator, 316-317, 384–388,
    450–452
Voltage-controlled oscillator (VCO),
    395–399, 410–413
Voltage divider network (FET), 221–224
Voltage (shunt) feedback, 237–241
Voltage follower, 188–191
Voltage gain:
    bipolar transistor amplifier, 134–142, 181
    cascode amplifier, 182–187
    composite voltage follower, 189–191
    Darlington amplifier, 191–192
    decibel (dB), 133, 284
    differential pair (FET), 201–205
    feedback amplifier, 235, 236, 242, 245,
        247
        open loop, 236, 239
    field-effect transistor amplifier, 142–147
    frequency-dependent (*see* Frequency
        response)
    operational amplifier, 280–284
        Bode plots for, 288–289
        common mode, 283
        complete amplifier, example, 294–295
        differential mode, 283
        frequency dependent (*see* Frequency
            response)
        gain-stabilized transconductance
            amplifier, 313
        open loop, 284
        operational transconductance amplifier,
            313

W

Waveform generator, 399–405
    pulsed sine wave, 403–404
    sawtooth, 400–404
    sinusoidal oscillator (*see* Oscillators)
    square wave (VCO), 397–399
    triangular wave (VCO), 397–399
Wave-shaping circuit, 399
Weighted-resistor D/A converter 455–456
Wide land operational amplifier, 298–308
Wien bridge oscillator, 270–271

Z

Zener breakdown, 26
Zener diode, 225, 297
Zero crossing detector, 449